MICROBIAL
BIODETERIORATION

ECONOMIC MICROBIOLOGY

Series Editor

A. H. ROSE

Volume 1. Alcoholic Beverages

Volume 2. Primary Products of Metabolism

Volume 3. Secondary Products of Metabolism

Volume 4. Microbial Biomass

Volume 5. Microbial Enzymes and Bioconversions

Volume 6. Microbial Biodeterioration

In preparation

Volume 7. Fermented Foods

Volume 8. Food Microbiology

ECONOMIC MICROBIOLOGY
Volume 6

MICROBIAL
BIODETERIORATION

edited by

A. H. ROSE

School of Biological Sciences
University of Bath,
Bath, England

1981

ACADEMIC PRESS

A Subsidiary of Harcourt Brace Jovanovich, Publishers

LONDON   NEW YORK   TORONTO   SYDNEY   SAN FRANCISCO

ACADEMIC PRESS INC. (LONDON) LTD.
24/28 Oval Road
London NW1

United States edition published by
ACADEMIC PRESS INC.
111 Fifth Avenue
New York, New York 10003

British Library Cataloguing in Publication Data

Microbial biodeterioration. (Economic
microbiology; v. 6)
1. Biodegradation
I. Rose, A. H.   II. Series
620.1'1223   QP517.B5 M 5
ISBN 0-12-596556-7

*Filmset by Northumberland Press Ltd, Gateshead, Tyne and Wear
Printed in Great Britain by Fletcher and Sons Ltd, Norwich*

CONTRIBUTORS

C. C. ANDREWS, Naval Weapons Support Center, Department of the Navy, Crane, Indiana 47522, U.S.A.

ROSAMUND M. BAIRD, Department of Medical Microbiology, St. Bartholomew's Hospital, Charterhouse Square, London EC1, England.

D. J. DICKINSON, Department of Pure and Applied Biology, Imperial College of Science and Technology, University of London, London SW7 2BB, England.

C. GENNER, Department of Microbiology, University College of Wales, P.O. Box 97, Cardiff CF1 1XP, Wales.

B. HAINES, British Leather Manufacturers' Research Association, King's Park Road, Moulton Park, Northampton NN3 1JD, England.

E. C. HILL, Department of Microbiology, University College of Wales, P.O. Box 97, Cardiff CF1 1XP, Wales.

R. E. KLAUSMEIER, Naval Weapons Support Center, Department of the Navy, Crane, Indiana 47522, U.S.A.

J. F. LEVY, Department of Pure and Applied Biology, Imperial College of Science and Technology, University of London, London SW7 2BB, England.

J. LEWIS, International Wool Secretariat, Ilkley, Yorkshire LS29 8PB, England. (Present address, Dinoval Chemicals Ltd., 54 Dundonald Road, London SW19 3PH, England.)

J. D. A. MILLER, Corrosion and Protection Centre, University of Manchester Institute of Science and Technology, P.O. Box 88, Manchester M60 1QD, England.

T. G. MITCHELL, Group Research and Development Centre, British-American Tobacco Company Limited, Regents Park Road, Southampton SO9 1PE, England.

A. H. ROSE, Zymology Laboratory, School of Biological Sciences, University of Bath, Claverton Down, Bath BA2 7AY, Avon, England.

P. C. STAUBER, Group Research and Development Centre, British-American Tobacco Company Limited, Regents Park Road, Southampton SO9 1PE, England.

ALICJA B. STRZELCZYK, Laboratory of Paper and Leather Conservation, Institute of Conservation and Restoration of Antiquities, Nicholas Copernicus University, 87–100 Toruń, Poland.

H. WINTERS, Department of Biology, Fairleigh Dickinson University, Teaneck, Rutherford, New Jersey 07070, U.S.A.

B. J. ZYSKA, Central Mining Institute, Plac Gwarków 1, 40–951 Katowice, Poland.

PREFACE

Previous volumes in this series have dealt with primary and secondary products of microbial metabolism, production of microbial biomass and of industrially important microbial enzymes and transformations which can be affected by microbes. All of these processes have in common that microbial activity is exploited for commercial purposes. The present volume is quite different in this respect in that it deals with unwanted microbial activity.

The bulk of the materials with which man deals in his day to day life are susceptible, at least under certain conditions, to attack by micro-organisms. There are exceptions, and these are principally man-made materials such as plastics, which are highly resistant, or entirely resistant, to attack by micro-organisms. The present volume deals with 'microbial biodeterioration', and it is essential to realize the restraints which have been placed on the contributions included in the volume. It is particularly important to distinguish between the terms bio-deterioration and biodegradation. Man has for long appreciated, not least in recent financially stringent years, that all materials and com-pounds made by living organisms are subject to decomposition by micro-organisms. This is the basis of cycling of elements in the biosphere. Degradation of naturally occurring compounds is therefore essential to man's existence on this planet. Biodeterioration, on the other hand, implies a change in the quality or value of a material which makes the material less useful in aesthetic or utilitarian terms. Biodeterioration can be brought about by a variety of living organisms, including insects. The contributions in the present volume are, however, restricted to biodeterioration processes brought about by micro-organisms on various classes of material.

However, the volume does not deal with all aspects of microbial biodeterioration, for there is no chapter on biodeterioration of foods and food products or of agricultural materials. It is planned to include these aspects in later volumes in the series.

Finally, I am again extremely grateful for the advice given to me during the preparation of this volume by my colleague, Alan Rayner.

June 1981   ANTHONY H. ROSE

CONTENTS

Contributors v
Preface to Volume 6 vii

1. History and Scientific Basis of Microbial Biodeterioration of Materials
ANTHONY H. ROSE

I. Introduction 1
II. What is Biodeterioration? 2
III. Physiological Aspects of Microbial Biodeterioration of Materials . . . 4
    A. Environmental Conditions and Biodeterioration 4
    B. Methods of Microbial Attack in Biodeterioration 8
IV. Combating Microbial Biodeterioration of Materials 9
    A. The Past 9
    B. The Present 10
    C. The Future 16
References 17

2. Wood
J. F. LEVY and D. J. DICKINSON

I. Introduction 19
II. Degradation of Wood 21
III. Biodeterioration of Wood 21
    A. Marine Borers 22
    B. Insects 22
    C. Micro-Organisms 23
IV. Factors Affecting Decay 28
    A. Wood Structure 28
    B. Moisture Relationships 31
    C. Nutrient and Nitrogen Relationships 33
V. Effects of Decay 34
    A. Soft Rots 35
    B. Wood-Rotting Basidiomycetes 36
VI. Mechanisms of Decay: Enzymes and Enzyme Action 37
VII. Where Decay Occurs 39
    A. Living and Dead Wood 40
    B. Infection of Wood in Standing Trees 40
    C. Post-Harvest Deterioration 42
    D. Packaged Timber in Transport 44
    E. Damage in Service 45
VIII. Colonization of Wood and the Effect of Wood Preservatives . . . 51
References 56

3. Stone
ALICJA B. STRZELCZYK

I. Chemical Air Pollution as an Agent Responsible for Deterioration . . .	61
II. Origin of the Microflora on Rocks and Stones	63
III. Role of Chemolithotrophic Bacteria in the Decomposition of Rocks and Stones .	65
A. Activity of Sulphur-Oxidizing Bacteria	66
B. Activity of Nitrifying Bacteria	70
IV. Role of Heterotrophic Bacteria and Fungi in the Destruction of Rocks and Stones .	71
V. Role of Algae in Destruction of Rocks and Stones	74
VI. Attempts at Combating Microflora on Stone Monuments	77
References.	78

4. Wool
J. LEWIS

I. Introduction	81
II. Fleece Damage Caused by Micro-Organisms	84
III. Bacterial Proteolytic Enzyme in Removal of Wool from Sheepskins . . .	89
A. The Sweating Process	92
B. Enzyme Depilation	96
C. Recovery of Wool from Skin Pieces	97
IV. Microbial Damage of Processed Wool	98
A. Conditions that Favour Development of Mildew on Wool	99
B. Detection of Mildew Damage	101
C. Testing of Textile Materials for Resistance to Microbial Attack . . .	103
D. Chemical Control of Mildew	110
E. Eradication of Mildew on Infected Goods	114
F. Characteristics of Selected Treatments	114
V. Damage of Domestic Wool Textiles by Detergents Containing Proteolytic Enzymes	116
A. Effects of Enzyme Detergents on Wool	116
B. Treatments to Inhibit Degradation of Wool in Enzyme-Detergent Solutions .	122
VI. Conclusion	127
VII. Acknowledgements	128
References.	128

5. Hides and Skins
BETTY M. HAINES

I. Introduction	131
II. Deterioration by Autolysis	133
III. Deterioration by Bacteria	136
IV. Curing	141
A. Salting	141
B. Drying	143
C. Dry-Salting	144
D. Storage	144
V. Alternative Methods of Cure	145

A. Biocides 145
B. Chilling 146
C. Freezing 146
VI. Deterioration by Moulds 146
References. 147

6. Metals
J. D. A. MILLER

I. Introduction 150
   A. Basic Principles of Electrochemical Corrosion 150
   B. Classification of Microbial Corrosion Processes 153
II. Corrosion by Concentration Cell Formation 155
   A. Mechanism 155
   B. Microbial Growth and Concentration Cell Corrosion in Industrial Situations . 156
   C. Control of Microbial Growth in Cooling Water Systems 159
III. Sulphuric Acid Corrosion 162
   A. Acid Production by Thiobacilli 162
   B. Corrosion by Thiobacilli 163
   C. Detection of Causative Organisms 164
IV. Aviation Fuel Tank Corrosion 165
   A. Historical 165
   B. The Fungus *Cladosporium resinae*, and Other Contaminants 167
   C. Mechanism of Corrosion of Integral Fuel Tanks of Subsonic Aircraft . . 171
   D. The Possibility of Fungal Growth in Supersonic Aircraft 173
   E. Prevention and Control 174
V. Corrosion by the Sulphate-Reducing Bacteria 179
   A. Historical 179
   B. The Sulphate-Reducing Bacteria 181
   C. Nature and Extent of Corrosion by the Sulphate-Reducing Bacteria . . 183
   D. Mechanism of Iron Corrosion by the Sulphate-Reducing Bacteria . . 187
   E. Corrosion of Other Metals by the Sulphate-Reducing Bacteria . . . 194
   F. Prevention and Control 195
   G. Detection of Sulphate-Reducing Bacteria 198
References. 199

7. Paintings and Sculptures
ALICJA B. STRZELCZYK

I. Introduction 203
II. Effect of Moisture on Microbial Deterioration of Paintings and Sculptures . 205
III. Development of Microbial Communities on the Surface of Paintings . . 207
IV. General Forms of Microbial Activity Affecting Paintings and Painted Sculptures . 209
V. Microbial Deterioration of Paintings on Wood and Painted Sculptures . . 210
   A. Decay of the Wooden Support 210
   B. Decay of Paint Layers 212
VI. Microbial Deterioration of Paintings on Canvas 215
VII. Microbial Deterioration of Paintings on Paper 218
VIII. Microbial Deterioration of Mural Paintings 220

     A. Development of Heterotrophic Micro-Organisms   220
     B. Development of Autotrophic Micro-Organisms.   222
  IX. Combating Micro-Organisms on Paintings and Sculptures   .   .   .   223
     A. Methods for Disinfecting Paintings and Sculptures   224
     B. Fungicides Used in Disinfection   226
References.   233

8. Tobacco
T. C. MITCHELL and P. C. STAUBER

    I. Introduction   235
   II. Types of Biodeterioration   237
  III. Barn Rots   240
  IV. Marketing, Storage and Shipping of Cured Leaf   246
    V. Manufacturing.   251
   VI. Distribution   253
  VII. Economic Aspects   254
VIII. Acknowledgements   256
References.   256

9. Fuels and Oils
C. GENNER and E. C. HILL

    I. Introduction   260
     A. High Volume Ration of Hydrocarbon to Water   260
     B. High Ration of Water to Hydrocarbon   260
   II. Fuels   261
     A. Conditions for Microbial Growth   261
     B. Types of Fuels Involved   261
     C. Consequences of Microbial Growth   263
     D. Growth Associated with Subsonic Aircraft   264
     E. Fuel Baffles.   269
     F. Supersonic Aircraft   269
     G. Detection of Microbial Infections in Fuel Systems   .   .   .   271
     H. Anti-Microbial Procedures   272
     I. Infections in Light Fuel Oils.   278
     J. Natural History of *Cladosporium resinae*   .   .   .   280
  III. Lubricating and Hydraulic Oils.   280
     A. Introduction   280
     B. Characteristics of Oils Involved   281
     C. Consequences of Microbial Growth   282
  IV. Soluble Oil Emulsions   288
     A. Ecological Aspects of 'Soluble' Oil Emulsions in Machine Tools.   .   288
     B. The Spoilage Process   291
     C. Biocides in Cutting Oils   292
     D. On-Site Test Methods.   293
     E. Physical Methods of Controlling Micro-Organisms   .   .   .   293

F. Rolling Mill Emulsions   294
G. Other Oil-in-Water Emulsions   294
H. Health Aspects of Emulsion Spoilage   295
V. Water-In-Oil Emulsions   300
VI. Conclusions   300
References   301

10. Latex Paints
HARVEY WINTERS

I. Introduction   307
II. Bacterial Attack   308
III. Biodeterioration of Exterior Paint Films on Wood Substrates   318
References   320

11. Rubber
BRONISLAW J. ZYSKA

I. Introduction   323
II. Interaction of Rubber with Micro-Organisms   325
A. Micro-Organisms of *Hevea* Latex   325
B. Micro-Organisms of Raw Natural Rubber   327
C. Microbial Resistance of Synthetic Rubbers   329
D. Microbiological Resistance of Compounding Ingredients   333
E. Microbiological Resistance of Vulcanizates   337
F. Microbial Deterioration of Certain Rubber Goods   347
III. Colonization of Rubber Products to Pathogenic Micro-Organisms .   .   .   367
IV. Protection Against Micro-Organisms   368
A. Preservation of *Hevea* Latex   368
B. Preservation of Natural Rubber   370
C. Protection of Rubber Products   371
V. Acknowledgements   379
References   380

12. Drugs and Cosmetics
ROSAMUND M. BAIRD

I. Introduction   387
II. Spoilage   388
A. Biodegradation of the Active Ingredient   390
B. Changes in Physical Properties   393
C. Aesthetic Deterioration   394
III. Health Hazard   395
A. Sterile Pharmaceuticals   395
B. Non-Sterile Pharmaceuticals in Cosmetics   397
IV. Assessment of the Risk   401
A. Incidence of Contamination   401
B. Hazard to the User   406

V. Sources of Contamination 408
   A. Contamination in Manufacture 408
   B. Contamination During Use 414
VI. Factors Influencing Microbial Survival 418
VII. Monitoring and Control 420
   A. International Standards 421
   B. Control During Manufacture 423
   C. Control During Use 424
VIII. Economic Aspects 424
References. 426

13. Plastics

ROBERT E. KLAUSMEIER and CATHERINE C. ANDREWS

I. Introduction 431
II. Plastics—An Introduction 432
III. Microbiology of Plastic Formulations 436
   A. Microbiology of Polymers 436
   B. Microbiology of Plasticizers 447
   C. Biochemistry of Plasticizer Degradation 456
IV. Preservatives and Antimicrobials. 459
V. Test Methods 463
VI. Summary 471
References. 472

Author Index 475
Subject Index 491

1. History and Scientific Basis of Microbial Biodeterioration of Materials

ANTHONY H. ROSE

Zymology Laboratory, School of Biological Sciences, Bath University, Bath, Avon, England

I. Introduction	. . . .	1
II. What is Biodeterioration?	. . . .	2
III. Physiological Aspects of Microbial Biodeterioration of Materials .	. . .	4
A. Environmental Conditions and Biodeterioration .	. . . .	4
B. Methods of Microbial Attack in Biodeterioration .	. . . .	8
IV. Combating Microbial Biodeterioration of Materials .	. . . .	9
A. The Past	. . . .	9
B. The Present .	. . . .	10
C. The Future .	. . . .	16
References .	. . . .	17

I. INTRODUCTION

Seminal to Man's early development, and the evolution of a primitive and later a more sophisticated civilization, was his ability to fashion and fabricate naturally occurring materials. Originally, there were three classes of such material, namely those used to make hunting weapons and similar implements, food materials, and materials employed, often after being subjected to some degree of fashioning, as covering materials. Since all of these were naturally occurring materials, they were *ipso facto* prone to microbial biodeterioration. A major factor in the development of civilization was Man's capacity, firstly, to recognize decay in naturally occurring materials, to avoid and prevent it where possible obviously empirically and, much later, to combat it preferably after understanding the causes of the deterioration process. During the past century, these three classes of material have been joined by a fourth, which has become available largely as a result of the development of a chemical industry. This fourth class includes materials that are chemically modified forms of natural materials, such as oils, and non-natural purely synthetic materials. The latter include industrial chemicals, and plastics materials which Man has invented and developed primarily with the aim of avoiding the economic losses that accompany decay and biodeterioration of naturally occurring materials.

II. WHAT IS BIODETERIORATION?

The word 'deterioration' is derived from the Latin verb *deteriorare* meaning 'to worsen'. In its Anglicized form, the verb has come to mean, and here I quote from the Third Edition of The Shorter Oxford English Dictionary, 'to lower in quality or value ... to degenerate'. It follows that, by biodeterioration, it is implied that biological agents are the cause of the lowering in quality or value, or of degeneration. In a series entitled *Economic Microbiology*, it is implied that the biological agents dealt with are micro-organisms. The nouns 'quality' and 'value', and the verb 'degenerate', infer an appreciation of the material, either in aesthetic or utilitarian terms. At this point, it is important to distinguish between deterioration and degradation. The latter term must encompass biological changes that we understand by biodeterioration, but the former term is restrictive in that it embodies the concept of a lowering in quality or value of an article or commodity that, after it has been subjected to biodeterioration, can retain a degree of quality or value such as to make it still appreciated and usable by mankind. In other words, implicit in the accepted use of the term 'biodeterioration' is that the articles and materials under consideration are, because they are inherently so or have been made so by Man, reasonably resistant to the action of microbes that may cause deterioration, always bearing in mind the environmental conditions under which the article or material is used by Man. Eggins and Oxley (1980) recently addressed themselves to the terms biodeterioration and biodegradation, and arrived at a similar basis for distinguishing them.

Two principal commodities which are subject to microbial attack and which have a significant place in Man's economy are agricultural products and food materials. Both of these are destined, ultimately and in large part, for human or animal consumption, so that they are not inherently resistant to microbial attack. Neither category or material is dealt with in the present volume. The microbiology of agricultural materials, and in particular the problem of post-harvest decay which can account for very considerable economic losses in many countries of the World, is covered in a number of recent review articles. The microbiology of foods has, over the years, received much greater attention, firstly because its control is of interest to a larger number of people than agricultural products, and also because a breakdown in control of the microbiological standards of food often has important public health implications. There are several texts which deal with food microbiology (Ayres *et al.*, 1980; Frazier and Westhoff, 1978; Jay, 1970). Volume 8 in the present series will contain chapters covering various ways in which unwanted microbial activity in foods can be avoided, and descriptions of some of the main groups of micro-organisms which are causative agents of food spoilage. Volume 7 includes accounts of desirable microbial activity in foods.

Because it affects articles and materials that Man has, usually at some considerable expense, designed and arranged to resist microbial attack, the subject of biodeterioration, and in particular that attributable to micro-organisms, has attracted a considerable body of scientific workers and, in turn, spawned a voluminous literature. International meetings of scientists interested in biodeterioration are held regularly under the auspices of The Standing Committee for International Biodeterioration Symposia (SCIBS) and published proceedings of these meetings are most valuable sources of information. The first meeting was held in Southampton, England, in September 1968, and reports of papers presented there were published in a volume edited by Walters and Elphick (1968). The second symposium was held in Lunteran, Holland, in September 1971, and the proceedings appeared in 1972 in a volume edited by Walters and Hueck-Van der Plas. Kingston, Rhode Island, U.S.A. was the venue for the third meeting in August 1975, the proceedings of which were published in 1976 (Sharpley and Kaplan, 1976).

The Bundesanstalt für Materialprüfing were hosts for the fourth meeting, which was held in Berlin-Dahlem, West Germany, in August 1978. The proceedings of this symposium have been published in a text edited by Oxley *et al.* (1980). The fifth symposium is scheduled to be held in Aberdeen, Scotland, in September 1981.

When discussing the literature on microbial biodegradation, the Biodeterioration Centre, housed in the University of Aston in Birmingham (address: St. Peter's College, Saltley, Birmingham B8 3TE, England) has a forefront position. This centre is supported financially by a consortium of British and international organizations, numbering near 20. The aims of the centre are to collate, sift and to disseminate information on biodegradation of all materials. The principal activity of the centre is to publish, and supply for a charge, information obtained from their store of data. They also produce three important publications. The best known one is the *International Biodeterioration Bulletin*, published quarterly, and containing original papers and review articles, as well as the Biodeterioration Society Newsletter. The *Bulletin* is now in its 16th volume, and it has as editor-in-chief H. O. W. Eggins. The centre also produces two other publications, namely *Biodeterioration Research Titles* and *Waste Materials Biodeterioration Research Titles* each providing about 2,000 references a year.

An important text dealing with microbial biodeterioration is that edited in 1975 by Gilbert for the Society for Applied Bacteriology. Although it is over 10 years old, the review by Walters (1971) is also valuable reading on the subject since it discusses the principles behind the use of techniques for combating deterioration of materials caused by microbes. Other reviews in this area appear in *Advances in Applied Microbiology*, *Developments in Industrial Microbiology* and in *Progress in Industrial Microbiology*.

III. PHYSIOLOGICAL ASPECTS OF MICROBIAL BIODETERIORATION OF MATERIALS

A. Environmental Conditions and Biodeterioration

The materials discussed in this volume are relatively resistant to attack by micro-organisms for either or both of two reasons. Firstly, many are made up of components that are virtually insoluble in water, or have a very low water solubility, with the result that potential microbial nutrients are not readily supplied by the materials. Secondly, the materials are often used under environmental conditions, such as shortage of water and nutrients and low temperatures, that in general restrict microbial growth. In the following paragraphs, there is a brief discussion of the environmental conditions often associated with the use of resistant materials, and of the manner in which these factors affect microbial activity.

1. Water Activity

Water accounts for between 80 and 90% of the weight of a micro-organism. All chemical reactions that take place in living organisms require an aqueous environment, and water must therefore be in the environment if the organism is to grow and reproduce. It must be, more-over, in the liquid phase, and this confines biological activity to temperatures ranging from around $-2°C$ (or lower in solutions of high osmotic pressure) to approximately $100°C$; this is known as the bio-kinetic zone. There are several reviews on the water relationships of micro-organisms which merit perusal. Corry's (1973) review, although some decade old, still is worth examination. The more recent review by Smith (1978) emphasizes the responses of individual classes of microbe to dry environments, whereas that from Dallyn and Fox (1980) stresses more practical aspects of studying microbial responses to water.

The water requirements of a micro-organism are expressed quantita-tively in the form of the water activity (a_w) of the environment or material. This is equal to p/p_0, p being the vapour pressure of the solution and p_0 the vapour pressure of water.

Values for a_w can be calculated using the equation:

$$\ln a_w = \frac{-vm\theta}{55.5}$$

where v = the number of ions formed by each solute molecule, m = the molar concentration of solute and θ = the molar osmotic coefficient, values for which are listed for various solutes in a number of textbooks. Water has an a_w value of 1.000; this value decreases when solutes are dissolved in water.

Micro-organisms can grow in media with a_w values between 0.99 and about 0.63. For any one organism, the important values within this range are the optimum and minimum a_w values. These have been determined for a number of micro-organisms, and they seem to be remarkably constant for a particular species and to be independent of the nature of the dissolved solutes. The general effect of lowering the a_w value of a medium below the optimum is to increase the length of the lag phase of growth and to decrease the growth rate and the size of the final crop of organisms. However, virtually nothing is known of the manner in which low water activity adversely affects micro-organisms. It has been suggested that water stress causes microbes to divert energy from growth to osmoregulation (Bernstein, 1963). Another suggestion (Christian and Waltho, 1962) is that low intra-cellular water potential inhibits enzyme activity in miro-organisms.

On the whole, bacteria require media of higher a_w value (0.99–0.93) than either yeasts or moulds. Staphylococci and micrococci charac-teristically have lower optimum a_w values in this range. With *Salmonella oranienburg*, the a_w value of the medium has important effects on the physiology of the bacterium; only at a_w values below 0.97 is proline required for growth of the bacterium. Moreover, in media lacking amino acids, accumulation of potassium ions by this bacterium increases to a maximum as the a_w value of the medium is lowered to 0.975, and then decreases as the water value is lowered further to 0.96. Yeasts also vary in the optimum a_w values required for growth, but the values for these organisms (0.91–0.88) are lower than those for the majority of bacteria. A few yeasts, e.g. *Saccharomyces rouxii*, can grow in media of a_w value as low as 0.73 and these are known as osmophilic yeasts. *Saccharomyces rouxii*, unlike non-osmophilic yeasts, contains glycerol and arabitol, and it is believed that these intracellular polyols in some way enable the yeast to grow in media with low water activities. Moulds are, in general, better able to withstand dry conditions than other micro-organisms, and for some strains, e.g. *Aspergillus echinulatus* (Moreau, 1979), the lower limit of a_w value is near 0.60. A few moulds, e.g. *Monascus bisporus*, have an upper a_w limit of approximately 0.97, and are described as xerophilic.

The water relations of materials discussed in this book differ consider-ably. Some, such as hides and skins, tobacco and drugs, and cosmetics, contain high concentrations of microbial nutrients, but microbial activity is minimized by maintaining the material in a dry state. At the opposite end of the scale are plastics, which are virtually insoluble in water and almost totally resistant to microbial attack. In between, there is a variety of situations, with much depending on the water contents of environments in which the materials are used. Fuels and oils are excellent carbon sources for growth of some microbes but, because of their extremely hydrophobic nature, they efficiently repel water. Wood and wool can also absorb and retain water, and they too provide good sources of nutrients. Stone does not, but it has the capacity to retain moisture for long periods. Materials specially manufactured to provide protective surfaces, such as paints (including those applied to paintings and sculptures) and rubber, are materials that rarely provide environments with water activities sufficiently high to permit microbial activity.

2. Availability of Nutrients

Many of the materials considered in this volume are resistant to microbial invasion because they do not readily provide a source of microbial nutrients. Microbes that colonize these materials rely on nutrients provided by a slow breakdown of compounds in the materials and by environmental agencies. It is easy to appreciate therefore that microbes colonizing the materials usually grow slowly and spasmodically. At most times, there is one or more nutrient limiting microbial growth on the material. Two of the major classes of nutrient are those that provide a source of carbon (and often of energy too) and nitrogen. Limitation of growth by lack of either or both of these classes of nutrient also has important effects on the physiology of the invading micro-organisms.

The importance of the carbon and energy source on growth of heterotrophic micro-organisms has been studied over many years (Mandelstam and McQuillen, 1973; Rose, 1976; Doelle, 1975). Particular attention has been given to the phenomenon of carbon catabolite repression, in which glucose or compounds formed by catabolism of glucose regulate the expression of a variety of genes (Paigen and Williams, 1970). In general, glucose or its catabolites, through the action of cyclic AMP at the gene level, prevent expression of genes that often are concerned with utilization of other carbon sources. Relief from carbon catabolite repression, a situation that frequently prevails in natural environments where resistant materials are used, can therefore permit a micro-organism to utilize a wider range of carbon-containing nutrients.

Ammonia catabolite repression, a phenomenon which was discovered only in the past decade, is in many ways similar to carbon catabolite repression in that the presence of ammonium ions prevents expression of genes (Aharanowitz, 1980). It is, however, a much less well understood phenomenon. Again, micro-organisms that colonize resistant materials in natural environments will, under most conditions, not be subject to ammonia catabolite repression.

Some micro-organisms, when grown under nutrient-deficient conditions which usually entails relief from carbon and ammonia catabolite repression, are able to form differentiated structures, such as spores (Ashworth and Smith, 1973; Smith and Berry, 1974; Peberdy, 1980). With some microbes, these differentiated structures are much more resistant to environmental stresses, such as heat and desiccation, and therefore constitute a type of survival kit for a micro-organism. Because of the nutritionally poor nature of environments in which resistant materials are used, formation of these resistant structures is generally favoured, and is an important factor in the survival and dissemination of the invading organisms.

B. Methods of Microbial Attack in Biodeterioration

Deterioration of a material can occur when that material encounters a moist atmosphere. The manner in which the deterioration process takes place can be considered under three headings.

Some materials, principally those that are maintained in a dry state such as drugs and cosmetics as well as tobacco, can, when moistened, provide a wide variety of microbial nutrients. It is imperative, therefore, that these materials be kept at all times in a desiccated, or near desiccated, condition. A similar situation obtains with oils which, when covered with a film of water, can be used as a source of carbon by a variety of micro-organisms (Levi et al., 1979).

Other materials, even when moist, do not provide a ready supply of nutrients, and micro-organisms can invade these materials only if they have the capacity to break down constituents to give utilizable nutrients. The capacity to break down these materials is invariably associated with the ability of an organism to elaborate depolymerizing enzymes, such as cellulases and proteinases. Volume 5 of this series deals with production of microbial enzymes, and several of the chapters in the volume survey the micro-organisms that are able to elaborate various classes of depolymerizing enzyme.

Finally, there are those materials that undergo deterioration because of the elaboration by an organism of a corrosive chemical. The principal chemical that causes deterioration of materials in this way is sulphuric acid, which is an end product of oxidation of sulphur-containing compounds by certain chemolithotrophic bacteria. This manner of attack is encountered in microbial biodeterioration of metals and of stone.

IV. COMBATING MICROBIAL BIODETERIORATION OF MATERIALS

A. The Past

Despite the fact that Man's appreciation of the causative role of micro-organisms in many biodeterioration processes extends back a century at the most, the history of preventing deterioration of materials by microbes is surprisingly long and varied (Block, 1979). This history can be divided into two eras without, as one might anticipate, there being a clear line of demarkation between the two.

Although the human body, dead or alive, can hardly be described as a resistant material, some of the earliest examples of combating microbial attack of perishable materials are seen in mummification. The principal substances used to preserve the deceased human body in ancient Egypt were natron, salt and resins, all of which were thought then to have magical and often life-giving qualities, and which were therefore assumed to be capable of retaining life in the mummy. Some centuries later, plant extracts, especially oils, began to be used to combat biodeterioration. Books of the Apocrypha in the Old Testament refer to using cedar oil as a preservative, as do the writings of Ovid and Horace. Alexander III, King of Macedon and usually known as Alexander the Great, ordered bridges to be treated with olive oil to prevent decay.

Fumes from burning materials, because they were pungent, later came to be used to combat biodeterioration. Burning sulphur (fire and brimstone) and burning wood were widely employed throughout the Middle Ages to prevent spread of micro-organisms, especially pathogenic ones. The states of Renaissance Italy suspected that ships were importing disease, as they probably were, and it was decreed by the health authorities in the the City of Venice in 1438 that ships' cargoes should be fumigated. Meyer (1962) described treatment of a letter with vinegar in the year 1485, the object of which was simply to preserve the missive. A few centuries later, considerable effort was taken to preserve ships' planking against both shipworm and microbial attack. The still familiar mercuric chloride, which had become available as a result of early developments in the chemical industry, is recorded as being first used as a biocide for this purpose in 1705; its use was patented in England in 1832. Two other salts which were similarly used were copper sulphate and zinc chloride, both patented in the 1830s (Block, 1879). Pressurized injection of wood with coal-tar creosote, a procedure which came to be known as the Bethell process, was patented in 1838 (Hunt and Garratt, 1953).

Although the 19th and early 20th Centuries saw antimicrobial compounds used increasingly in agriculture, and particularly to combat bacterial diseases of man and animals, progress in the application of these compounds in the preservation of inert materials, except possibly timbers, was decidedly slow. World War II, however, changed the situation, and ushered in the second era in Man's combating of microbial degradation of resistant materials. Introduction into tropical areas of the World of a wide range of military equipment, extending from canvas covers to electronic valves, gave a completely new dimension to our understanding of the extent to, and speed at, which microorganisms can destroy what were hitherto regarded as inert materials. It was during this period that it first came to be appreciated that aircraft and other liquid fuels can become contaminated with microbes. Large government-sponsored research programmes, especially in the United States, were promptly set in motion, and as a result a huge new variety of chemical compounds came into use to combat microbial attack of inert materials. This era has extended to the present day, when the industry concerned with preventing microbial attack of these materials is large and flourishing.

B. The Present

Today, the principal stratagem used to combat microbial deterioration of materials continues to be application of antimicrobial chemicals, which are usually referred to as *biocides*. Biocides have been subdivided into five major groups (Sharpell, 1980), namely health-related biocides (antiseptics, disinfectants), food and food-related preservatives, crop and plant protectants, cosmetic and skin-care biocides, and industrial biocides. With one exception (drugs and cosmetics), the materials covered in the present volume are protected with biocides included in the last category.

Sharpell (1980) goes on to divide industrial biocides into two classes. The first class includes biocides that are used as an adjunt in an industrial process, as with incorporation of biocides into machine-cutting fluids. Here, biocides are added to water-based materials which are prone to bacterial as well as fungal contamination. Prominent among bacteria that contaminate these materials are species of *Bacillus*, *Brevibacterium*, *Enterobacter*, *Flavobacterium*, *Proteus* and *Pseudomonas* (Sharpell, 1980). Fungal contaminants found include species of *Aspergillus*, *Fusarium* and *Penicillium*. These microbes can cause a variety of changes to properties of the material, ranging from production of foul odours to formation of slime which can block nozzles. They are often able to split emulsions.

The second category includes biocides that are incorporated into a finished product, such as wood veneers, hides, plastics and rubber products. The majority of these products contain little or no water, and the principal contaminating microbes are therefore fungi. Species commonly found on these materials include members of the genera *Aspergillus* and *Penicillium* along with *Aureobasidium pullulans*, *Chaetomium globosum* and *Streptoverticillium rubrireticuli* (Sharpell, 1980).

A biocide, to be suitable for a particular application, must fill many criteria. Not only must it be effective in the material which it is designed to protect, but it must also be long lasting, reasonably cheap and stable. Importantly, however, it must be passed as toxicologically acceptable in natural environments, and microbes must be slow in developing resistance to it. The last two considerations have been very important in the recent development of biocides. Early on in the application of industrial biocides, resistance by micro-organisms was soon encountered, and the natural move was to switch to another biocide. However, the room for manoeuvre has, in recent years, been severely curtailed as many biocides have been withdrawn after they have been shown to have a degree of persistence and toxicity that is unacceptable

Table 1

A list of some industrial biocides. Derived from Sharpell (1980) and reproduced with the permission of the author and the Society for Industrial Microbiology

Composition	Per cent	Trade names	Application
2-[(Hydroxymethyl)amino]-2-methylpropanol		Troysan 192 (Troy)	Latex paints, resin emulsions
3,5-Dimethyl-tetrahydro-1,2,5,-2H-thiadiazine-2-thione		Biocide N-521 (Stauffer)	Leather, paint, glue
Hexahydro-1,3,5-tris-(2-hydroxyethyl)-5-triazine	78	Grotan® (Lehn & Fink) Onyxide® 200 (Onyx)	Cutting oils and diluted coolants
2-n-Octyl-4-isothiazolin-3-one	45	Skane® M-8 (Rohm & Haas)	Latex and oil-based paints, in-can paint preservative
2-n-Octyl-4-isothiazolin-3-one	5	Kathon® LM (Rohm & Hass)	Fabrics
2-n-Octyl-4-isothiazolin-3-one	8.1	Kathon® LP (Rohm & Haas)	Wet processing of hides
5-Chloro-2-methyl-4-isothiazolin-3-one	10.1	Kathon® 886 MW (Rohm & Haas)	Wood veneer, cutting fluids and coolants, paste, paper and paperboard
2-Methyl-4-isothiazolin-3-one	13.9		
2[(Hydroxymethyl)amino]ethanol		Troysan 174 (Troy)	Paints, resin emulsions
Hexachlorodimethyl-sulphone		Stauffer® N-1386 (Stauffer)	Industrial emulsions
Diiodomethyl p-tolyl-sulphone		Amical® 48 (Abbott)	Paint
Para-Chlorophenyl diiodomethyl-sulphone		Amical® 77 (Abbott)	Paint
2,4,5,6-Tetrachloroisophthalonitrile		Nopcocide® N-96 (Nopco)	Latex paints
Poly [oxyethylene (dimethyliminio) ethylene (dimethyliminio)] ethylene dichloride	60	Busan® 77 (Buckman)	Cutting fluids
3-Iodo-2-propynyl butylcarbamate	40	Troysan Polyphase Anti-mildew (Troy)	Interior and exterior coatings
1,1'-(2-Butenylene) bis-(3,5,7-triaza-1-azoniaadamantane chloride)		Cosan® 265 (Cosan)	Latex paints, resin emulsions, adhesives
10,10'-Oxybisphenoxarsine 5% in a polymeric resin carrier		Vinyzene® SB-1 (Ventron)	Polyvinylchloride, polyurethane, and other polymeric compositions
10,10'-Oxybisphenoxarsine in various non-volatile plasticizer carriers		Vinyzene® BP series (Ventron)	Film and sheeting, extruded plastics, plastisols, moulded goods, organosols, fabric coatings

Table 1 (cont.)

Composition	Per cent	Trade names	Application
Solubilized copper 8-quinolinolate		Cumilate® series (Ventron)	Wood and wood products, glues and adhesives, paper products
Rosin amine D-pentachlorophenate		Cumimene® series (Ventron)	Paper, textiles, rope, emulsion systems
2,2'-Methylenebis-(4-chlorophenol)		Cuniphen® series (Ventron) G-4 (Dichlorophene Givaudan)	Textiles, rubber products, hoses
6-Acetoxy-2,4-dimethyl-m-dioxane		Giv-Gard DXN® (Brand of Dimethoxane) Socci® 7370 (Ventron)	Metal-working fluids, textile lubricants, polymer emulsions, other aqueous emulsions
Ortho-Phenylphenol		DOWICIDE® 1 ANTI-MICROBIAL	Metal-working fluids, polishes, adhesives, gums, latexes, textiles
Ortho-Phenylphenate (sodium-o-phenylphenate tetrahydrate)		DOWICIDE® A ANTI-MICROBIAL (Dow)	
Pentachlorophenol	88	DOWICIDE® EC-7 ANTIMICROBIAL (Dow)	Wood preservation
2,3,4,6-Tetrachlorophenol	12		
Sodium pentachlorophenate		DOWICIDE® G-ST ANTIMICROBIAL (Dow)	Paper and paper products leather, hides, drilling muds
1-(3-Chloroallyl)-3,5,7-triaza-1-azoniaadamantane chloride with sodium bicarbonate		DOWICIL® 75 PRESERVATIVE (Dow)	Adhesives, metal-working fluids, latex paints, textile emulsions, water-based coating formulations
2,2-Dibromo-3-nitrilopropionamide (DBNPA) (20%, 10%, or 5%) in poly(ethylene glycol)		Dow Antimicrobials 7287 8536 and XD-8259 ANTIMICROBIAL (Dow) Biosperse® 240, 244 (Drew)	Metal-working fluids, oil recovery
1,2-Dibromo-2,4-dicyanobutane		Tektamer® 38 (Kerck)	Aqueous paints, latex emulsions, joint cements, adhesives
2-(4-Thiazolyl)-benzimidazole		Metasol TK 100® Merck TK-50° 50% (Merck)	Paint

Table 1 (cont.)

Composition	Per cent	Trade names	Application
1,2-Benzisothiazolin-3-one	30	Proxel® CRL (ICI)	Adhesives, latex, paper coatings, aqueous emulsions
4-(2-Nitrobutyl)morpholine 70%; 4,4′-(2-ethyl-2-nitrotrimethylene) dimorpholine	20	Bioban P-1487® (IMC)	Metal-working fluids, petroleum production, jet fuels
Tris(hydroxymethyl) nitromethane		Tris Nitro® (IMC)	Oil-in-water emulsions
Sodium dimethyldithiocarbamate	27.6	Vancide® 51 (Vanderbilt)	Paper and paperboard, cotton fabrics, paste, wood veneer, cutting oils
Sodium 2-mercaptobenzothiazole	2.4		
Zinc dimethyldithiocarbamate	87.0	Vancide® 51Z (Vanderbilt)	Adhesives, paper and paperboard, textiles
Zinc 2-mercaptobenzothiazole	7.5		
N-Trichloromethylthio-4-cyclohexene-1,2-dicarboximide		Vancide® 89 (Vanderbilt)	Polythylene, paint, paste, rubber,and rubber-coated products
N-(Trichloromethylthio)phthalimide		Fungitrol® 11 (Tenneco)	Nonaqueous paints and caulking compounds
2-(Thiocyanomethylthio)benzothiazole	30	Busan® 30 (Buckman)	Wood
Potassium N-hydroxymethyl-N-methyldithiocarbamate	32	Busan® 52 (Buckman)	Water-thinned colloids, emulsion resins, emulsion paints, waxes, cutting oils, adhesives
Sodium 2-mercaptobenzothiazole			
2-(Thiocyanomethylthio)benzothiazole	32	Busan® 74 (Buckman)	Paint films
2-Hydroxypropyl methanethiosulphonate			
Hexahydro-1,3,5-triethyl-s-triazine		Vancide® TH (Vanderbilt)	Cutting oils, synthetic rubber latexes, adhesives, latex emulsions
Benzylbromoacetate		Merbac-35® (Merck)	Paint raw materials (cellulose, casein)
Tetrahydro-3,5-dimethyl-2H-1,3,5-thiadiazine-2-thione and blends		Metasol® D3T (Merck) Biosperse® 210 (Drew)	Clay slurries, adhesives, glues, latex emulsions, casein, titanium slurries
5-Hydroxymethoxymethyl-1-aza-3,7-dioxabicyclo-	24.5	Nuosept® 95 (Tenneco)	Latex paints

Table 1 (cont.)

Composition	Per cent	Trade names	Application
5-Hydroxymethyl-1-aza-3,7-dioxabicyclo(3,3,0)octane	17.7		
5-Hydroxypoly methyleneoxy (74% C_2, 21% C_3, 4% C_4, 1% C_5) methyl-1-aza-3,7-dioxabicyclo(3,3,0)octane	7.8		
Organic mercurials		Various	Paints
Ethyl p-hydroxybenzoate		Ethyl Parasept® (Tenneco and others)	Adhesives, starch and gum solutions, inks, polishes, latexes, other emulsions
Methyl p-hydroxybenzoate		Methyl Parasept	
Propyl p-hydroxybenzoate		Propyl Parasept	
Butyl p-hydroxybenzoate		Butyl Parasept	
Sodium salt of 2-mercaptobenzothiazole		NX-84 (Tenneco)	Adhesives, textiles, paper, rug backings, waxes
Chlorethylene bisthiocyanate	10	Cytox® 3810 (American Cyanamid)	Water systems, emulsions
Zinc 2-pyridinethiol-N-oxide	20	Ormadine® 645 (Olin Corp.)	Water-based latex paints and emulsions, joint cement, polyvinylchloride acrylic water-based paints, polyvinyl acetate latex, adhesives, styrene-butadiene, water-based latex paints
5,4′-Dibromosalicylanilide	36		
3,5,6′-Tribromosalicylanilide	36		
Other brominated salicylanilides	7		
Zinc 2-pyridinethiol-1-oxide powder aqueous dispersion	95 48	Zinc Omadine (Olin Corp.)	Aqueous based metal-working fluids, plastics
Sodium 2-pyridinethiol-1-oxide powder aqueous solution	90 40	Sodium Omadine (Olin Corp.)	Aqueous-based metal-working fluid systems, vinyl, latex emulsions for short-term, in-can inhibition of bacterial growth
2,2′-Oxybis-(4,4,6-trimethyl-1,3,2-dioxaborinane)-2,2′(1-methyltrimethylenedioxy)-bis-(4-methyl-1,3,2-dioxaborinane)	95	Biobor® JF (U.S. Borax and Chem. Corp.)	Hydrocarbon fuels, boat and ship fuel and marine storage, home heating fuel

with modern-day environmental legislation. Specifications in this legislation obviously vary from one country to another. Here, the literature is most comprehensive for the United States, and Engler (1980) has written an informative article on United States regulations pertaining to use of industrial biocides.

The number of biocides retailed commercially is huge. C. H. Kline and Co. Inc. of Fairfield, New Jersey, U.S.A. recently compiled a list (Galvanek et al., 1979), and a list presented by Sharpell (1980), and compiled in mid-1979, is reproduced in derived form in Table 1. Again, information on individual types of biocide is most complete for the United States where, some ten years ago, phenolic industrial biocides were the commonest in use, followed by halogenated compounds, inorganic biocides, quaternary ammonium derivatives, organosulphur compounds and organometallics (Marouchoc, 1979). The order may now well have changed somewhat, particularly following the imposition of stricter environmental legislation. Marouchoc (1979) reviewed phenolic industrial biocides, and Petrocci et al. (1979) did likewise for quaternary ammonium biocides. Legislation in the United States has done much to restrict application of mercury-containing biocides, but their application in protective coatings has been reviewed by Machemer (1979).

The number of tests required to evaluate a biocide must, today, seem also limitless. In the United States, several organizations are involved, including the Association of Official Agricultural Chemists (AOAC), the American Association of Textile Chemists and Colorists (AATCC) and the American Society for Testing and Materials (ASTM), together with the Environmental Protection Agency (EPA). Sharpell (1979) has provided an informative account of the manner in which the ASTM develops protocols for testing antimicrobial agents.

C. The Future

Both Block (1979) and Johnson (1980) have recently peered into the crystal ball to see what the future holds for industrial biocides. Neither was very hopeful of a healthy growth certainly in the early 1980s, both stressing the disincentive nature of much of the recently introduced legislation both in the United States and in other countries. Nevertheless, Johnson (1980) referred to an estimated growth of 4–6% per annum, based on weight of industrial biocides. It might be argued that, in the short term with continued economic difficulties in many countries of the World, there will be an increased need for industrial biocides as materials are replaced less frequently. In the long term, one strategy must be to attempt to manufacture materials that inherently are less prone to microbial biodeterioration. Meanwhile, there will doubtless be modifications made to individual biocides such as to render them less toxic. At least two suggestions have appeared which might decrease unwanted persistence of toxic biocides. One is that a biocide may be immobilized, so that some control can be exercized on its access to the commodity that requires protection. Another is that a chemical self-destruct mechanism might be built into biocide formulations. There will undoubtedly be fewer new biocides introduced onto the commerical market in the foreseeable future. However, there will certainly be no shortage of problems for the economic microbiologist concerned with protecting resistant materials against microbial attack.

REFERENCES

Aharonowitz, Y. (1980). *Annual Review of Microbiology* **34**, 209.
Ashworth, J. M. and Smith, J. E., eds. (1973). *Symposium of the Society for General Microbiology* **23**, 1.
Ayres, J. C., Mundt, J. O. and Sandine, W. E. (1980). 'Microbiology of Foods'. W. B. Freeman, San Francisco.
Bernstein, L. (1963). *American Journal of Botany* **50**, 360.
Block, S. S. (1979). *Developments in Industrial Microbiology* **20**, 81.
Christian, J. H. B. and Waltho, J. A. (1962). *Journal of Applied Bacteriology* **25**, 369.
Corry, J. E. L. (1973). *Progress in Industrial Microbiology* **12**, 73.
Dallyn, H. and Fox, A. (1980). *In* 'Microbial Growth and Survival in Extremes of Environment'. Society for Applied Bacteriology Technical Series No. 15 (G. W. Gould and J. E. L. Correy, eds.), pp. 129–139. Academic Press, London.
Doelle, H. W. (1975). 'Bacterial Metabolism', 2nd edition. Academic Press, New York.
Eggins, H. O. W. and Oxley, T. A. (1980). *International Biodeterioration Bulletin* **16**, 53.
Engler, R. (1980). *Developments in Industrial Microbiology* **21**, 117.
Frazier, W. C. and Westhoff, D. C. (1978). 'Food Microbiology', 3rd edition. McGraw-Hill, New York.
Galvanek, T. E., Payne, R. A. and Stelka, J. P. (1979). 'Biocides U.S.A. 1979'. C. H. Kline and Co. Inc., Fairfield, New Jersey.
Gilbert, R. J., ed. (1975). 'Microbial Aspects of the Biodeterioration of Materials'. Academic Press, London.
Hunt, G. M. and Garratt, G. A. (1953). 'Wood Preservation'. McGraw-Hill, New York.
Jay, J. M. (1970). 'Modern Food Microbiology'. Van Nostrand, New York.
Johnson, R. L. (1980). *Developments in Industrial Microbiology* **21**, 141.

Levi, J. D., Shennan, J. L. and Ebbon, G. P. (1979). *In* 'Microbial Biomass' (A. H. Rose, ed.), pp. 361–419. Academic Press, London.

Machemer, W. E. (1979). *Developments in Industrial Microbiology* **20**, 25.

Mandelstam, J. and McQuillen, K. (1973). 'Biochemistry of Bacterial Growth', 2nd edition. Blackwells Scientific Publications, Oxford.

Marouchoc, S. R. (1979). *Developments in Industrial Microbiology* **20**, 15.

Meyer, K. K. (1962). 'Disinfected Mail'. The Gossip Printery Inc., Holton, Kansas.

Moreau, C. (1979). 'Moulds, Toxins and Food' (translated and edited by M. O. Moss). John Wiley and Sons, Chichester.

Oxley, T. A., Allsopp, D. and Becker, G., eds. (1980). *Biodeterioration. Proceedings of the Fourth International Symposium, Berlin*. Pitman, London.

Paigen, K. and Williams, B. (1970). *Advances in Microbial Physiology* **4**, 251.

Peberdy, J. F. (1980). 'Developmental Microbiology'. Blackie, Glasgow.

Petrocci, A., Clarke, P., Merianos, J. and Green, H. (1979). *Developments in Industrial Microbiology* **20**, 11.

Rose, A. H. (1976). 'Chemical Microbiology. An Introduction to Microbial Physiology', 3rd edition. Butterworths, London.

Sharpell, F. H. (1979). *Developments in Industrial Microbiology* **20**, 73.

Sharpell, F. H. (1980). *Developments in Industrial Microbiology* **21**, 133.

Sharpley, J. M. and Kaplan, A. M., eds. (1976). *Proceedings of the Third International Biodegradation Symposium*. Applied Science Publishers Ltd., Barking.

Smith, D. W. (1978). *In* 'Microbial Life in Extreme Environments' (D. J. Kushner, ed.), pp. 369–380. Academic Press, London.

Smith, J. E. and Berry, D. R. (1974). 'An Introduction to the Biochemistry of Fungal Development'. Academic Press, London.

Walters, A. H. (1971). *Progress in Industrial Microbiology* **10**, 179.

Walters, A. H. and Elphick, J. S., eds. (1968). '*Biodeterioration of Materials, Volume 1*'. Applied Science Publishers Ltd., Barking.

Walters, A. H. and Hueck-Van Der Plas, E. H., eds. (1972). '*Biodeterioration of Materials, Volume 2*'. Applied Science Publishers Ltd., Barking.

2. Wood

J. F. LEVY and D. J. DICKINSON

Timber Technology Section, Department of Botany, Imperial College, University of London, London, England

I. Introduction	.	19
II. Degradation of Wood	.	21
III. Biodeterioration of Wood	.	21
A. Marine Borers	.	22
B. Insects	.	22
C. Micro-Organisms	.	23
IV. Factors Affecting Decay	.	28
A. Wood Structure	.	28
B. Moisture Relationships	.	31
C. Nutrients and Nitrogen Relationships	.	33
V. Effects of Decay	.	34
A. Soft Rots	.	35
B. Wood-Rotting Basidiomycetes	.	36
VI. Mechanisms of Decay: Enzymes and Enzyme Action	.	37
VII. Where Decay Occurs	.	39
A. Living and Dead Wood	.	40
B. Infection of Wood in Standing Trees	.	40
C. Post-Harvest Deterioration	.	42
D. Packaged Timber in Transport	.	
E. Damage in Service	.	45
VIII. Colonization of Wood and the Effect of Wood Preservatives	.	51
References	.	56

I. INTRODUCTION

It is common knowledge that wood and the timber derived from it can decay. It is equally freely expressed that wood is the one structural material used extensively by man, that can be regarded as a renewable resource. What is not so well known is the amount of timber and wood products imported annually into the United Kingdom. In 1979, this amounted to £2,754,000,000 (Campbell, 1980); or over 90% of the total wood used in the United Kingdom. To offset and lower this enormous burden on the annual national balance of payments, two principal lines of action may be attempted; an increase in the forest area in the United Kingdom, as outlined by the Centre for Agricultural Strategy in a recent report (Anon, 1980); the second is the minimization (if not elimination) of decay of timber in service. The outcry from all sides, following the publication of the Centre for Agricultural Strategy Report 6, against any increase in forest area is a clear indication of the problems to be overcome before any of the recommendations in the report can be put into effect. Prevention of decay by choice of durable wood species, good design detail or addition of toxic chemicals to wood has been practised from time immemorial. The railway system in the United Kingdom has run on rails laid on preservative-treated wooden sleepers almost since railways began; the United Kingdom wood-preservation industry is a world leader in the science, technology and practice of wood preservation; the British Wood Preserving Association celebrated its golden jubilee in 1980.

Prevention of decay or deterioration of wood requires an understanding of the processes involved, which includes the organisms, the factors affecting their activity, what they do to the wood cell walls, their possible degradation mechanisms and enzyme action, from the standing tree in the forest to the final timber product in service, where the onset of decay occurs.

The carbon and nitrogen cycles are examples of the recycling processes that occur in nature by which complex organic remains are broken down into their simplest forms which serve as nutrients to be utilized by other plants and subsequently by other animals in the course of their growth and development. Wood in the form of twigs, branches and stems of forest trees which have fallen on the forest floor are simply part of these recycling processes. In attempting to remove some of the useful wood from the natural cycle and to utilize it as timber in his tools, buildings and structures, man has learned, often to his cost, that micro-organisms do not readily give up their food source. Wood, being aesthetically pleasing, light, strong, easy to work and, at least in former days, readily to hand, is used in a wide variety of environments and still remains one of the most useful structural materials available. Since, however, it remains a potential food source for micro-organisms a complete appreciation of its degradation is essential if it is to be used to its full effectiveness.

II. DEGRADATION OF WOOD

Although the biodeterioration of wood is the most important of the means by which wood in service is degraded, it should not be forgotten that other means exist. Wood can be destroyed partly or completely by fire, and chemical or mechanical agencies. Fire will ultimately convert the wood to carbon dioxide and water with a relatively small residual ash. It should never be forgotten that authorities, such as the Food and Agriculture Organization of the United Nations and the International Union of Forest Research Organizations, have estimated that about half the World's forest crop is burnt for fuel or firewood. Fire can also be the limiting factor in nature, in areas where fires are common.

Wood can also be destroyed by both acids and alkalis, some species being less susceptible than others. The years immediately following the Second World War, saw the introduction of alkaline detergents while draining boards in domestic kitchens were being made of wood. The interaction between the two produced a black gelatinous mass which was not appreciated by housewives.

Destruction of wood by mechanical means requires little explanation, although it is not always appreciated that abrasion can accentuate deterioration of timber during the course of decay. As much as 50% of log volume can be lost by sawing, resawing, planing, sanding and moulding into components for timber products.

III. BIODETERIORATION OF WOOD

Wood can be destroyed by many types of organism and when considering the biological agencies involved, it is not only the micro-organisms alone, but also higher forms which are associated in one way or another with micro-organisms.

A. Marine Borers

Marine borers, such as shipworm (e.g. *Teredo navalis*) or gribble (*Limnoria* spp.), can be very destructive to wooden boats and wooden structures in the sea, such as wharves, jetties, breakwaters and groynes. The literature is inconclusive as to whether it is necessary for some degree of fungal decay to occur before the larvae can settle and bore beneath the surface, or whether the organisms will burrow into sound timber. Nevertheless, some fungal decay does appear to encourage successful colonization and there is a symbiotic flora in the gut (Oliver, 1962; Bletchly, 1967; Jones and Eltringham, 1968).

B. Insects

Two main groups of insects are destroyers of wood, namely members of the Isoptera or Termites and wood-boring members of the Coleoptera or wood-boring beetles. In both groups, the insects are frequently associated, in some form or other, with bacteria or fungi, either as gut flora, food for nuturing young larvae or nymphs; or partially to decay the wood before infestation by the animals.

Termites are widespread throughout the tropics and sub-tropics and comprise a very large number of species which may be divided into two main groups, namely dry wood termites and subterranean termites. The latter group contains species with a well-defined colonial existence which include in part of their nest a region known as a 'fungal garden' where fragments of wood are brought by the adults to develop fungal decay in the moist ground. Newly hatched larvae are brought to feed on the fungi until they are of sufficient size and age to fend for themselves (Anon, 1969; Harris, 1971; Hickin, 1963).

Of the wood-boring beetles, five groups are noteworthy in the United Kingdom, namely ambrosia beetles, the common furniture beetle, the death watch beetle, the house longhorn beetle, and powder post beetles. Some show symbiosis with micro-organisms. Ambrosia beetles are widely distributed throughout the world. The adults burrow into dead branches, fallen trees and logs to lay their eggs. They carry fungal spores with them which brush off their bodies as they move through the galleries. The spores germinate and produce a fungal mycelium

on which the young larvae feed after hatching and which causes discolouration and partial decay of the wood around the galleries. The
adults, after pupation are infected with fungal spores by the time they
emerge from the wood (Baker, 1963; Baker and Kreger-Van-Rij,
1964; Graham, 1967). Norris (1966, 1976) has shown that the ambrosia
beetle, *Xyleborus sharpi* contains a symbiotic gut flora as well as the type
of symbiosis described by Baker (1963) and Francke-Grosmann (1966a
and b).

Baker and Bletchly (1966) calculated that larvae of the common
furniture beetle (*Anobium punctatum*), after burrowing through wood,
appear to contain more nitrogen than could be accounted for by the
quantity obtained from the volume of wood that passed through
their gut (Bletchly, 1969). They suspected the presence of microbial
symbionts, some of which might have been capable of fixing atmospheric nitrogen, but were unable to prove it (Baker *et al.*, 1970).

Larvae of the death watch beetle (*Xestobium rufo-villosum*) were shown
by Fisher (1940) to continue to burrow slowly through sound oak wood
for 10 years without any sign of pupation but, if the oak was partially
decayed by a wood-rotting fungus, pupation occurred after about two
years and the adult beetles would emerge three years after egg-laying
(Fisher, 1941; Hickin, 1963). The powder post beetle (*Lyctus brunneus*),
on the other hand, like the house longhorn beetle (*Hylotrupes bajulus*),
is a sapwood borer. Wilson (1933) and Parkin (1940) demonstrated
that the larvae were destroying the wood to reach the starch food
reserves of the tree, stored in the ray parenchyma cells. Whether or
not symbiosis exists in these two groups has not been demonstrated.
Wood-boring insects must, therefore, receive a mention in any account
of microbial deterioration of wood, since many of them exhibit a form
of symbiosis with bacteria or fungi or both, without which they are
unable to complete their life cycle.

C. Micro-Organisms

Marine borers and insects are large by comparison with the size of a
softwood tracheid or hardwood fibre and, by grinding or biting their
way through wood, these organisms form tunnels or galleries many
times the diameter of a single cell, in contrast to micro-organisms which
act directly on the individual components of the wood cell wall. Micro-

organisms use extracellular enzymes to destroy the wood cell walls *in situ* and are thus acting on a molecular scale. Their activities have, therefore, to be studied at much higher magnifications and can be divided into two phases, that of passive colonization and of active destruction of part of the wall. The colonization phase involves many micro-organisms from a wide range of taxonomic groups. As will be discussed later, individual species may act in different ways under different circumstances and a pattern of progression, if not succession and association, only becomes apparent when they are assigned to a series of ecological niches or physiological groups (Clubbe, 1980a, b). Clubbe recognizes six such groupings one of which is further sub-divided into two, as follows: bacteria, primary moulds, stainers, soft rots, wood-rotting basidiomycetes (white rots and brown rots) and secondary moulds. These groups require definition.

1. Bacteria

Clubbe (1980a, b) records that mainly Gram-positive rods are present in exposed wood, but many other cultural forms have been noted. Rossell *et al.* (1973), in a literature review relating to the presence, action and interaction of bacteria in wood, suggest that identification of isolates is no easy task since characters can vary *in vivo* and *vitro*. They note that *Bacillus* spp. have been found frequently (McCreary *et al.*, 1965) particularly *B. asterosporus* (Hejmanek and Rypacek, 1954), *B. cereus* and *B. megaterium* (Ellwood and Ecklund, 1959; Knuth, 1964), *B. macerans* (Ellwood and Ecklund, 1959), *B. subtilis* and *B. pumulus* (Knuth, 1964). *B. polymyxa* was found in the greatest frequency by Knuth (1964), MacPeak (1963), Greaves (1966), Ellwood and Ecklund (1959) and Knuth and McCoy (1962).

Gram-positive rods of *Corynebacterium humiferum* have been reported and those of *Clostridium* spp. have been found to ferment cellulose by Smyth and Obold (1930), and Norman and Fuller (1942). *Clostridium omelianski* was identified by Karnop (1967), Imschenezki (1959) and Liese and Karnop (1968).

Sarcina spp. have been found by Ellwood and Ecklund (1959), *Micrococcus* spp. by Lutz *et al.* (1966) and those of *Staphylococcus* spp. by Knuth (1964).

Members of the genus *Vibrio* have been found to utlize cellulose (Siu, 1951; Norman and Fuller, 1942). The minor genus *Celivibrio* also utilizes cellulose (Smyth and Obold, 1930). Hawker *et al.* (1960) and Enebo (1951) found it in soil. *Desulphovibrio desulphuricans* was isolated from pine by Knuth and McCoy (1962). Greaves (1970) identified a *Spirillum* sp. Many Gram-negative rods, for example of *Xanthomonas* spp. (McCreary *et al.*, 1965) and *Serratia* and *Flabovacterium* spp. (Greaves, 1970), have been found either to inhabit wood or to utilize cellulose. *Pseudomonas* spp. have been found by McCreary *et al.* (1965), Kelman and Cowling (1965), Lutz *et al.* (1966) and Greaves (1970), and *Pseudomonas synxanthaceae* was isolated by Knuth (1964), although he considered it to be unimportant, and *Pseudomonas solancearum* by Kelman and Cowling (1965). *Cellulomonas* spp. were found to attack cellulose (Smyth and Obold, 1930; Freeman *et al.*, 1948). *Aerobacter aerogenes* and *Aerobacter cloacae* were recognized by Knuth (1964).

Clubbe (1980a, b) includes actinomycetes in this group and cites species of *Streptomyces* which often show a random distribution. *Micromonospora* sp. was found by Harmsen and Nissen (1965) and rods of *Cytophaga*, an intermediate between bacteria and myxobacteria, are well known for utilizing cellulose (Siu, 1951; Rogers, 1962). The activities of these organisms in relation to wood decay is under active investigation at the present time (Baecker and King, 1980a, b; Cavalcante and Eaton, 1980; Cavalcante, 1981).

Banks and Dearling (1973), Boutelje and Bravery (1968), Dunleavy *et al.* (1973), Greaves and Levy (1965), Harmsen and Nissen (1965), Levy (1971), Liese (1970), McQuire (1970), Rossell *et al.* (1973) and Shigo (1966), have all presented evidence for the action of bacteria on the woody cell wall and in particular the membrane of the crossfield pits and bordered pits. Clear evidence has been presented to show that bacteria, by pectinase or cellulase activity or a combination of both, can make holes through or totally destroy the pit membranes. Most of the evidence has been accumulated from ponded wood(i.e. wood stored in water), but Levy (1975) has reported that a similar phenomena occurs at the ground line in timber in ground contact.

In ponded wood, the increase in permeability brought about by the partial or total destruction of the pit membranes has been shown by Dunleavy and McQuire (1970) and Banks and Dearling (1973) to be rapid. The process in wood in ground contact is slower, but opening up the pit membranes, gaseous exchange becomes easier to a greater depth into the wood and this may have two marked effects. Firstly, the conditions inside the wood become less anaerobic and therefore suitable for fungal growth, and secondly there are now open pathways for fungal hyphae to pass through walls from cell to cell whilst effecting passive colonization of the wood.

Sharp and Millbank (1973) demonstrated that nitrogen fixation can occur in wood in ground contact. Bacteria isolated from wood have also been shown to fix atmospheric nitrogen at some distance beneath the surface layers of the wood. Whether these bacteria take an active part in destruction of the pit membranes is not known, but wood in ground contact for periods of days, weeks or years has been shown to contain nitrogenase activity associated with nitrogen fixation.

Levy (1975) also suggested that bacteria may be either synergistic or antagonistic organisms, and Levy and Greaves (1966) were clearly of the opinion that an association between soft-rot fungi and bacteria caused a more rapid destruction of softwood in underground mines.

The activities may be summarized as follows. (a) A progressive breakdown of the pit membranes in sapwood. This increases the permeability of the wood, allows gaseous exchange and free movement of water from lumen to lumen and gives access to fungi not capable of destroying cell wall or pit membrane (Levy, 1973, 1975). (b) Fixation of atmospheric nitrogen, which increases progressively from an exposed surface into wood as colonization proceeds (Baines and Millbank, 1976; Levy et al., 1974; Baines et al., 1977). (c) Antagonism or synergism with fungal colonists (Levy, 1975). (d) As a result of their colonization, bacterial cells and their metabolic products will be present in the wood which is thus changed by their mere presence. (e) Bacteria can also destroy the wood cell wall in both sapwood and heartwood (Levy and Greaves, 1966), especially under anaerobic conditions such as in wooden piles (Boutelje and Bravery, 1968). This phenomenon has been extremely difficult to demonstrate in the laboratory (Schmidt, 1981), although T. Nilsson (personal communication) claims to have at least one bacterial isolate capable of attacking the wood cell wall.

2. Primary Moulds

This group is comprised of fungi that do not appear able to degrade the wood cell wall and yet are usually the first fungal colonizers of wood. They can be regarded as being closely akin to Garrett's sugar fungi (Garrett, 1951, 1955). Their food source is likely to be sugars or simple carbohydrates present in parenchyma cells of the sapwood or brought by water movement to other parts of the wood. Being unable to utilize the wood cell wall, they can only penetrate into wood through natural openings, such as cut cells or through openings made by other organisms, such as the open pits after bacteria have destroyed the pit membranes. Any organism capable of fast growth and acting as a 'sugar fungus' is represented in this group. Phycomycetes, ascomycetes and Fungi Imperfecti are all included.

3. Stainers

This group is characterized by pigmentation of the hyphal walls, which causes discolouration of infected sapwood. This is sometimes blue in colour and is known as bluestain or sapstain (Kaarik, 1974, 1980). They colonize the ray parenchyma cells of the sapwood utilizing the cell contents and stored food reserves of the tree as a nutrient source. Hyphae of these fungi are able to penetrate through the wood cell walls in a characteristic manner, passing from one ray to another in a horizontal tangential direction through all the intervening cells and cell walls. Penetration through the cell wall is by means of a fine constriction of the hypha. On making contact with the wall, the hyphal tip swells to form a transpressorium (Liese and Schmid, 1964), the constricted hypha grows through the wall and on emerging into the lumen of the adjacent cell resumes its normal shape and size. This group is comprised of ascomycetes and Fungi Imperfecti and many species also appear capable of causing soft rot (Krapivina, 1960; Karkanis, 1966).

4. Soft Rots

This group of fungi is also composed of ascomycetes and Fungi Imperfecti. They degrade wood by penetrating into the middle layer of the secondary wall (the S_2 layer) of the wood cell wall and forming, therein, chains of cavities (Corbett, 1965). They may also act as stainers or degrade the cell wall by direct lysis from hyphal contact from the lumen (Corbett's Type II attack), but their characteristic feature is their ability to form cavities inside the wall (Corbett's Type I attack). It is this feature which causes them to be grouped together (Kaarik, 1974, 1980). Soft-rot fungi are widespread, but their economic importance, as will be seen later, depends on their ability to colonize and destroy wood under certain circumstances where the other main types of wood-rotting fungi are inhibited.

5. Wood-Rotting Basidiomycetes

This group comprises the main wood-rotting fungi and can be divided into two sub-groups, namely the white rots and the brown rots. These were characterized by the former group being capable of completely degrading both holocellulose and lignin, whereas the latter only slightly modify lignin but destroy holocellulose. White rots bleach the wood as decay progresses, whereas the brown rots leave the lignin as a brown, brittle, friable residue (Campbell, 1952; Bavendamm, 1928). Within the last ten years, observations in the scanning electron microscope have shown up other characteristic differences between the white and brown rots principally in the way in which they destroy the cell wall (Bravery, 1971, 1975; Nasroun, 1971; Crossley, 1980), which will be described and discussed later.

6. *Secondary Moulds*

This group comprises those organisms that can be shown to possess an active cellulase system by production of a clearance zone when grown on ball-milled cellulose agar, but do not appear capable of utilizing cellulose in the undecayed wood cell wall. They have been isolated from wood that is undergoing active decay by soft, white or brown rot fungi, by Clubbe (1978, 1980a, b) who suggested that they were only capable of utilizing the cell-wall cellulose after partial breakdown of the wall by other fungi. The cellulose used by the secondary moulds might have been a true nutritional excess or the results of competition between the two groups of organism for partially decayed substrate. Clubbe (1980) names *Trichoderma viride* and *Gliocladium roseum* as the predominant secondary moulds.

IV. FACTORS AFFECTING DECAY

A. Wood Structure

Wood can be described as a structure rather than a material with a very marked anisotropy, due to elongation of cells parallel to the axis of growth in the stem of the tree or branch and a marked localized lateral elongation of cells in a radial direction by formation of the rays. This forms the basis of the anatomy of wood, the axial elongation of the fibres, vessels or tracheids forming the 'grain' of the wood. Thus the term 'along the grain' refers to the axis of the stem and 'across the grain' in the radial direction to movement along or parallel to the rays. Lateral movement 'across the grain' in a tangential direction is at right angles to the rays. The term 'end grain' denotes the transverse face of a log or stem cut at right angles to the axis or 'grain' of the wood.

This orientation of grain was shown by Corbett (1963, 1965) to have a marked effect on the rate of colonization of wood by fungi, movement along the grain being faster than movement across the grain in a radial direction, which was itself faster than that across the grain in a tangential direction. Grain orientation also has a marked effect on movement of water through timber, as will be discussed later (Baines and Levy, 1979).

Commerical timbers are divided into softwoods and hardwoods. Softwoods are all gymnosperms, and are very uniform in structure being constituted of tracheids with parenchyma restricted almost entirely to rays and resin ducts. They are, therefore, composed of cells with a complete cell wall, the only access from lumen to lumen being through cell walls or pit membranes. Their air-dry weight ranges from about 350 kg/m^3 to 720 kg/m^3. Hardwoods, on the other hand, are all angiosperms and are therefore constituted of fibres, vessels and parenchyma as well as some tracheids. The vessels consist of vessel elements which are either open ended (simple perforation plates) or at best have the open grill-like structure of scalariform perforation plates. These elements are joined end-to-end to form vessels so that access from lumen to lumen along the grain is virtually unrestricted, except where tyloses may be present, and can form an open pathway or network, for the passage of liquids or organisms. The combinations and permutations possible with the four basic cell types make the hardwoods very diverse in structure and properties, their air-dry weight ranging from 80 kg/m^3 for balsa to 1,250 kg/m^3 for lignum vitae. In some species like Oak (*Quercus robur*), the early wood vessels are very much larger in diameter than those of the late wood, and such woods are called 'ring-porous hardwoods'. Where there is little or no difference in vessel size and distribution, the species, e.g. birch (*Betula pendula*), are known as 'diffuse porous hardwoods'. In addition, the hemicellulose component of the cell wall will differ between softwood and hardwood and as Lewis (1976) has demonstrated, certain wood-rotting basidiomycetes may be able to destroy the one in preference to the other.

Wood, when first formed, performs the triple function of water conduction, mechanical strength and storage of the reserve food supplies of the tree, and is called the sapwood. After some years, the inner portion of the wood becomes converted progressively into heartwood, so that a mature tree will have a central heartwood core surrounded by a relatively thin outer region inside the cambium of sapwood. Heartwood formation involves removal of the reserve food materials stored in the ray parenchyma, and impregnation of the cell walls by additional materials which are collectively known as 'extractives'. These extractives often provide the dark colouration typical of such woods as mahogany, walnut or ebony. In the heartwood of some species, these extractives are toxic in nature and impart a natural durability as, for example, western red cedar, teak or jarrah. In these species, there will be a considerable difference in the ability of micro-organisms to destroy the sapwood and heartwood. The sapwood will be susceptible to decay with a readily available supplementary source of nutrient in the ray parenchyma and the absence of toxic extractives in the cell walls. The heartwood on the other hand, could be highly resistant to decay with the presence of toxic extractives in the cell walls and no extra sources of nutrients. The sapwood of most species is perishable, but the range of natural durability of heartwood will vary from species to species and sometimes also within a species, where a fast-grown tree of a particular sort may have less durable heartwood than a slow-grown specimen of the same species. Natural durability is often evaluated by burying stakes to half their length in soil and finding the time it takes for the wood to decay sufficiently at the ground line so that it breaks off; a species is said to be perishable when this occurs in less than five years and very durable when it remains unbroken for more than 25 years.

One other characteristic varies considerably between species and this is termed 'treatability'. This term represents the ease or difficulty with which a liquid can be impregnated into wood. Some species, such as *Pinus radiata* sapwood, take up water and wood preservatives very easily, whereas others can hardly be penetrated. In modern commercial practice, the natural durability and treatability of a particular species are important properties for effective utilization in a situation with a high decay hazard. High natural durability may be acceptable, but a timber of low durability can only be used if it can be treated effectively with a toxic chemical and this means reasonable treatability.

B. Moisture Relationships

Wood is widely described in the literature as being a hydroscopic material which will take up an equilibrium moisture content with its surroundings. The force with which water is held in the wood is then in equilibrium with the force with which water is held in its immediate surroundings, be it soil, air, plaster or brickwork. In the living tree, the outer sapwood is the main pathway for water movement from roots to leaves and any wood cut from this part of the tree at that time will have a high moisture content and is said to be in the 'green' state. The water it contains will be in the cell lumen where it is called 'free water' and in the cell walls where it is known as 'bound water'. As a piece of wood dries from the green state, the free water is lost and there is a theoretical state where all of the free water has gone and only the bound water remains. This theoretical point is known as the 'fibre saturation point'. As this piece of wood dries still further, the bound water is lost and the cell walls contract causing shrinkage to the wood as a whole. The amount of water present in the piece of wood is expressed as the percentage moisture content ($\%$ m.c.) which is the weight of water present in the wood expressed as a percentage of the oven-dry weight of the wood itself. The fibre saturation point is about 30% m.c. in most woods; air-dried wood will be about 14–18% m.c., and wood in a centrally heated building will be about 8–10% m.c.

For fungal decay to occur, the moisture content must be above 20%. If wood is kept below that moisture content, fungi cannot cause decay, which is why a house with good design details protects the wood in its structure and contents by isolating the wood from external sources of moisture, be it rising damp, a leaky roof or internal condensation. If this cannot be ensured as, for example, in the case of window joinery, then naturally durable species or treatable species to which a wood preservative has been effectively applied should be used. Uptake of water from wet surroundings may be as liquid water or as water vapour and, equally, loss of water to drier surrounds may be in either state.

The expression 'equilibrium moisture content' may give the impression that water in wood is in a static condition. This may sometimes be so, but there are many situations where the equilibrium moisture content at any one part of the wood may remain constant and yet the water itself is moving. Take the example of a post in the ground. One can assume with confidence that the basic anatomical structure of the wood is similar throughout its length. Yet, as soon as the post is erected, a vertical zonation of microhabitats quickly becomes apparent. Four moisture zones can be found (Levy, 1968). (a) Conditions of intermittent wetting and drying at the top of the post. (b) Permanent low water content above the ground. (c) Permanent high water content below ground level. (d) Excess water-producing anaerobiosis where the bottom of the post reaches the water table. These four zones will each attract a different variety, range and activity of micro-organisms, and this will progressively increase the difference between each zone. Some decay may occur in time at (a), whereas virtually no decay apart from surface discolouration will occur at (b). In both cases, although there is sufficient oxygen to support fungal growth, the short supply of water has prevented it from occurring. At (c) and (d) there will be plenty of water, but oxygen becomes the limiting factor. At the ground line there is both oxygen from above and water from below and it is here that the greatest fungal activity takes place. In more anaerobic regions below the ground only bacteria are usually present.

Baines and Levy (1979) demonstrated that a stick of wood with the grain parallel to its long axis will act as a wick if there is a water gradient between its two ends. They quantified this wick action by measuring the rate of water movement through sticks of Scots pine under different rates of evaporation. Baines (1981) has also shown that in the case of the post in the ground already referred to, the equilibrium moisture contents of the four moisture zones along its length will remain constant but nevertheless water will move from zone (d) to zone (c) and will evaporate about the boundary between zones (c) and (b). In this region materials dissolved in or carried by the water will be deposited where evaporation takes place. This is the groundline region where the most active decay occurs and, if the materials deposited are nutrients, it would appear that the activity may well reflect the fortunate circumstances that provide conditions that are wet enough and yet sufficiently aerobic to support fungal growth and provide continuous supply of additional nutrients to sustain it.

C. Nutrient and Nitrogen Relationships

Cowling (1965), discussing the nutritional aspects of wood-inhabiting fungi, pointed out the importance of the low content of nitrogen normally present in wood. He also considered the availability of cellulose in cell walls, and how far other wall constitutents prevented its breakdown. Cowling and Merrill (1966) have also discussed the nitrogen relationships of wood, ways in which it might be accumulated or conserved by the fungal colonists and its role in wood deterioration.

Greaves and Levy (1965) demonstrated the importance of readily available stored food reserves and possibly residual protoplast in the parenchyma of the rays of the sapwood for initial colonization of wood. They showed that, in laboratory experiments with monocultures of fungi and bacteria, the ray parenchyma was usually the first tissue to be substantially colonized before initiation of decay of cell walls. Levy and Olofinboba (1966), investigating the reasons for rapid colonization and subsequent discolouration of the sapwood of freshly felled logs of tropical hardwood species, showed a similar importance for the axial parenchyma in initial colonization of the wood.

King et al. (1974) have shown that water-soluble nutrients can be moved through wood and deposited near the surfaces of cut timber in normal drying and seasoning processes. Evaporation from the surface causes a movement of water and materials in solution towards the surfaces, and the solutes remain as food sources which encourage scavenging micro-organisms to colonize wood (King et al., 1976). Although, initially, wood is a nitrogen-poor environment for basidiomycetes, the nitrogen content can increase by microbial and physical effects. Sharp and Millbank (1973), Levy et al. (1974), Baines and Millbank (1976) and Baines et al. (1977) have demonstrated that nitrogen fixation can occur in wood providing it is wet enough. Baines and Millbank (1976) showed the progressive penetration of nitrogen-fixing organisms into wood from an exposed face probably via the ray parenchyma.

Uju (1979) carried out an investigation of the nitrogen relationships of wood in ground contact by adding a solution of a soluble inorganic nitrogenous salt to sterile soil into which sterile sticks of Scots pine (*Pinus sylvestris*) had been half buried. He showed that an increase in nitrogen content took place not only in the half of the stick below soil level but also in the half of the stick projecting above the soil level. He subsequently demonstrated this increase in nitrogen in the wood above ground level in a variety of conditions. He produced evidence to show that some of the increase must have been due to movement of soluble nitrogen, since it could not be correlated with the presence of any organism (Uju *et al.*, 1981). King *et al.* (1981) in studies of the total nitrogen balance of wood in soils demonstrated increases of the nitrogen content and implicated the transfer from the soil by the colonizing micro-organisms.

V. EFFECTS OF DECAY

The gross effect of the action of wood-rotting organisms is to lower the strength of wood by destruction of the main wall component responsible for this property, namely cellulose. All three groups of fungi (soft, white and brown rots) are capable of this. Wood decayed by each type may be distinguished by its own characteristics. For example, soft rot produces a pronounced softening of the surfaces when wet and a recognizable small square cracking of the surface when dry; white rot tends to bleach the wood; brown rot leaves a brown, brittle, friable residue.

At higher magnifications, the effects on cell walls are seen to be equally characteristic, soft-rot fungal hyphae forming chains of cavities in the S_2 or middle layer of the secondary cell wall, white rots forming bore holes through the wall and eroding the wall from the lumen forming grooves or channels in which the hyphae lie, brown rots producing an extracellular enzyme that diffuses into the wall from hyphae in the lumen and destroys the cellulose without the hypha penetrating into the wall. Of the other groups of organism, only bacteria and stainers appear to have an effect on the cell wall which has already been described. Corbett (1963, 1965) proposed the term 'micro-morphology' for these types or patterns of destruction of the cell wall. The micro-morphology of the three rot types has given rise to a series of hypotheses as to the way in which their enzymes may act and recent work with the electron miscroscope, substantiated by biochemical investigation, has gone a long way to suggest mechanisms for decay (Crossley, 1980) and possible relationships that might exist between the microdistribution of wood preservatives in the wall layers and the decay mechanisms of fungi (Dickinson and Levy, 1979).

A. Soft Rots

The micromorphology of these organisms had been illustrated by Schacht (1863) in the middle of the last century, and much later by Bailey and Vestal (1937). Their economic importance as wood-decay organisms was not, however, appreciated until Findlay and Savory gave an account of their activities in 1950. Savory (1954a, b, 1955) published a series of papers which established the term 'soft rot' and demonstrated that these fungi were responsible for degradation of the timber fill in water-cooling towers which had hitherto been thought to be due to chemical breakdown. He reported on aspects of the physiology and life history of the organisms and pointed out that they became established where there was little competition from other wood-rotting micro-organisms.

Savory (1954a) showed that soft-rot fungi grew in the S_2 layer of the secondary cell wall, forming chains of cavities with pointed ends which were arranged in a helix parallel to the orientation of the cellulose microfibrils. Corbett (1963, 1965) was the first to observe that a short side branch of a hypha lying in the cell lumen penetrated the S_3 layer perpendicular to the wall and, on reaching the S_2 layer (or having passed into the S_2 layer of the adjacent cell wall), appeared to turn at right angles to lie parallel to the cellulose microfibrils and at the angle, branched at 180°, to form what she termed a 'T-branch'. A cavity was formed by wall lysis round the hyphal branches, forming the cross of the T and the hypha then increased in girth. Fine hyphal strands, termed 'proboscis' hyphae (Corbett 1963, 1965), form the initial extension growth from the distal ends of the hypha in the cavity (Crossley and Levy, 1977; Levy, 1980).

Leightley (1977), Leightley and Eaton (1977) and M. Hale (personal communicaton) have studied development of proboscis hyphae and cavity formation by time-lapse photography. The proboscis hypha develops from the tip of a cavity hypha, and continues penetration through the wall in the same helix as the cavity hypha, in a direction parallel to the cellulose microfibrils. After a period of growth, further extension of this proboscis hypha ceases while it increases in girth, and a new cavity forms round it. After some time, extension growth begins anew by development of another proboscis hypha, and the process is repeated time and again. A hypha in a cavity appears to be capable of forming a new T-branch and so of setting up a new helical chain of cavities. The cavity itself is straight-edged even when viewed under a transmission electron microscope (Findlay, 1970). This suggests that the action of enzymes in forming the cavity is very exact and at no time do they diffuse freely into the wood cell wall beyond the sharp edge of the cavity. The pointed ends of the cavity and the general orientation of the cavities, with respect to the cellulose microfibrils, have given rise to much speculation. Preston (1979) and Crossley (1980) have suggested possible mechanisms. With time for development of more helices of cavities, they begin to coalesce and, as Zainal has shown (1975), the whole of the S_2 layer can be removed with only a thin layer of the S_3 separating the cavity from the lumen.

Corbett (1963, 1965) described two micromorphological forms of decay by soft-rot fungi. Cavity formation she termed 'Type 1 decay'; the other type, 'Type 2 decay', was formed when a hypha came to lie on the inner surface of the wall and a trough was formed by lysis of the wall in the immediate vicinity of the hypha. This 'lumen erosion' trough has been described by others, notably Zainal (1975) and Crossley (1980), who, with the aid of transmission electron microscopy, have given further details of the development of this form of degrade.

B. Wood-Rotting Basidiomycetes

Until Findlay and Savory (1950) first suggested the economic importance of soft rot, Basidiomycetes were regarded as the only fungi capable of causing severe economic degradation of timber in service. Significant micromorphological differences between the white rots and the brown rots were not suspected until Bravery (1971, 1972, 1975, 1976) published a series of scanning electron photomicrographs. His observations were confirmed by Nasroun (1971).

1. White Rots

Bravery (1971, 1972, 1975, 1976) showed the hyphae of the white-rot fungus *Coriolus versicolor* penetrated into the cell lumen and, like the hypha of soft rots in Corbett's type-2 decay, came to lie on the inner surface of the cell wall. Lysis occurred along the hyphal contact with the wall forming a trough with a central ridge on which the hypha rested. New troughs were formed in association with each new branch formed from the original hypha, and by repeated branchings, the troughs coalesced to form an apparent total erosion of the wall. This erosion developed from the lumen progressively through the S_3 and S_2 layers. Where the trough was formed parallel to the cellulose microfibrils its edges were smooth, but where it cut across their orientation the edges were ragged where the ends of the cellulose microfibrils projected (Bravery, 1971; Levy, 1980). The formation of troughs or channels with well defined edges a relatively short distance from the hypha suggests that there could be some restriction to free movement of extracellular enzymes involved in the lysis. This phenomenon is shown to a greater or lesser degree by other white rots; but T. Nilsson (personal communication) has also noted effects similar to those normally associated with brown rots.

2. Brown Rots

Bravery (1971) showed that although the hyphae of brown rots came to lie on the inner surface of the cell wall, no troughs or channels were formed, neither did the hypha penetrate into the walls and form chains of cavities inside. Instead, the wall surface and the hyphae change very little but the inner layers of the wall are heavily degraded, becoming brown, friable and disorganized (Bravery, 1971; Nasroun, 1971; Crossley, 1980; Levy, 1980). It gives the impression that in this type of fungus, the enzymes were capable of diffusing some distance away from hyphae and penetrating into cell walls, where the cellulose and hemicellulose is destroyed, leaving the lignin largely unaltered as residue. Fukazawa et al. (1976), using a fluorescent stain to show up breakdown products of wood decay, have confirmed the micro-morphological patterns for the three groups of wood-rotting fungi by showing that the enzyme activity of the soft and white rots is restricted within the cavity or trough, whereas that of the brown rots appears to be diffused throughout the wall.

VI. MECHANISMS OF DECAY: ENZYMES AND ENZYME ACTION

Montgomery (1980) described wood as an ideal high-calorie carbon source, which suffers from the disadvantage that it is highly polymeric (Mandels and Reese, 1957; Norkans, 1967) and thus highly insoluble. In order to be ingested as a food source by a micro-organism, it has to be broken down into small water-soluble molecules. Since wood is also a very stable mixture of polymers it will take a very long time to break down unless a suitable catalyst is present. For a fungal enzyme to act as such a catalyst, it must be capable of acting outside the fungal cell or else it cannot make contact with the wood. Montgomery (1980) makes the point that such an extracellular enzyme must, therefore, work in an uncontrolled environment outside the fungal cell and has to meet certain requirements, namely: (1) to be able to penetrate into the fine structure of the cell wall it should be small in size; (2) to be efficient it should show long-lasting activity; (3) to facilitate movement and non-rejection it should be ionically suited to its environment. Although the latter two points may be true for extracellular enzymes of wood-rotting fungi, the first would suggest the possibility that one explanation for the observed micromorphological differences could be enzyme size, those of the soft and white rots being too large to penetrate into the wall, whereas those of brown rot are small enough to do so. For this to be so, the brown-rot enzymes would have to be smaller than 4.0 nm in any one dimension, and no evidence exists to substantiate this, or indeed to suggest that any difference in size exists between cellulases of all three wood-rotting types. Montgomery (1980) sums up by stating that all wood-rotting micro-organisms use extracellular enzymes which are too large to penetrate into the native (or unaltered) cell wall.

Koenigs (1974, 1975) has shown that chemical breakdown of wood by a water–ferrous iron system was very similar to that of brown-rot decay and quite unlike soft-rot or white-rot decay. Highley (1978) concluded that there is strong evidence for brown rots being different from other groups of fungi in using a non-protein oxidative mechanism to break down cellulose in wood, and gave indications of no retention of the enzymes to hyphae. Green (1980) and Green et al. (1980) gave evidence to suggest a high degree of retention between enzyme and hyphae in the case of soft and white rots, but no retention by brown rot. Washing with an acetate buffer and a $\beta(1-3)$ glucanase lowered enzyme retention to a low level and produced an active cell-free extract. This suggested the possibility that the enzyme might be attached to the fungal cell wall or an extracellular mucilage if it was present. Crossley (1980), using the transmission electron microscope, demonstrated the presence of a substance of very similar conformation to mucilage associated with hyphae of a soft rot and a white rot in the process of wood decay. He found no such substance associated with brown-rot hyphae in the process of wood decay.

The presence of mucilage associated with the hyphae of those fungi shown to have a high retention of enzyme to hyphae, and the absence of mucilage in those fungi with no retention of enzyme, suggests the possibility that the two phenomena are associated with enzyme being retained by or within the mucilage. Formation of discrete cavities by hyphae of soft-rot fungi and troughs by white-rot hyphae suggest that limitation of the lysis zone is due to retention of the enzyme to the mucilage round the hyphae. In the absence of mucilage, brown-rot enzymes are free to move and the possibility that these fungi may also be able to use a non-protein catalyst to break up the wall structure could provide a mechanism for unrestricted diffusion of their enzymes throughout the cell wall. Crossley (1980) has postulated mechanisms for the three types of micromorphology shown to exist in the process of wood decay. These are based on the concept of retention of enzymes by mucilage where this is present, non-protein oxidative mechanism for brown-rot activity and a detailed consideration of the fine structure of the wood cell wall. The way in which the lignin is circumnavigated by the brown rots and broken down by the white rots is not yet clearly understood. At the time of writing it is felt that a clearer picture of the biochemistry of the processes is immanent which should give a better indication of how the cell wall complex is broken down.

VII. WHERE DECAY OCCURS

Once wood has been formed in the growing tree, it can be subject to microbial degradation at any time, whether as sapwood or heartwood, in the living tree. It may decay as a dead standing tree or a fallen tree, as a log after the tree has been felled, while lying in the forest awaiting extraction, during extraction to the mill for conversion into timber, during transport as a log or milled timber, during seasoning, in the merchant's yard after seasoning, at the building or manufacturing site. It may also decay in service as a finished product, be it solid wood as railway sleepers, transmission poles, fence posts, roof trusses, floor boards, external joinery, furniture and shipbuilding or reconstituted as glulam beams, plywood, chipboard and other board materials. The critical factor in every case is the moisture content for, as Montgomery (1980) points out, since the wood substrate is some finite distance (however small that may be) from the hypha, the enzyme must travel between the hypha and the substrate. This requires the presence of at least a film of water, and means that dry wood, with a moisture content less than 20% m.c., is immune to microbial attack.

A. Living and Dead Wood

In standing trees, the sapwood is considered to be much more resistant
to decay than the heartwood. Hollow trees are not uncommon in old
woodlands where the dead heartwood has been destroyed but the living
sapwood has responded to invading organisms and overcome them.
Shigo (1968) has described the response of the tree to injury and
microbial invasion which has the effect of cutting off the wood present
at the time of injury from the subsequent newly formed wood which
remains free from infection and decay. This is seen as a discolouration
of wood formed at the time of injury. New wood formed after this
remains free from discolouration and subsequent invasion and decay by
fungi. The result of these events is healthy, blemish-free sapwood
surrounding stained or decayed innerwood. Such a course of events
can take up to 50 years to develop fully. The reaction of living trees
to decay organisms and its consequences have been reviewed by Mercer
(1980).

On the other hand, once the tree has been felled and the living cells
in the sapwood die, the wood becomes an inert substrate which makes
no response of any sort to invading organisms. The difference between
sapwood and heartwood then changes completely, the sapwood becoming highly susceptible to decay because of the stored food reserves and
residual protein in the parenchyma and an absence of toxic extractives
in the walls. The heartwood, without stored food reserves and often
containing toxic extractives, is usually more resistant. Sheffer and
Cowling (1966) have reviewed the factors that influence the susceptibility and resistance of timber to decay.

B. Infection of Wood in Standing Trees

As well as demonstrating the fundamental difference in decay patterns
between living trees and converted timber, the decay in standing trees
is of major importance in the subsequent commercial use of timber.
In many instances, infected timber finds its way into use with major
internal decay patterns that markedly affect its performance in service.
In large-dimension timbers, decay in storage gives rise to 'hidden decay'
which only becomes evident when the timber is resawn (Roff et al.,

1974). In the United Kingdom, this has always been a problem with large dimensional timbers, particularly those imported from North America. In many woods, the infection may continue and decay may proceed during the service life of the timber. An example is *Poria carbonica*, which causes considerable internal damage to transmission poles in service in North America where the fungus continues to decay the poles in service. It has also been the cause of decay in the structural members of water-cooling towers in the United Kingdom above the mist eliminators. This fungus must have been present in the timber before construction of the towers, being a North American fungus not normally found in the United Kingdom (Dickinson, 1980). Further examples are the decay of the famous flagpole at Kew Gardens by the same fungus, and the presence of *Polyporus sulphureus* in the oak timbers of H.M.S. *Victory*. In the past, such infections have been associated with so-called over-mature trees. Current forestry practice has also contributed to the decay and infection of standing timber. In the United Kingdom, butt-rot of conifers caused by *Heterobasidion annosum* formerly called *Fomes annosus* is a classic example of such a problem, and serves well to illustrate how correct practice and prevention of infection is the first line of defence in protecting commercial timbers.

Heterobasidion annosum is the most serious form of disease in British forests (Anon, 1967). It normally enters the tree through the root, penetrates the heartwood and grows up the stem to damage the timber as a butt-rot. Modes of entry can also be provided by stem and root wound caused by extraction damage. Minimizing such damage is essential in modern forestry practice if good-quality uninfected timber is to be produced. By far the most important source of infection is via cut stumps or during thinning. The fungus first grows in the stump down into the roots and then into the roots of neighbouring standing trees. These trees may infect their neighbours. The rate of development of the rot and the age at which it occurs vary considerably from species to species and from site to site. In exceptional cases, attacks of a 20-year-old crop may be infected as far as 3 metres up the stem in the form of stain and 1.5 metres in the form of decay. Generally speaking, if infection is present throughout the crop at an early age, serious damage may be expected by the time the crop is felled and ready for conversion into commercial timber. As when preventing infection by damage to standing trees during extraction of logs, infection through cut stumps can also be avoided by good forestry practice (Anon, 1967). In pine crops, *H. annosum* is often

checked when stumps are naturally infected by *Phlebia gigantea* formerly
called *Peniphora gigantea*. Rishbeth (1952) was able to show that inoculation with spores of *P. gigantea* gave good results in controlling infection
by *H. annosum*. *Phlebia gigantea* is now available commercially as a
suspension of spores and is used widely to control *H. annosum* (Greig,
1976). Fungi such as *Polyporus shweinitzii* and *Armillaria mellea* are also
important in causing decay in standing commercial timbers in British
forests but *H. annosum* is responsible for roughly 90% of decay in conifers
in Great Britain (Anon, 1967) much of this degrade in standing commercial timbers can be eliminated and prevented by good forestry
practice.

C. Post-Harvest Deterioration

In considering post-harvest deterioration, two distinct problems must
be dealt with. In log-form and in large-dimension timbers, drying may
take many months. In such cases, development of internal decay is very
important. Much of this decay may be a continuation of infection
carried over from the standing tree but, in all instances, green timber
which is held for any length of time will be at risk to post-harvest
infection by decay organisms. It is often desirable to hold large quantities of saw logs in order to ensure a continuous supply at mills. Also,
at certain times, large quantities of valuable saw logs become available
after widespread windblows. Such logs have to be stored in such a way
as to prevent deterioration. A common practice is to 'waterstore' such
timbers. Under these very wet conditions, normal decay organisms
cannot survive, and logs can be stored for long periods without apparent
deterioration. Water storage of green poles is a further example of
preventing decay by creating conditions unsuitable for growth of wood-
rotting fungi. Water-stored timber, however, is not completely immune
to microbiological degradation. Under such conditions, bacterial degradation of the pit structures of the sapwood can occur. This destruction
of the pits of species such as Scots pine causes a phenomenon known
as 'over absorbency'. Owing to removal of the pit membrane, the
sapwood is rendered very porous. In Scots pine, this can have both
beneficial and detrimental effects. In round timber likely to be creosoted
for use as transmission poles, this increased permeability removes the
risk of subsequent 'bleeding' of preservative after treatment. In joinery
timbers, however, where moisture uptake is a vital part of the decay

process, this absorbent timber can cause a much more rapid wetting-up of the timber. In treated joinery, there is a further problem due to excessive uptake of the commonly used organic solvent preservatives. Such timber does not dry off after treatment and leads to major painting problems during manufacture of the joinery unit. In refractory species, there has been considerable interest in recent years in using the 'ponding' process to open up the pits, increase the treatability of the wood, and make subsequent treatment with preservative possible (Dunleavy and McQuire, 1970; Dunleavy *et al.*, 1973; Fowlie, 1981). Control of such a process could make utilization of home-grown spruce for transmission poles a viable proposition in various parts of the United Kingdom.

As soon as any tree is felled, the sapwood is readily colonized by a wide range of fungi. The earliest colonizers are usually those fungi that obtain nourishment from residual food materials in the sapwood rather than those which degrade the cell walls. This situation is almost exactly paralleled in the infection of standing trees through wounds (Shigo, 1968) and the colonization of sound timber when placed in the ground (Levy, 1980) or when it becomes exposed to wetting in a situation such as external window joinery (Carey, 1980). Some of these early colonizers are stainers, which give rise to a blue or grey discolouration in the sapwood.

The sapwood of imported pine from Baltic regions, home-grown *Pinus sylvestris* and Southern pines serve well to illustrate the problem in reference to commercial timber imported into the United Kingdom. Softwood timbers converted from small trees now contain large proportions of sapwood. This is of major importance as regards decay problems in service but, as regards blue-stain, this high proportion of sapwood and the speed with which the fungus colonizes the wood have created a major problem in insuring clean blemish-free timber. Although the 'sapstain' fungi do not appreciably lower the strength of the wood, the unsightliness of severely stained timber can decrease its sale value by as much as 20% (Savory, 1971).

Sapstain is a problem at two stages after felling. In the log form, stain readily enters from areas where the bark is damaged due to the activity of bark-boring beetles. Such infections initially lead to wedge-shaped growths which rapidly spread and cause internal staining of the sapwood. This problem can be relatively easily avoided by preventing excessive bark damage and converting logs as soon as possible after felling or by water storage. The risk of log stain is much greater in summer and is dependent on the temperature. In Northern Scotland, for instance, stain in logs does not appear to be a major problem provided they are converted within two or three months. On Thetford Chase, which is in eastern England, a similar time period gives rise to major problems. In southern pines, such as the Portuguese maritime pines, staining is a major problem and logs are often timed to arrive at mills within days of felling to avoid logstain. Logstain, therefore, can again be controlled by good practice and management. But the problem at present is increasing due to the introduction of automatic felling machines. In Sweden, a 60% increase in logstain has occurred due to mechanization (B. Henningson, personal communication) necessitating the use of prophylactic treatments on logging machines in the forest.

Once the log is converted into planks, the problem of blue-stain greatly increases. Normal air-seasoning practices often prove inadequate to control it. In such cases, kilning without delay provides an answer, but timber is often held for some time before kilning. In mills producing large quantities of sawn softwood, some sort of chemical treatment is carried out as the timber comes from the saw. This treatment need only be superficial and adequate to control colonization during the seasoning and storage period.

D. Packaged Timber in Transport

Established timber-handling practices in conjunction with relatively mild fungicidal treatments have minimized blue-stain in imported timbers, but the establishment of timber-packaging systems for sawn timber led to an increase in the incidence of staining (Savory, 1970). Packaged timber has also led to a considerable increase in the incidence of incipient decay in imported timber. In countries exporting timbers in package form, control of *Phlebia gigantea* is considered more important than even controlling the blue-stain problem, and development work to find new replacement anti-stain chemicals now always include assessment of these chemicals in the control of organisms such as *P. gigantea* which cause a decay problem, e.g. in the treatment of packaged *Pinus radiata* from New Zealand, and of packaged Maritime pine from Portugal (Dickinson, 1980). The problem of decay in packaged timber can carry over into timber in service. Incipient decay in the packaged timber develops into serious decay because the timber is put into use without being seasoned and in a situation where it is not allowed to dry out for a prolonged period.

E. Damage in Service

1. Timbers Inside Buildings

It has been estimated (Scobie, 1980) that the cost of repairing damage caused by wood-rotting fungi in buildings in the United Kingdom amounts to £3 million per week or about £150 million per annum. None of this damage could have occurred if the timber had been kept dry, which may initially be a design problem. However, given time and poor or infrequent maintenance, water ingress is possible in many forms, such as rising damp, poor sub-floor ventilation, condensation and leaks in the plumbing, gutters, rainwater pipes and roofs. The first principle in eradication or curative treatment is to find the source of water ingress which is raising the moisture content of the wood and eliminate it. Then comes the problems of drying out the building, which may be easy in one case and very difficult in another, but which has to be overcome if the outbreak of decay is to be stopped and remedial measures taken.

Insect damage to timber in buildings is widespread throughout the country, but as far as microbial degradation is concerned the matter usually resolves itself into whether the outbreak is due to what is popularly called 'Dry Rot' or 'Wet Rot'. Much has been written about these forms of decay (Cartwright and Findlay, 1958; Anon, 1964). Anon (1974, 1978), and Dickinson (1980) have defined and explained the terms.

In the past, the term 'dry rot' was used to describe any type of decay in buildings, but in more recent times, it has been restricted to decay of wood caused by *Serpula lacrimans*. This fungus leaves the timber as a dry, brown, brittle, friable mass which on drying has cracked into deep fissures both along and across the grain, to give the so-called cuboidal cracking, and which can usually be crumbled to powder between finger and thumb. The term 'wet rot' is used to describe decay in buildings caused by any other fungus and often occurs under wetter conditions. In many cases, the mycelium is not visible and this may, in former times, have led to the belief that moisture alone was the cause of decay, hence the term wet rot. This absence of visible mycelium is by no means true of all of the fungi causing wet rot and on occasions, as Dickinson (1980) points out, outbreaks of wet rot with an obvious mycelium have often been confused with dry rot by inexperienced surveyors.

It is important to distinguish between the two rots, because *S. lacrimans* is more difficult to eradicate and the measures taken in remedial treatment have, therefore, to be much more stringent and, inevitably, more costly.

a. *Dry rot. Serpula lacrimans* rarely, if ever, occurs away from human habitation. The fungus often becomes established in timber in contact with damp masonry which takes a long time to dry out. This all too often, can cause re-infection of replacement timbers. The fungus can tolerate alkaline conditions in walls, and its mycelium can be found penetrating through porous material, such as coarse bricks and crumbling mortar and is commonly found between the plaster and brick work of walls. This can lead to an extensive spread of the fungus. *Serpula lacrimans* is often quoted as being able to colonize dry areas, but normally spread will only occur in wet conditions. The fungus produces substantial strands behind the advancing hyphal front and Jennings (1980) has given an account of their function and physiology. Their function within buildings has often been misunderstood, but they obviously contribute to the survival and success of this organism in an environment where the food source may not be continuous, but separated by other inert materials. It is a widely held belief that unlike wet rot, it cannot be controlled by drying alone. It can be controlled in this way, but in practical terms a wet wall may take a very long time to dry sufficiently to cease sustaining the fungus with water and, unless this can be done quickly, the fungus will remain viable.

In conditions of high humidity, such as a wet cellar, the fungus can produce spectacular masses of fine cotton-wool like mycelium. Droplets of water (lacrymations) occur over the mycelium which gives the fungus its specific name. In drier situations, the mycelium is much sparser and is reduced to a greyish felted thin layer which may have patches which are coloured vividly yellow or lilac. The strands are grey in colour. The brown colour of the rotted wood is due to the residual lignin residues left by the fungus, which is a brown rot. The wood splits into cuboidal pieces by deep cross-cracking. The fruit bodies vary from flat pancakes to irregular brackets, depending on their position. The fruit bodies produce an enormous number of spores over a long period of time with the result that the spores, which are individually bright orange in colour, assume a rust-red colour *en masse* visible to the naked eye.

b. *Wet rot.* The decay fungus most commonly found in buildings is not *S. lacrimans*, but *Coniophora puteana* which is thought to be responsible for up to 90% of the decay within buildings (Dickinson, 1980). It is commonly called the 'cellar fungus'. In the early stages of decay, the wood surface darkens in colour, splitting is normally along the grain with small cross-checking; however, decay can also resemble the cuboidal cracking normally associated with dry rot. *Coniophora puteana* can often cause internal decay leaving a relatively thin layer of sound wood at the surface. Little or no mycelium is usually visible but, when present, it often develops into fine dark brownish-black strands. The fruit body is a thin 'skin' coloured olive-green to dull-brown and rarely found in buildings.

A number of other wood-rotting fungi are also found in buildings (Findlay, 1951). *Poria vaillantii* affects wood in a similar way to *S. lacrimans*, whereas *Paxillus pannoides* which requires very moist conditions resembles *C. puteana* in the way in which it breaks down the timber. *Poria xantha* and *Lenzites trabea* are often associated with decay of timber in roofs. In recent years, the widespread construction of flat roofs has given rise to a great deal of decay of wood in these structures due to leakage and condensation problems.

2. *Timbers Exposed Externally, But Not in Ground Contact*

This type of exposure is typified by external cladding and window and door joinery in buildings. Here again, the problem is essentially of the ingress of water to the wood and its retention there for sufficient time to give rise to decay. In the United Kingdom, the traditional timber for window joinery was heartwood of Scots pine (or red deal or baltic redwood). This species is still used today, but whereas, formerly when the trees were over two feet in butt diameter it was possible to cut off all traces of the perishable sapwood with a relatively small loss of log volume, today the tree is less than a foot in butt diameter and contains little or no durable heartwood. The tree species is the same, but durable heartwood has been replaced by perishable sapwood and, over the last 20 years, a problem that has been close to a national disaster has arisen where one did not exist before.

At the same time, changes in design and manufacturing techniques have produced window and door frames that allow water to penetrate into the joints and then along the grain of the wood. In the paint systems used today, lead primer has been replaced by other materials. Savory and Carey (1975) showed that decay of external doors was often initiated by decay of dowels in the joints when a perishable hardwood was used without first being treated with a wood preservative. This increase in the incidence of decay in external joinery has stimulated a great deal of work in the United Kingdom, principally at the Princes Risborough laboratory of the Building Research Establishment. Tack (1968) showed in a survey in south-east England that at least half of the window frames examined had one or more of the lower joints moist enough for decay to occur. This survey and subsequent experimental work (Anon, 1969) clearly demonstrated that once moisture had penetrated into joinery the rate of drying out through a normal three-coat paint system was very slow. Other surveys (Soane, 1978; Beech and Newman, 1975) confirm the incidence of decay in untreated joinery.

At first, much of the decay was attributed to *Coniophora puteana*, but subsequent examinations and surveys revealed that *C. puteana* was in fact rare in window joinery and doors. Canadian hemlock, from Soane's (1978) survey has yielded a range of basidiomycetes. One of the most common was the white rot *Phellinus contigua* followed by *Coriolus versicolor*, and species of *Hyphoderma*, *Hyphodantia*, *Sistotrema* and *Stereum* were also identified. *Poria placenta* was also identified from two of the doors. This must have been the result of infection of packaged timber from Canada as previously mentioned, since this fungus does not occur naturally in the United Kingdom (Savory and Carey, 1975). In a systematic study of window frames from various sites, species of *Phellinus* were again shown to be abundant at one site with *Coniophora puteana* again being absent (Savory and Carey, 1976). The situation appears to result from establishment of moist conditions in the joinery components which are stabilized by the presence of the paint film which prevents drying. Under these conditions, certain white-rot fungi become established and cause extensive decay.

It was recognized that this was possibly the climax of an ecological sequence, and that other organisms might have been important in the early stages. Carey (1980), in an extensive survey of exposed painted simulated joinery components, has shown that an ecological sequence of colonization by micro-organisms exists starting with bacteria, moulds and stainers and climaxing in the final decay by basidiomycetes. Several of these early colonizers have been shown able to increase the absorbency of the timber from the joint in a manner similar to the ponding effect already described (Robins, 1978). This undoubtedly is a major contributory factor in the colonization and decay process.

Similar attention has also been focussed on the microbiology of joinery finishes such as paints and varnishes. Dickinson (1971) showed that *Aureobasidium pullulans* is a very common fungus in joinery and that once under the decorative finish was capable of erupting through it and therefore giving points of entry for water which would subsequently have difficulty in evaporating through the otherwise intact paint film. There has been much discussion as to whether or not this fungus is capable of penetrating the paint film directly and then causing subsequent eruption. Recently, F. Mendes (personal communication) demonstrated direct penetration in at least one type of varnish and paint system. This allowed colonization of the wood and subsequent eruption back through the surface film.

Work is continuing in this area in order to understand the importance of these fungi, but it seems likely that organisms such as *A. pullulans* are capable of degrading surface finishes possibly by direct penetration, allowing wetting up of the timber underneath and possibly detoxifying the fungicide tributyl tin oxide leading to colonization by decay organisms. These recent studies have underlined the importance of establishment of stable wet conditions in joinery. This has been recognized by the preservative industry, and most companies now offer water-repellent versions of joinery preservatives. Such systems and alternative decorative finishes are currently under investigation in a study following up the work of Carey (1980). Results of long-term trials in the U.S.A. indicate that water-repellent treatments alone may be just as effective as when included with a fungicide (Feist and Miraz, 1978).

3. Constructional Timber Other Than in Buildings

Outside buildings, it is recognized that timber will become wet in service and, generally, some provision is taken to ensure a reasonable service life. Constructional timber used in marine work and in the timber fills of water-cooling towers is of particular interest. Here the timber is permanently wet, and wood-rotting basidiomycetes are of little consequence, their place being taken by the soft rots which assume economic importance. In even wetter conditions of permanent anaerobic water-logging, soft-rot fungi give way to bacteria which may very slowly degrade cell walls. Boutelje and Bravery (1968) demonstrated the importance of anaerobic bacteria in deterioration of wooden pile foundations supporting a Swedish building.

In heavy construction work, mechanical abrasion can accentuate deterioration of timber during the course of decay. In the sea, structures are likely to be subjected to wave action and tidal surges, where the water may contain sand or shingle which will rapidly abraid any areas weakened by decay. In such exposures, the timber must not only be durable against micro-organisms but also strong enough to withstand abrasion.

4. Timber in Ground Contact

The greatest hazard will be in ground contact where the soil will act as a reservoir of both water and micro-organisms that have evolved to break down plant remains. Posts and poles are used widely for fences, and in telephone and electricity supply services. The posts may often be shaped with the sapwood removed, and here the natural durability of the heartwood is important. The timber is partly buried in the soil and, as already discussed, is liable to decay at the ground line, where oxygen, water and nutrients are readily available for soil micro-organisms. If heartwood durability is low, then it must have a high treatability rating to enable effective wood preservation to be carried out. Round poles will contain an outer layer of sapwood which, being permeable, can be treated with a wood preservative to bring it up to the durability rating of the durable heartwood. Since the effective strength of a pole depends on maintaining the strength characteristics of the outer layers rather than the core, it is important to ensure durability in the outer band. The United Kingdom Post Office claims protection of creosote-treated poles for 90–120 years.

Railway sleepers are also in ground contact. The treatment of permeable pine sapwood by creosote over the last 150 years has been one of the success stories of wood preservation, and sleepers have been successfully treated to prevent microbial deterioration during the mechanical life of the timber.

Since the last war there has been a steady move towards the use of water-borne salt preservatives, which have given excellent performance in pines. In recent years there have been some problems in the ground contact situation. These have normally been associated with soft-rot in treated hardwoods, particularly some hardwood species in the tropics. Aspects of these interactions between wood preservatives and fungi are dealt with later.

VIII. COLONIZATION OF WOOD AND THE EFFECT OF WOOD PRESERVATIVES

Studies on colonization by micro-organisms in wooden stakes in ground contact, and in simulated window joinery out of ground contact, serve to illustrate differences in these interactions between treated and un-treated hardwoods and softwoods. The reactions of the colonizing micro-organisms to these habitats can best be appreciated by regarding each wood species as a new substrate, and wood treated with a wood preservative as another type of substrate which is composed of wood and chemicals.

Clubbe (1980a, b) set up an experiment with Scots pine and birch stakes half-buried in the ground, half of which were treated with a 1% solution of a copper chrome arsenate-type wood preservative. Carey (1980) used a simple L-joint to simulate the bottom joint in a window frame, sealed at each end and painted with a typical three-coat paint system. Some of her L-joints were treated with a 1% solution of tributyl tin oxide, some with a 5% pentachlorophenol solution and some were untreated. Clubbe's (1980a, b) stakes were carefully orientated with respect to grain, being cut from sapwood so that one face was the outer tangential surface, one the inner tangential surface and the other two approximately radially orientated.

Both Clubbe (1980a, b) and Carey (1980) instituted a system of taking samples for isolation of micro-organisms on to selective media from a set pattern of points. This was an important criterion since the sampling points became an objective system rather than the more usual subjective isolations only from places where decay was suspected. Butcher (1971) has given a very clear appreciation of the problems of analysing fungal floras in wood. In the ground contact situation, bacteria were the first colonizers of untreated wood, followed by primary moulds and stainers. Wood-rotting fungi were not isolated for the first three months of exposure after which soft-rot fungi and wood-rotting basidiomycetes were isolated. After six months, soft-rot fungi declined in frequency of isolation, and wood-rotting basidiomycetes became the dominant member of the microflora, with only the secondary moulds as attendant species. However, in preservative-treated wood, the new substrate (both pine and birch) showed a major difference in fungal population. Wood-rotting basidiomycetes had been totally eliminated (the only group for which this occurred) and soft-rot fungi became the dominant member of the microflora. In Scots pine, these organisms, in spite of being in abundance, did not appear to degrade the wood cell walls, even after 18 months exposure. With birch, however, the treated wood as well as being heavily colonized by soft-rot fungi was seen to have heavy degradation of fibre walls (Clubbe, 1980a, b). Carey (1980), showed a similar progression of fungal types in the L-joints exposed out of ground contact, with a similar elimination of wood-rotting basidiomycetes but not soft-rot fungi in the preservative-treated wood. These results from natural exterior exposure of a preservative-treated and untreated softwood and hardwood confirmed the findings of experiments *in vitro* in the laboratory which had used both soil burial and monocultures of representatives of the three rot types.

In an attempt to understand the differences in performance between the hardwoods and softwoods, Sorkhoh and Dickinson (1976) subjected copper chrome arsenate-treated and untreated birch and Scots pine to monocultures of soft-rot, white-rot and brown-rot fungi in laboratory experiments. Their results demonstrated a fundamental difference between decay in treated and untreated wood, and between treated hardwood and treated softwood, at normal commercial levels of biocide. Both species of the untreated wood were decayed by soft-rot fungi and wood-rotting basidiomycetes. The treated material, however, performed quite differently. The treated softwood was completely protected, but the treated hardwood was protected only against the wood-rotting basidiomycetes, the fibres being attacked by soft-rot fungi although the rays were protected. Microdistribution studies of the preservative elements showed considerable differences between the two test-timber species.

Dickinson *et al.* (1976) showed that the treated softwood was evenly treated with the preservative components throughout the tissues. Analysis of the cell-wall layers of the softwood showed an even distribution across the cell wall, with relatively high levels within the S_2 layer of the wall. In the hardwood, the preservative components were unevenly distributed, with accumulations in the rays. The S_2 layers of the fibres were poorly treated but there was an accumulation of preservative on the lumen surfaces, a fact previously noticed in sycamore (Dickinson, 1974). When considering the way in which white- and brown-rot fungi attack the cell wall via the S_3 layer from the lumen, it is easy to see how a preservative deposited on the lumen surface might inhibit such fungi. However, in the case of soft-rot fungi attacking hardwood fibres, once entry has been gained into the poorly treated S_2 region of the wall, decay can proceed normally, since this is the usual mode of soft-rot attack. In contrast, in softwood, with even distribution and good treatment of the S_2 layers of the tracheids, protection from all fungi was achieved.

Differences between the effectiveness of preservative treatment in pine and in certain hardwoods has, thus, suggested differences in the microdistribution of the wood preservative within the wood. This has been shown to be of three types: (a) differences between the total amount of preservative in each tissue type (e.g. vessels, rays or parenchyma in hardwoods); (b) differences in the proportions of the salts in a mixed salt preservative in each tissue type; or (c) differences in the amount of preservative present in each layer of the cell wall. Previous work (Petty and Preston, 1968; Greaves, 1972, 1974; Dickinson, 1974; Dickinson et al., 1976) had shown the cell-wall layers of all cell types in Scots pine to be penetrated by a copper chrome arsenate-type of wood preservative, both as regards the total amount of preservative and the proportions of the three constituent elements. Birch, on the other hand, showed a lower treatment of the ray parenchyma than in Scots pine, but sufficient to prevent soft-rot decay, whereas the birch fibres had little or nothing in the S_2 layer and were heavily degraded by soft-rot fungi (Sorkhoh and Dickinson, 1976; Dickinson et al., 1976).

Greaves and Levy (1978) and Levy and Greaves (1978), using a scanning electron microscope, examined the pattern of distribution of a copper chrome arsenate wood preservative in 11 species of wood. They presented evidence to show that each timber had its own pattern of distribution through the tissue types which ranged from the comparative evenness of *Alstonia scholaris* on the one hand to the very uneven distribution in *Flindersia pimenteliana*. Dickinson and Sorkhoh (1976) and later Drysdale (1979) suggested that although the differences in distribution of total preservative and its disproportionate distribution in the various tissues present in each wood species went some way to explain differences in the performance of the treated wood, it was not the complete story. Further micro-analysis of ultrathin sections in the transmission electron microscope, to search for differences in the distribution of the preservative and the proportions of its constituent parts within the layers of the cell walls, could supply the answer. Hulme and Butcher (1977) and Butcher and Drysdale (1978) showed that, in laboratory tests, hardwood species had a greater susceptibility to soft-rot than softwood species and that control could be effected by increasing the loading of copper in each wood species.

Drysdale (1979) carried out an examination of four hardwood species compared to a softwood Scots pine (*Pinus sylvestris*). She showed that, although some timber species were more susceptible to soft-rot than others, control in the laboratory could be achieved by increasing the gross loading with copper chrome arsenate. Analysis of ultrathin sections of all treated wood species showed that in no case did an increased loading of copper chrome arsenate show any increase in the preservative components in the wall layers, particularly the S_2. *Alstonia scholaris*, which appeared from standard toxicity tests to be as effectively treated as the pine, showed an uneven distribution in the wall layers, with a high deposition of copper chrome arsenate on the lumen–wall surface, low in the S_2 layer and high in the middle lamella–primary wall complex. Observations from pit apertures showed a high concentration of copper chrome arsenate in the pit membrane which decreased in passing along the middle lamella in any direction away from the pit membrane, to increase again on approaching another pit.

This suggested that, in this timber, penetration of wood preservative solution occurs via the cell lumen, preservative is deposited on the cell wall, and enters the wall through the pit membrane and thence into the middle lamella. If this is so, then effective protection in *A. scholaris* could be due to total encapsulation of the wall of each cell, so that a chance penetration of soft-rot fungus into the wall limits its attack to that cell alone. Birch (*Betula pendula*) also showed substantial deposition in the pit membrane, but little or no penetration along the middle lamella (Drysdale *et al.*, 1980).

The earlier hypothesis supposed that wood-preservative solutions penetrated the cell lumina, thereby coating the inner surface of the cell wall and penetrating into the wall through the S_3 layer. The higher loading in the middle lamella–primary wall region was regarded as an accumulation of preservative in that region as the preservative penetrated from lumina. This meant that, in some species, there was less retention in the S_2 than in the middle lamella and, in a species like birch, little penetration from the lumen, which suggested some barrier to penetration at the S_3 layer. Drysdale's observations of the possible penetration into the middle lamella from the pit membrane with its attendant encapsulation of the cell in a preservative-treated sheath puts a new concept forward, and explains why others have found high loadings at the S_3 border and middle lamella and low loadings in the S_2 (Dickinson and Sorkhoh, 1976; Peters and Parameswaran, 1980). High loadings in the S_2 may not be necessary in *A. scholaris*, since any soft rot-type organism which does manage to pass through a minute gap in the preservative coating to the lumen wall surface will be trapped in the S_2 layer by treatment of the middle lamella.

The effect of wood preservatives on colonizing microfungi is, therefore, complex. So far, no clear-cut explanation of the differences in the mechanisms of action of these toxic chemicals to soft-rot fungi and to wood-rotting basidiomycetes has been attempted. What constitutes the 'tolerance' of soft-rot fungi to copper chrome arsenate-treated wood? Some interesting aspects of microbial ecology remain to be studied.

The work described in this chapter has involved many people in many parts of the world and the exciting observations of recent years have been the result of careful observation and experimentation in the past which has outlined the colonization of the organisms, their effect on the cell wall and the effect on the organisms of water and fungicides. The future use of wood depends on our being able to conserve the one renewable natural structural material, so that it is not wasted by bio-deterioration. To do this, the natural durability of wood will have to be increased to give the required service life. This means a full under-standing of how microbial deterioration of wood comes about so that an effective treatment can be administered which will give the required induced durability without using environmentally unacceptable chemi-cals. Much still remains to be done, but its importance in terms of the national economy should be evident to all. As much as £150 million per annum are required to repair fungal decay in building and £2,754 million per annum are required to pay for the import of wood and wood products! These are surely incentives enough so far as the United Kingdom alone is concerned.

REFERENCES

Anon (1964). 'Timber Pests and their Control'. Timber Research and Development Association. London (new edition in press).

Anon (1967). *Forestry Commission Leaflet* No. 5.

Anon (1969). *Technical Note No. 44.* Princes Risborough Laboratory, B.R.E. H.M.S.O., London.

Anon (1974). 'Timber Preservation'. Timber Research and Development Association/ British Wood Preserving Association, London.

Anon (1978). British Wood Preserving Association, Leaflet No. 1.

Anon (1980). 'Strategy for the U.K. Forest Industry'. Centre for Agricultural Strategy, Report No. 6.

Baecker, A. A. W. and King, B. (1980a). *In* Proceedings of the 4th International Bio-deterioration Symposium, Berlin 1978. (T. A. Oxley, G. Becker and D. Allsopp, eds.). Pitman Publishing, London.

Baecker, A. A. W. and King, B. (1980b). *International Research Group of Wood Preservation* Document IRG/WP/1116.

Bailey, I. W. and Vestal, H. R. (1937). *Journal of the Arnold Arboretum* **18**, 196.

Baines, E. F. (1981). Ph.D. Thesis: University of London.

Baines, E. F. and Levy, J. F. (1979). *Journal of the Institute of Wood Science* **8**(45), 109.

Baines, E. F. and Millbank, J. W. (1976). *Material und Organismen*, Beiheft **3**, 167.

Baines, E. F., Dickinson, D. J., Levy, J. F. and Millbank, J. W. (1977). *Record of the 1977 Annual Convention of the British Wood Preserving Association*, p. 33.

Baker, J. M. (1963). *In* 'Symbiotic Assocations'. 13th Symposium of the Society for General Microbiology, pp. 232–265. Cambridge University Press.

Baker, J. M. and Bletchly, J. D. (1966). *Journal of the Institute of Wood Science* **3**(17), 53.

Baker, J. M. and Kreger-Van Rij, N. J. W. (1964). *Antonie von Leeuwenhoek* **30**, 433.

Baker, J. M., Laidlaw, R. A. and Smith, G. A. (1970). *Holzforschung* **24**, 45.

Banks, W. B. and Dearling, T. B. (1973). *Material und Organismen* **8**, 39.

Bavendamm, W. (1928). *Zentralblatt für Bakteriologie* **76**, 172.

Beech, J. C. and Newman, P. L. (1975). *Building Research Establishment, Current Paper* CP 97/75 DOE.

Bletchly, J. D. (1967). 'Insect and marine borer damage to timber and woodwork'. H.M.S.O., London.

Bletchly, J. D. (1969). *Journal of the Institute of Wood Science* **3**(22), 43.

Boutelje, J. B. and Bravery, A. F. (1968). *Journal of the Institute of Wood Science* **4**(20), 47.

Bravery, A. F. (1971). *Journal of the Institute of Wood Science* **5**(30), 13.

Bravery, A. F. (1972). Ph.D. Thesis: University of London.

Bravery, A. F. (1975). *In* 'Biological Transformation of Wood by Micro-organisms' (W. Liese, ed.). pp. 129–142. Springer-Verlag, Berlin.

Bravery, A. F. (1976). *Material und Organismen*, Beiheft **3**, 331.

Butcher, J. A. (1971). *Material und Organismen* **6**, 209.

Butcher, J. A. and Drysdale, J. A. (1978). *Material und Organismen* **13**, 187.

Campbell, J. (1980). Policy for Forestry. Letter, *The Times* No. 60570 (8.3.80), p. 13.

Campbell, W. G. (1952). *In* 'Wood Chemistry' (L. E. Wise and E. C. Jahn, eds.) vol. 2, chapter 27. Reinhold Publishing Corporation, New York.

Carey, J. K. (1980). Ph.D. Thesis: University of London.

Cartwright, K. St. G. and Findlay, W. P. K. (1958). 'Decay of Timber and its Prevention'. H.M.S.O. London (2nd Edn.).

Cavalcante, M. S. (1981). Ph.D. Thesis: Portsmouth Polytechnic.

Cavalcante, M. S. and Eaton, R. A. (1980). *International Research Group on Wood Preservation* Document IRG/WP/1110.

Clubbe, C. P. (1978). *International Research Group on Wood Preservation* Document IRG/WP/186.

Clubbe, C. P. (1980a). *International Research Group on Wood Preservation* Document IRG/WP/1107.

Clubbe, C. P. (1980b). Ph.D. Thesis: University of London.

Corbett, N. H. (1963). Ph.D. Thesis: University of London.

Corbett, N. H. (1965). *Journal of the Institute of Wood Science* **3**(14), 18.

Cowling, E. B. (1965). *In* 'Cellular Ultrastructure of Woody Plants' (W. A. Cote, ed.), pp. 341–368. Syracuse University Press.

Cowling, E. B. and Merrill, W. (1966). *Canadian Journal of Botany* **44**, 1533.

Crossley, A. (1980). Ph.D. Thesis: University of London.

Crossley, A. and Levy, J. F. (1977). *Journal of the Institute of Wood Science* **7**(43), 30.

Dickinson, D. J. (1971). *Record 1971 Annual Convention of the British Wood Preserving Association*, p. 151.

Dickinson, D. J. (1974). *Material und Organismen* **9**, 21.

Dickinson, D. J. (1980). *In* 'Decomposition by Basidiomycetes' (J. Frankland and J. Hedger, eds.), British Mycological Society, Symposium No. 3, Cambridge University Press (in press).

Dickinson, D. J. and Levy, J. F. (1979). *Record of the 1979 Annual Convention of the British Wood Preserving Association*, p. 33.

Dickinson, D. J. and Sorkhoh, N. A.A. H. (1976). *SEM/76* **2**, 549.

Dickinson, D. J., Sorkhoh, N. A. A. H. and Levy, J. F. (1976). *Record of the 1977 Annual Convention of the British Wood Preserving Association*, p. 25.

Drysdale, J. A. (1979). D.I.C. Thesis: Imperial College, London.

Drysdale, J. A., Dickinson, D. J. and Levy, J. F. (1980). *Material und Organismen* **15**, 287.

Dunleavy, J. A. and McQuire, A. J. (1970). *Journal of the Institute of Wood Science* **5**(26), 20.

Dunleavy, J. A., Moroney, J. P. and Rossell, S. E. (1973). *Record of the 1973 Annual Convention of the British Wood Preserving Association*, p. 127.

Ellwood, E. L. and Ecklund, B. A. (1959). *Forest Products Journal* **9**, 283.

Enebo, L. (1951). *Physiologie Plantarum* **4**, 652.

Feist, W. C. and Miraz, E. A. (1978). *Forest Products Journal* **28**, 31.

Findlay, G. W. D. (1970). Ph.D. Thesis: University of London.

Findlay, W. P. K. (1951). *Transactions of the British Mycological Society* **34**, 35.

Findlay, W. P. K. and Savory, J. G. (1950). *Proceedings of the VIIth International Botanical Congress, Stockholm*, p. 315.

Fisher, R. C. (1940). *Annals of Applied Biology* **27**, 545.

Fisher, R. C. (1941). *Annals of Applied Biology* **28**, 244.

Fowlie, I. M., Hutchison, G. O. and Oxley, T. A. (1981). *Record of the 1981 Annual Convention of the British Wood Preserving Association* (in press).

Francke-Grossmann, H. (1966a). *In* 'Symbiosis, its Physical and Biochemical Significance', vol II. Academic Press, London.

Francke-Grossmann, H. (1966b). *Material und Organismen*, Beiheft **1**, 503.

Freeman, G., Baillie, A. J. and McInnes, C. A. (1948). *Chemistry and Industry* (May), 279.

Fukazawa, K., Imagawa, H. and Doi, S. (1976). *Research Bulletin of the College Experimental Forests, College of Agriculture, Hokkaido* **33**, 101.

Graham, K. (1967). *Annual Review of Entomology* **12**, 105.

Garrett, S. D. (1951). *New Phytologist* **50**, 149.

Garrett, S. D. (1955). *Transactions of the British Mycological Society* **38**, 1.

Greaves, H. (1966). *Material und Organismen*, Beiheft **1**, 61.

Greaves, H. (1970). *Holzforschung* **24**, 6.

Greaves, H. (1972). *Material und Organismen* **7**, 277.

Greaves, H. (1974). *Holzforschung* **28**, 193.

Greaves, H. and Levy, J. F. (1965). *Journal of the Institute of Wood Science* **3**(15), 55.

Greaves, H. and Levy, J. F. (1978). *Holzforschung* **32**, 200.

Green, N. B. (1980). *Journal of the Institute of Wood Science* **8**(47), 221.

Green, N. B., Dickinson, D. J. and Levy, J. F. (1980). *International Research Group on Wood Preservation* Document IRG/WP/1111.

Greig, B. J. W. (1976). *European Journal of Forest Pathology* **5**, 286.

Harmsen, L. and Nissen, T. V. (1965). *Nature, London* **206**, 319.

Harris, W. V. (1971). 'Termites: their Recognition and Control', 2nd edn. Longman, London.

Hawker, L. E., Linton, A. H., Folkes, B. F. and Carlile, M. J. (1960). 'An Introduction to the Biology of Micro-organisms'. Arnold, London.

Hejmanek, M. and Rypacek, V. (1954). *Ebenda* **360**, 225–237.

Hickin, N. E. (1963). 'The Insect Factor in Wood Decay'. Hutchinson, London.

Highley, T. L. (1978). *Material und Organismen* **12**, 161.

Hulme, M. A. and Butcher, J. A. (1977). *Material und Organismen* **12**, 223.

Imschenezki, A. H. (1959). 'Mikrobiologie der Cellulose'. Akademie–Verlag, Berlin.

Jennings, D. (1980). *In* 'Decomposition by Basidiomycetes' (J. Frankland and J. Hedger, eds), British Mycological Society, Symposium No. 3. Cambridge University Press (in press).

Jones, E. B. G. and Eltringham, S. K. (1968). 'Marine borers, Fungi and Fouling Organisms of Wood'. OECD Paris.

Kaarik, A. (1974). *International Research Group on Wood Preservation* Document IRG/WP/125.

Kaarik, A. (1980). *International Research Group on Wood Preservation* Document IRG/WP/199.

Karkanis, A. G. (1966). M.Sc. Thesis: University of London.

Karnop, G. (1967). Ph.D. Thesis: University of Hamburg.

Kelman, A. and Cowling, E. B. (1965). *Phytopathology* **55**, 148.

King, B., Oxley, T. A. and Long, K. D. (1974). *Material und Organismen* **9**, 241.

King, B., Oxley, T. A. and Long, K. D. (1976). *Material und Organismen*, Beiheft **3**, 264.

King, B., Mowe, G. and Smith, G. M. (1981) *Record of the 1981 Annual Convention of the British Wood Preserving Association* (in press).

Krapivina, l. G. (1960). *lensoi zhurnal* **3**, 130. (CSIRO Australian translation No. 5329, 1961).

Knuth, D. T. (1964). *Dissertation Abstracts* **25**, No. 2175.

Knuth, D. T. and McCoy, E. (1962). *Forest Products Journal* **12**, 437.

Koenigs, J. W. (1974). *Wood and Fibre* **6**, 66.

Koenigs, J. W. (1975). *In* 'Cellulose as a Chemical and Energy Resource' (R. C. Wilke, ed.), Biotechnology and Bioengineering Symposium. No. 5, pp. 151–159. John Wiley and Sons, N.Y.

Leightley, L. E. (1977). Ph.D. Thesis: C.N.A.A., Portsmouth Polytechnic.

Leightley, L. E. and Eaton, R. A. (1977). *Record of the 1977 Annual Convention of the British Wood Preserving Association*, p. 221.

Levy, J. F. (1968). *In* 'Biodeterioration of Materials' (A. H. Walters and J. J. Elphick, eds.), pp. 424–428. Elsevier, Amsterdam.

Levy, J. F. (1971). *Timber Trades Journal* **227**, 71.

Levy, J. F. (1973). *British Wood Preserving Association News Sheet* **130**, 1.

Levy, J. F. (1975). *In* 'Biological Transformation of Wood by Micro-organisms' (W. Liese, ed.), pp. 64–73. Springer-Verlag, Berlin.

Levy, J. F. (1980). *In* 'Decomposition by Basidiomycetes' (J. Frankland and J. Hedger, eds.), British Mycological Society Symposium No. 3. Cambridge University Press (in press).

Levy, J. F. and Greaves, H. (1966). *British Wood Preserving Association News Sheet* **68**, 1.

Levy, J. F. and Greaves, H. (1978). *Holzforschung* **32**, 209.

Levy, J. F. and Olofinboba, M. (1966). *British Wood Preserving Association News Sheet* **80**, 1.

Levy, J. F., Millbank, J. W., Dwyer, G. and Baines, E. F. (1974). *Record of the 1974 Annual Convention of the British Wood Preserving Association*, p. 3.

Lewis, P. F. (1976). *Material und Organismen*, Beiheft **3**, 113.

Liese, W. (1970). *Record of the 1970 Annual Convention of the British Wood Preserving Association*, pp. 81–97.

Liese, W. and Karnop, G. (1968). *Holz als Roh-und Werkstoff* **26**, 202.

Liese, W. and Schmid, R. (1964). *Phytopathologische Zeitschrift* **51**, 385.

Lutz, J. F., Duncan, C. G. and Sheffer, T. C. (1966). *Forest Products Journal* **16**, 23.

McCreary, M., Cozenda, B. and Shigo, A. L. (1965). *Phytopathology* **55**, 129

Macpeak, M. D. (1963). *Western Pine Association Research Note* **2**, 116.

McQuire, A. J. (1970). Ph.D. Thesis: University of Leeds.

Mandels, M. and Reese, E. T. (1957). *Journal of Bacteriology* **73**, 269.

Mercer, P. (1980). *In* 'Decomposition by Basidiomycetes' (J. Frankland and J. Hedger, ed.), British Mycological Society Symposium No. 3. Cambridge University Press.

Montgomery, R. A. P. (1980). *In* 'Decomposition by Basidiomycetes' (J. Frankland and J. Hedger, eds.), British Mycological Society Symposium No. 3. Cambridge University Press.

Nasroun, T. A. H. (1971). D.I.C. Thesis: Imperial College, London.

Norkans, B. (1967). *Advances in Applied Microbiology* **9**, 91.

Norman, A. G. and Fuller, W. H. (1942). *Advances in Enzymology* **2**, 239.

Norris, D. M. (1966). *Material und Organismen*, Beiheft **1**, 523.

Norris, D. M. (1976). *Material und Organismen*, Beiheft **3**, 479.

Oliver, A. C. (1962). *Journal of the Institute of Wood Science* **2**(9), 32.

Parkin, E. A. (1940). *Journal of Experimental Botany* **17**, 364.

Peters, G. A. and Parameswaran, N. (1980). *Wood Science and Technology* **14**, 81.

Petty, J. A. and Preston, R. D. (1968). *Holzforschung* **22**, 174.

Preston, R. D. (1979). *Wood Science and Technology* **13**, 155.

Rishbeth, J. (1952). *Forestry* **25**, 41.

Roff, J. W., Cserjesi, A. J. and Swan, G. W. (1974). Canadian Forest Service Publication No. 1325.

Robins, C. S. (1978). Third Year Undergraduate project: Imperial College, London.

Rogers, H. J. (1962). *In* 'The Bacteria' (R. Y. Stainer and I. C. Gunsalus, eds.), vol. II. Academic Press, London.

Rossell, S. E., Abbot, E. M. and Levy, J. F. (1973). *Journal of the Institute of Wood Science* **6**(32), 28.

Savory, J. G. (1954a). *Journal of Applied Bacteriology* **17**, 213.

Savory, J. G. (1954b). *Annals of Applied Biology* **41**, 336.

Savory, J. G. (1955). *Record of the 1955 Annual Convention of the British Wood Preserving Association*, p. 3.

Savory, J. G. (1970). *Timber Lab. News*. No. 6.

Savory, J. G. (1971). *International Biodeterioration Bulletin* **7**, 91.

Savory, J. G. and Carey, J. K. (1975). *Timber Trades Journal* **295** (5171), 12.

Savory, J. G. and Carey, J. K. (1976). *Record of the 1976 Annual Convention of the British Wood Preserving Association*, p. 3.

Schacht, H. (1863). *Jahrbucher für Wissenschaftliche Botanik* **3**, 442.

Schmidt, O. (1981). Proceedings of the International Symposium on Wood and Pulping Chemistry, Stockholm 1981—The 19th Ekman Days.

Scobie, D. (1980). *Timber Trades Journal* **312** (5403), 21.

Sharp, R. F. and Millbank, J. W. (1973). *Experientia* **29**, 895.

Sheffer, T. C. and Cowling, E. B. (1976). *Annual Review of Phytopathology* **4**, 147.

Shigo, A. L. (1966). *United States Forest Service Research Paper* NE-47.

Shigo, A. L. (1968). *Phytopathology* **58**, 1493.

Siu, R. G. H. (1951). 'Microbial Decomposition of Cellulose'. Reinhold, New York.

Soane, G. E. (1978). *British Wood Preserving Association News Sheet*, No. 151, 1.

Sorkhoh, N. A. A. H. and Dickinson, D. J. (1976). *Material und Organismen*, Beiheft **3**, 287–293.

Smyth, H. F. and Obold, W. I. (1930). 'Industrial microbiology'. Bailliere, Tindall and Cox, London.

Tack, C. H. (1968) *Building* **214**, 135.

Uju, G. C. (1979). Ph.D. Thesis: University of London.

Uju, G. C., Baines, E. F. and Levy, J. F. (1981). *Journal of the Institute of Wood Science* **9**(49), 23.

Wilson, S. E. (1933). *Annals of Applied Biology* **20**, 661.

Zainal, A. S. (1975). Ph.D. Thesis: University of London.

3. Stone

ALICJA B. STRZELCZYK

Laboratory of Paper and Leather Conservation, Institute of Conservation and Restoration of Antiquities, Nicolaus Copernicus University, Torun, Poland

I. Chemical Air Pollution as an Agent Responsible for Deterioration . . .	61
II. Origin of the Microflora on Rocks and Stones	63
III. Role of Chemolithotrophic Bacteria in the Decomposition of Rocks and Stones .	65
A. Activity of Sulphur-Oxidizing Bacteria	66
B. Activity of Nitrifying Bacteria	70
IV. Role of Heterotrophic Bacteria and Fungi in Destruction of Rocks and Stones .	71
V. Role of Algae in Destruction of Rocks and Stones	74
VI. Attempts at Combating Microflora on Stone Monuments	77
References	78

I. CHEMICAL AIR POLLUTION AS AN AGENT RESPONSIBLE FOR DETERIORATION

The problem of microbial decomposition of rocks and stone objects of art has two contrasting aspects, although the process itself is essentially the same. It is Man's approach to the problem that is different in each case. The first aspect of the problem is related to the participation of micro-organisms in the weathering of natural rocks, releasing mineral compounds contained in the rocks, and to all processes involved in soil formation as a whole. From the point of view of Man as a soil exploiter, these are useful initiating soil-forming processes, which expand our supplies of arable land. These processes are, at the same time, conducive to the ageing of the Earth's crust; disintegration of rocks followed by formation of a fertile soil.

The other aspect of the problem is microbial disintegration of stone as a building material and/or as a raw material in creative art. Disintegration of stone in this context is a highly harmful process causing irreparable loss in our cultural heritage each year. This is all the greater if the objects undergoing weathering and microbial deterioration are of artistic value. The parts most liable to destruction are the outsides of stone elements, which of course determine their artistic expression. These include sculptures, external walls of buildings, portals, monuments and many other elements of architecture.

The different attitude of man towards rock weathering on one hand and of deterioration of stone objects of art on the other are reflected in the names of these two essentially identical processes; namely weathering of rocks as a positive phenomenon of the process of soil formation, and corrosion of stone, an undesirable phenomenon of deterioration of stone objects (Domasłowski, 1966; Ciach, 1967).

In spite of the conceptual separation of these two processes, investigations on the nature of each of them are complementary, evidence for which comes from, among others, Krumbein's (1972, 1973) studies pointing out the great resemblance in the mechanisms of deterioration of rocks and stone objects.

The rate of decomposition of these materials depends on a number of agents in their environment. Decomposition of rocks and stones is known to proceed at a considerable rate in industrialized areas, where the air contains large concentrations of acidic chemicals (Domasłowski, 1966), and at a much lower rate in areas with chemically pure air. Przedpelski (1957), 20 years ago, wrote that stone objects of art in industrial cities deteriorated at a rate 10 times that of similar objects in the country. Pochon (1964) denies the claim that stone monuments in the country should deteriorate at a slower rate than those in the town. It seems that in the 1980s it will no longer be possible to speak of any differences in the degree of chemical pollution between town and country.

Industrial works, densely dotted over the area and which emit smoke laden with all sorts of chemical compounds, contribute greatly to the deterioration of stone monuments, the outsides of buildings, portals and other elements of architecture. Smoke and gas pollutants of the air include solid particles in the form of soot and tar-like substances dispersed in the air, as well as sulphur dioxide, sulphuric acid vapours, nitric oxide, aldehyde, sulphates and chlorides. The concentrations of air pollutants range from several thousand to several hundred thousand $\mu g/m^3$ of air (Commins, 1962).

Among chemical air pollutants, the most active in deterioration of stone objects are those gases that readily form acid solutions with water. Particularly destructive for stone is sulphur dioxide, which forms in the course of combustion of coal containing 2–8% sulphur. A considerable proportion of the sulphur (60–80%) passes into the atmosphere as sulphur dioxide (Ciach, 1967). This becomes oxidized and, combined with water, yields sulphuric acid, which penetrates inside the stone and corrodes it. When precipitating with rain and solid particles, it enriches the soil as sulphur in its oxidized form. The amount of sulphur dioxide, which in industrialized areas is about 60 mg/m^3 of air, becomes multiplied by the number of cubic metres of air with which the exposed stone objects come into contact in the course of their long lives. The same coefficients must be considered in estimating the sulphur content in soils.

Pochon (1964) claims that the total amount of sulphur-containing compounds in soil amounts to 200–1000 kg/ha. This concentration is considerably increased by polluted precipitation. Thus the sulphur cycle in nature is augmented by additional amounts of this element, exerting a considerable influence on the whole of man's environment and, consequently, also on the condition of stone objects of art (Strzelczyk, 1967).

II. ORIGIN OF THE MICROFLORA ON ROCKS AND STONES

The main source of contamination of rocks and stones is the surrounding soil, which contains large numbers of bacteria, fungi, streptomycetes and algae. The abundance of these micro-organisms in soil depends on its content of organic compound. Certain constituents of cultivated soils are known to increase their fertility and therefore the number of micro-organisms. Soils in towns are fertilized by sources such as sewage, animal excrements and organic remains. These supplements activate microbial processes in soil. The number of bacteria in soil according to Burges and Raw (1967) reaches thousands to several hundred million cells per gram of soil. In very fertile soils, the number of bacteria may reach as much as billions of cells per gram. Apart from bacteria, there are large numbers of streptomycetes, fungi and algae.

All of these micro-organisms arrive at the surface of stones shortly after leaving their natural habitats, and, together with organic particles, they constitute the contamination of rocks and stones. Some rocks are rich in organic substance because of the way in which they have evolved. After extraction from quarries, the organic content of stone greatly increases. Organic compounds can penetrate to a depth of several centimetres inside the stone (Krumbein, 1973). The microflora colonizing the surfaces of rocks and stone includes auto- and heterotrophic organisms. Webley and his coworkers (1963) claim that the microflora inside and on the surface of stone are identical with soil microflora. Walsh (1968) in his work on ecological considerations of biodeterioration states: 'organisms concerned in biodeterioration are usually a selection of those found in the soil'. Considering, however, the specific properties of stone, including its lower water permeability and lower oxygen supply compared with soil, it should be presumed that differences in the microflora living on stone and in soil are quantitative rather than qualitative.

The microflora living on rocks and stones constitutes a complex community of organisms capable of carrying on processes of rock weathering and stone decomposition, while exhibiting interrelationships, and being involved in decomposition of organic material. Thus, in time, stone surfaces become colonized by communities whose development depends on the availability of nutrients and, to an even greater degree, on temperature and humidity. These communities develop in microniches in fissures, cracks and hollows of rocks and monuments. Their presence may at times be beneficial to the stone surfaces in that they may form a cover, so helping to regulate humidity balance in the stones. This refers primarily to the thick crusts formed under certain conditions by algae and lichens.

On more exposed stone surfaces, it is not possible for a thick layer of organisms to be maintained because it is washed away by rainwater. These surfaces, as well as those adjoining them, may serve as habitats for chemolithotrophic organisms, predominately algae and bacteria. These cause formation of crater-like tubules on the rock and stone surface, in which water, and not infrequently acidic products of microbial metabolism, accumulate, exerting a continuous lytic effect on the rock and deepening the already existing pits.

The microflora living on rocks and stones is capable of penetrating and living inside them as well. This has been confirmed by Myers and

McCready (1966), who found that the bacterium *Serratia marcescens* labelled with radioactive phosphorus was able to penetrate inside sandstone to a depth of about 36 cm in 84 hours at normal atmospheric pressure, and inside limestone to a depth of about 8 cm. This contradicts the opinions of other authors who state that bacteria are not capable of migrating inside rocks to any considerable depth because of the minute size of the pores and the absence of sufficient pressure to help penetration.

Another source of micro-organisms in stone is groundwater infiltrating from the soil to inefficiently waterproofed outside walls of buildings or socles of stone monuments. The rate at which water rises in stone depends on the size of the contact area of the non-waterproofed stone with soil, as well as on the drying capacity of the front of the damp area in the stone. Water passing up from the ground carries with it soluble components from the soil and the atmosphere, as well as huge numbers of micro-organisms. The penetration of water and bacteria is facilitated by the presence in the water of considerable amounts of dissolved carbon dioxide, which is a weak acid. Micro-organisms living inside the stone are those that find adequate conditions of humidity, oxygen and food supply.

III. ROLE OF CHEMOLITHOTROPHIC BACTERIA IN THE DECOMPOSITION OF ROCKS AND STONES

Among micro-organisms that get on to the surface and inside stone, an unquestionable part in the processes of degradation of stone objects is played by chemolithotrophic bacteria, are principally sulphur and nitrifying bacteria. The role of these micro-organisms, in spite of their great activity, is still underestimated, or even denied by some authors (Kieslinger, 1966) who are not concerned with soil microbiology and do not realize how intensive microbial processes are in soil. On the other hand, the reports of enormous amounts of chemical compounds (sulphur compounds included) emitted by chemical works into the atmosphere suggest that all the blame for degradation of stone monuments should be put on them. As evidence of the destructive effect of gases from the atmosphere on stone, we know that only 50–100 years ago, when air pollution was lower, the rate of degradation of stone was not so rapid as it is now. Also, at that time the amounts of sulphurous compounds that got from the air into the soil and formed substrate for bacteria participating in transformation of sulphur compounds in soil and stone were smaller.

Studies carried out by Pochon *et al.* (1951), Thiebaud and Lajudie (1963), Krumbein and Pochon (1964), Pochon (1964, 1966) and Pochon and Jaton (1968) have demonstrated the importance of bacteria of the genus *Thiobacillus* in 'plate' deterioration of stone objects. Kauffmann's (1960) paper speaks of the role of nitrifyers in the process of stone decomposition. Krumbein's further studies of the problem (Krumbein, 1966, 1968, 1973) unquestionably point out the role of sulphur-oxidizing and nitrifying bacteria in processes of stone decomposition. In his detailed ecological studies on the part played by microorganisms in stone and rock deterioration, Krumbein (1972, 1973) points to the fact that in cities the activity of the autotrophic microflora is largely stimulated by the enzymically very active heterotrophic microflora, which lives on refuse and garbage characteristic of these areas. Heterotrophs supply the sulphur bacteria and nitrifying bacteria with energy substrates (hydrogen sulphide and ammonia). Moreover, according to this author, there is a clear causal relation between the occurrence of sulphur bacteria and nitrifying bacteria and the symptoms of 'plate' and superificial deterioration of stone.

Krumbein expresses the opinion that, under urban conditions, a differentiation between stone and rock deterioration effected by chemical agents and that due to chemolithotrophic bacteria is extremely difficult. He points out that the greatest accumulations of stone-deterioration symptoms are observed in microniches such as hollows in stone monuments where water evaporation is obstructed and where the richest accumulations of the autotrophic and heterotrophic microflora are found. The products of microbial activity and the signs of deterioration are often the same as those due to purely chemical corrosion of rocks and stones.

A. Activity of Sulphur-Oxidizing Bacteria

The destruction of stone objects varies in form. The most sensitive to microbial deterioration are limestones and sandstones with a calcareous binder, though purely siliceous rocks are also liable to this kind of corrosion. Plate deterioration is particularly frequent and a definite role in this is ascribed to bacteria that oxidize sulphur-containing compounds. This kind of deterioration affects stone elements of buildings, monuments and sculptures imperfectly isolated from the ground and continually moistened by ground water. These objects are generally prevented from drying quickly or from being washed by rainwater. Their surface is darkened, sometimes nearly black, with the tars, soot and other air pollutants that have settled on them. It generally forms a tight crust, which cracks easily and falls off. At times, this crust is so tight and solid that it looks like entirely sound stone surface until it is mechanically injured. Beneath the crust, there is usually a layer of powdered stone devoid of binder. In the case of sandstone with calcareous binder, this is sand. In an advanced stage of deterioration

Fig. 1. Photograph of the townhall in Toruń, Poland showing the loss of upper layers from a stone sculpture. Photograph taken by J. Wolski.

the crust falls off the stone surface, thus ruining the surface form of the object of art, and the exposed deeper layer of stone becomes deteriorated in its turn. This results in deformation of the object and strips it of its artistic form (Fig. 1).

The mechanism of this process as presented by Pochon (1964) is as follows. Firstly, the soil in cities and industrialized areas is continually being enriched with oxidized and reduced sulphur compounds (sulphur dioxide, hydrogen sulphide) through the agency of falling solid particles and rain. Oxidized sulphur compounds then become rapidly reduced in wet soil by autotrophic anaerobic bacteria such as *Desulfovibrio desulfuricans*. These bacteria are very common in soil, as demonstrated by Pochon (1964). They are particularly abundant in wet soil near

Fig. 2. The Eskens' portal in Toruń, Poland showing deterioration of sandstone. Photograph taken by J. Wolski.

damp walls. They utilize oxygen obtained from reduction of sulphates for oxidizing hydrogen, and this supplies them with primary energy necessary to reduce carbon dioxide. Ground water, infiltrating stone, carries with it a rich soil microflora, a certain percentage of which are sulphur bacteria. Moreover, it is rich in carbon dioxide of microbial origin, which facilitates penetration of micro-organisms inside the stone. Most bacteria, fungi, streptomycetes and other micro-organisms are not capable of colonizing the subsurface layers of stone, but these layers have been found to be particularly rich in ammonifiers and sulphur-oxidizing bacteria such as *Thiobacillus thiooxydans*. The latter organisms use reduced sulphur compounds (sulphur, sulphides, thiosulphates, tetrathionates) and oxidize them to sulphuric acid. In the process, energy is released which is utilized by bacteria as primary energy. It has been found that, in the subsurface regions of deteriorated stone, *Thiobacillus thiooxydans* rises from 11,400 to 1,400,000 cells/g of stone. The sulphuric acid, as it is being formed, then reacts with calcium

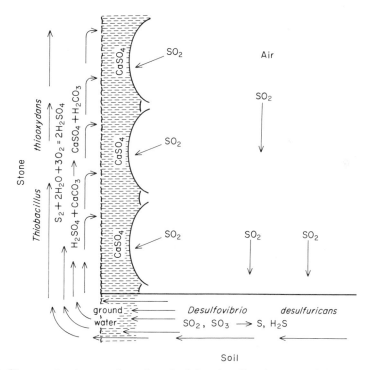

Fig. 3. Diagram showing transformation of sulphur in soil and stone, and the mechanism of stone deterioration.

carbonate in limestone and in the calcareous binder of sandstone. In the process, calcium sulphate is formed, and this in solution migrates towards the stone surface, where it is precipitated and binds two molecules of crystalline water $(CaSO_4 . 2H_2O)$. While so doing, it blocks the pores of the stone surface and, since its volume is 100% greater than that of the dissolved calcium carbonate, it tends to burst them. Moreover, gypsum, which is formed in the process, has a thermal expansion about five times that of calcium carbonate (Penkala, 1961). All of these agencies contribute to formation of crust and to its swelling. The crust has considerable weight, exceeding that of the calcium carbonate.

Similar marks of deterioration formed on the high parts of buildings also contain large numbers of *Th. thiooxydans*. The deteriorated areas are always near places liable to become damp (Fig. 2).

Figure 3 shows a diagram of the transformations of sulphur in soil and deteriorated stone.

The mechanism of deterioration caused by direct action of sulphur dioxide is much the same. Krumbein (1973), however, points out that chemical agents cannot be the only ones causing stone deterioration. If it were so, deterioration would not be confined to certain areas, but would affect large areas exposed to polluted air.

B. Activity of Nitrifying Bacteria

Within the last two decades, extensive evidence has accumulated on the part played by nitrifying bacteria in surface decomposition of stone and rock, though their occurrence and activity depend on other biological agents (Kauffmann, 1960; Krumbein, 1968, 1972). The bacteria oxidize ammonia to nitric and nitrous acid. They have been found to be particularly active in environments surrounding bird urine, excrement or protein substances vigorously decomposed by ammonifiers. A. B. Strzelczyk (unpublished work) isolated, from the ground course of St. John's basilica in Torun, nitrifying bacteria which, when inoculated on silica gel with enamel calcium carbonate, gave dissolution zones of a size never before observed in these bacteria isolated from cultivated soils.

The effect of nitrifying bacteria on limestone involves formation of calcium nitrate which, being readily soluble in water, is either washed away or crystallizes on the stone surface to form characteristic white spots described by Gargani (1968) on mural paintings as 'whitening of nitrates'. Superficial pits and craters are formed in the stone but no sulphur-oxidizing bacteria are found there. Indeed, the total numbers of bacteria are rather low, and there is a preponderance of nitrifyers and aerobic nitrogen-fixing bacteria (Krumbein and Pochon, 1964). Krumbein (1973) gives examples of intense activity of nitrifying bacteria (about 10,000 cells/g of limestone) resulting in formation on limestone of white snow-like efflorescences. Nitrates also exert a destructive effect on stone by changing their volume during transition from the dry to the wet state. Krumbein (1973) has demonstrated a causal relation between the number of nitrifying bacteria, the nitrate content in stones and the degree of destruction. He obtained similar results while studying damages caused by sulphur-oxidizing bacteria. The groups of bacteria in question are capable of decomposing carbonate rock, carbonate mortars, concrete, cement, as well as minerals and rocks.

IV. ROLE OF HETEROTROPHIC BACTERIA AND FUNGI IN DESTRUCTION OF ROCKS AND STONES

A number of writers referred to in this review believe that the activity of sulphur-oxidizing and nitrifying bacteria is accompanied by a very abundant heterotrophic microflora, among which an important part is played by ammonifying micro-organisms. Pochon (1964) found, in all three zones of plate ('en plaque') deterioration (the external (gypsum) one, the powdered one and the sound or internal one) considerable numbers of these bacteria (2×10^5–2×10^6 cells/g). Webley and his coworkers (1963) found that stone layers showing a smaller degree of deterioration contained fewer heterotrophic micro-organisms.

Stones and rocks situated in the open air are liable to become covered with dust and infiltrated with rain- and groundwater which, in a short time, results in their being saturated with considerable amounts of organic substance sufficient for the development of a rich heterotrophic microflora, which utilizes all available organic constituents in the stone. It has been found that concentrations of organic substance of 10–1 μM are enough for micro-organisms to live and reproduce. Acids of bacterial origin promote dissolution of rocks and stones and the release of a number of elements in the process. Henderson and Duff (1963) found that the most active in stone decomposition are non-sporulating Gram-negative rods. They were capable of vigorously breaking up silicates due to the considerable amounts of 2-oxogluconic acid they produced.

A particular role is attributed to bacterial slimes, which are acidic. Many stone-, rock- and soil-dwelling bacteria are slime-producing organisms. Slimes contribute to a large extent to the dissolving of rocks and stones and to rendering many cations (Fe, Mn, Al, Si and others) accessible to plants. Abundant occurrence of these bacteria may also affect the permeability of rock by expanding their capillaries as a result of the organic acids and carbonic acid they produce.

As revealed by Krumbein (1966), heterotrophic bacteria prefer stones of a neutral pH value. Krumbein (1973) expressed the opinion that the conditions prevailing in big cities, associated with all sorts of pollution, stimulate development of heterotrophic microflora. This is also promoted by milder ambient temperature in built-up areas.

Among rock- and stone-dwelling heterotrophs that promote weathering, mention must be made of the fungi. Their important role has been pointed out by Webley and his coworkers (1963), Henderson and Duff (1963) and other authors (Krumbein, 1972, 1973; Hueck van der Plas, 1968). They have noted that the fungi are not only capable of living on the surface of stones and rocks, but also contribute to their decomposition. Among the most active in this respect are, according to Henderson and Duff (1963), members of the genera *Botrytis*, *Aspergillus*, *Cephalosporium*, *Spicaria*, *Fusarium*, *Hormodendrum*, *Mucor*, *Penicillium* and *Trichoderma*. These fungi have been observed to dissolve the following silicate rocks: aprophylite, olivine, vermiculite, saponite and wollastonite. Kaolonite, labradorite, orthoclase and talc among others were not attacked. It has been found that, in the process of decomposition of silicate rocks, free silica is released, adding to the acidity of the medium in which the fungi live. Silica may help to dissolve phosphate rocks. The neutral pH value of culture media containing 4% glucose, on which these fungi were cultured, was lowered to 1.8 for *Aspergillus niger*, to 2.2 for *Spicaria* sp. and to 2.6 for *Penicillium* sp. Chromatographic studies showed that the fungi produced many organic acids responsible for dissolution of rocks and stones. Thus, in 50 ml of culture medium containing 4% glucose the following acids were found: in cultures of *A. niger*, critric acid (23.7 mg), oxalic acid (322.0 mg), fumaric acid

(29.5 mg); in cultures of *Spicaria* sp., acetic acid (19.3 mg), oxalic acid (25.9 mg), formic acid (8.3 mg); in cultures of *Penicillium* sp., citric acid (65.3 mg); in cultures of another strain of *Penicillium* sp., citric acid (73.6 mg).

Calcium-containing rocks (augite, talc, wollastonite) according to these authors, were rather weakly dissolved by oxalic acid of fungal origin, but much more vigorously by bacterial 2-oxogluconic acid. On the whole, however, the amount of silica released was greater in fungal than in bacterial cultures.

Bassi and Chiatante (1976) studied the capacity of fungi to attack marble surfaces, as well as the effect of pigeon excrement. The reason for undertaking these investigations was deterioration of the outside of Milan Cathedral, which is submitted to intensive fouling by pigeon excrement. Considering what is known about the part played by hetero-trophic micro-organisms in stone destruction, it was presumed that pigeon excrement might stimulate decomposition of marble by fungi. The authors studied samples of deteriorated marble from the outside of the Cathedral, and tested the effect of fungi on sound marble from the Candoglia quarry in the presence and absence of sterilized pigeon excrement. They reached the following conclusions. Firstly, from deteriorated marble fragments from the outside walls, members of the following species were isolated, *Mucor hiemalis*, *Fusarium oxysporum*, *Aspergillus clavatus*, *Aspergillus fumigatus*, *Aspergillus repens*, *Aspergillus ustus*, *Penicillium cyclopium* and two strains of bacteria related to *Diplococcus* sp. and *Bacillus megaterium*. Secondly, in cultures supplemented with sterilized pigeon excrement, fungal development was incomparably better than on non-enriched media. Thirdly, four of the fungal strains studied proved to be capable of lowering the pH value of the liquid medium they grew in to pH 3.6. Fourthly, already after 20 days of culture the polished marble surface which served as a substrate for the fungi showed pits, streaks, depressions and loss of smoothness. Finally, the authors put forward the opinion that the agent responsible for deterioration of marbles on the outside of Milan Cathedral are fungi, whose development is stimulated by pigeon excrement. They do not deny, however, the participation in these processes of the components of bird excrement and air pollution. No doubt a considerable part is also played by heterotrophic bacteria.

Many lower fungi, particularly aspergilli, are xerophilic, and their spores germinate at low relative humidity. These may cause uninterrupted deterioration of the outside of the Cathedral during periods of summer drought. During the rest of the year, the climate prevailing in this region of Italy ensures sufficient humidity and a temperature well within the range for activity of fungi and bacteria.

Krumbein (1973) did not find any deterioration of stone from pigeon excrement that formed thick crusts. It seems that the reason for the discrepancy in observations on this subject lies in the different environmental conditions in stones, leading to formation of niches characterized by specific conditions and distinct microclimates.

V. ROLE OF ALGAE IN DESTRUCTION OF ROCKS AND STONES

The activity of algae on rocks and stones is vigorous and obvious. In the first place, they foul the surface on which they develop. They form stains which vary in colour from light green to dark grey–green, and by so doing blur the clarity of the form of objects. Sometimes they entirely mask their surface and make engravings on it unreadable.

The richest development of algae is noted under conditions of nearly full saturation of the rock with water. As the substrate dries out, the number of living algal cells decreases rapidly. The first to die are individuals most sensitive to desiccation. Only seldom, however, are all members of a species simultaneously destroyed. As the rock becomes wet again, algal communities revive. On neglected stone surfaces and damp rocks, endolytic (forming canals) and chasmolytic (living in chasms) algae develop (Krumbein, 1973).

Algae have been found to produce and secrete a variety of metabolic products, among which predominate organic acids, including lactic, oxalic, succinic, acetic, glycoleic and pyruvic acids (Round, 1966; Hueck van der Plas, 1968). These acids either directly dissolve rock and stone components or increase their solubility, for example by the reaction:

$$CaCO_3 + H_2CO_3 \rightleftharpoons Ca (HCO_3)_2$$

Acidic calcium carbonate, which forms in the reaction, is 100-fold more readily soluble in water than $CaCO_3$ (Domasłowski, 1975). After evaporation of water, calcium bicarbonate becomes decarbonized to form calcium carbonate. In this case, algal acid metabolic products stimulate the migration of salt in the stone. Moreover, the change in solubility of rock and stone constituents alters the diameter of the capillaries in stones, the chemical composition of the water filling these capillaries, the coefficient of thermal expansion of the stones and many other parameters that increase the sensitivity of stone to physical weathering.

Algae secrete products of assimilation besides organic acids. Among the former are proteins, which are chelating agents and contribute to dissolution of phosphate rocks (Provasoli, 1958). They also secrete sugars, growth-promoting substances and antibiotics. These substances are responsible for formation over the surface of the algae of a thin layer of bacteria which utilize these compounds and are resistant to toxic substances secreted by the algae (Bell *et al.*, 1974). These bacteria, which are epiphytic, participate together with the algae in decomposition of the stone. Among the properties of algae which make them independent of the environment is their ability to fix atmospheric nitrogen, a property that is particularly well developed in blue–green algae (Provasoli, 1958). This feature is the reason for rather uneconomical use of amino acids and proteins by these organisms.

Many algae show a marked sensitivity to the pH value of the environment. Acid siliceous rocks (granites, gneisses) are inhabited by silicotrophic algae (mostly green algae) and alkaline rocks (limestones, marbles) by calcitrophic algae (mostly blue–green algae). There are, however, genera of micro-organisms for which the pH value of the substrate is not growth-limiting. For example, *Gleocapsa sanguinea* grows as often on granites, staining them red, as on limestones, but then the stains are violet in colour.

Growth of algae depends to a large extent on the availability of various elements, the most important of which are calcium and magnesium (Provasoli, 1958). This is the reason why the algae show special preference for calcareous rocks, walls and plasters.

An example of this phenomenon is the invasion of algae in the Lascaux Caves in the department of Dordogne in France (Lefevre, 1974). The caves were discovered in 1940. Over an area of 1500 m³ their walls are covered with rock painting and drawings showing animal scenes executed in earth pigments. They are a unique monument, whose age is estimated at 15,000 years. Shortly after their discovery, the caves were opened to visitors. It is estimated that they were visited daily by an average of 500–600 persons. This resulted in a build-up of humidity and carbon dioxide which, combined with the lighting that had to be put in, caused an invasion of algae on the rock drawings.

Microbiological studies showed infection of the air and walls of the caves with soil bacteria, fungi and streptomycetes, which may have had an indirect effect on spreading of algae. The author reports the presence of blue–green algae, green algae and diatoms, but does not give a full list of algal species that inhabited the inside of the Lascaux Caves. He only mentions *Chlorella* sp., *Stichococcus bacilliaris*, *Chlorobotrys* sp., *Brecteacoccus* sp. and *Palmelococcus* sp.

Thiebaud and Lajudie (1963) found, on the ancient buildings at Saint-Cloud, *Oscillatoria* and *Gleocapsa* species of the blue–green algae, *Chlorella* and *Pleurococcus* of the green algae, and *Navicula* and *Fragilaria* of the diatoms.

Table 1

Algae isolated from ancient stone. From Sokoll (1977)

Tribe	Algae found on:	
	Limestone	Sandstone
Chlorophyta	*Stichococcus bacilliaris*, *Chlorella vulgaris*, *Chlorella elipsoidea*, *Ulothrix punctata*	*Stichococcus bacilliaris*, *Chlorella vulgaris*, *Chlorella elipsoidea*
Cyanophyta	*Oscillatoria pseudogeminata*, *Oscillatoria terebriformis*, *Oscillatoria subtilissima*, *Phormidium lignicola*, *Lyngbia martensiana*, *Lyngbia aerugineocoerulea*, *Oocystis parva*, *Oocystis marssoni*, *Gleocapsa helvetica*	
Chrysophyta	*Botryochloris minima*, *Fragilaria* sp.	*Botryochloris minima*

Sokoll in my laboratory studied algae occurring on stone monuments and plasters. She isolated a number of algae, which are listed in Table 1. The algae isolated by Sokoll (1977) and Czerwonka (1976) were similar to those referred to in the literature as lithophytes. Shields and Durrel (1964) found the rock-dwelling algae *Pleurococcus* sp. and *Trentophila aurea* growing on hard, dry and shaded rocks; softer rocks were inhabited by the blue–green algae *Gleocapsa* sp., *Tolypothrix byssoides* and *Stigonema minutum*. On sheltered isolated surfaces, species of *Scytonema* and *Calotrix* develop. *Stigonema* species are characteristic of siliceous rocks, whereas *Schizothrix* and *Symploca* species are confined to calcareous rocks. *Fragilaria capucina* is one of the few diatoms inhabiting rock surfaces.

Other organisms known to develop on rocks and neglected stone surfaces are lichens. These organisms grow as the result of temporary symbiosis of aerophytic algae and fungi. They may affect rock decomposition in different ways. Under some conditions, a thick layer of lichens may protect the stone against frequent changes in humidity, which are a severe threat to the durability of stone. They form an insulating layer on the rock surface maintaining on it a steady microclimate. In most cases, however, these organisms help rain and condensation water to accumulate on the stone surface. They also corrode rock surface by excreting organic acids, most commonly oxalic acid.

VI. ATTEMPTS AT COMBATING MICROFLORA ON STONE MONUMENTS

As already pointed out, ancient stone situated in the open air is exposed to bacteria, fungi, algae and other organisms. In the course of conservatory treatment, it is frequently subject to mould growth. This may happen when prolonged treatment is necessary to remove salts saturating the stone. The treatment involves saturation of the stone with water and wrapping it in a thick layer of wet cotton wool. The wrappings often become overgrown with fungi, in which case disinfection is necessary. To inhibit the development of infection, the desalinating wrappings need to be sprayed with a solution of $0.3–0.5\%$ p-chloro-m-cresol in ethanol. This treatment disinfects the wrappings and prevents infection from being transferred on to the stone and to other uninfected objects in the laboratory.

The only way of combating bacteria and fungi living on stone objects standing in the open is by improving hygienic conditions, i.e. cutting them off from a source of water and cleaning their surface. Preventing access of water stops the development of micro-organisms, stops the supply of nutrients and inhibits the circulation of salt in stone. There are several ways of drying stone objects. The choice among them depends on the situation of the stone, the inflow and kind of water (ground or condensation) and on the temperature of the stone surface. The task is so difficult that it takes years to improve humidity conditions in the object. The problem appears in its most striking form in caves, crypts and archaeological excavations, where condensation water, which is the most troublesome factor, is conducive to microbial attack. Such conditions combined with long-duration lighting promote development of algae on stone surfaces. However, it has been found that the process of deterioration of stone objects stops after they have been taken into a museum.

Treatment with disinfectants to destroy the microflora developing on stone objects in the open has so far failed to give satisfactory results. All chemical compounds are subject to washing out by rainwater, inactivation by ultraviolet radiation, oxidation and decomposition by adapted micro-organisms. This is the conclusion from the author's long experience in the field of disinfection of stone objects.

In the literature, references on combating algae on stone objects cannot be found. Fitzgerald's (1971) valuable and interesting work on algicides contains no hint on combating aerophytic algae. On the other hand, most of the algicides included in the list of chemicals used for combating micro-organisms destroying various materials compiled by Hueck van der Plas (1966) are soluble in water. These compounds cannot therefore be used for preventive protection of stone objects.

We have found that some commercial preparations are fully effective for immediate combating of algal infection on stone. The durability of their effect, however, depends on the amount of precipitation. Disinfected fragments of monuments remain free from algal growth for some time, but algae then reappeared. This secondary growth contained communities that originally were not predominant (Czerwonka, 1976; Sokoll, 1977; Staszewska, 1977). Among the reviving algae were typical aerophytes, namely species of *Stichococcus* and *Protococcus*.

The best results in our work have been obtained by spraying or brushing stone surfaces with a 3% ethanol solution of Lastanox TA (a commercial preparation made by Lahema, Brno, Czechoslovakia), or a 3% ethanol solution (suspension) of Afalon 50 (marketed by Farbwerke Hoest AG, West Germany; a urea preparation), or a 3% solution of Atrazin 50 (Fixn. Pest Control Ltd., G.B., a derivative of atrazin).

Work on finding means of durably protecting stone objects against algal attack is being continued. To achieve this target, it is necessary to apply toxic and durable preparations, which would not show any noxious effect on stone monuments.

REFERENCES

Bell, W. H., Lang, J. M. and Mitchell, R. (1974). *Limnology and Oceanography,* **19**, 833–839.

Bassi, M. and Chiatante, D. (1976). *International Biodeterioration Bulletin* **12**(3), 73–79.
Burges, A. and Raw, F. (1967). 'Soil Biology'. Academic Press, London.
Ciach, T. (1967). Biblioteka Muzealnictwa i Ochrony Zabytków, B, XIX, 27–52, Warszawa.
Commins, B. T. (1962). *Research, London*, 421–425.
Czerwonka, M. (1976). M.Sc. Thesis. N. Copernicus University, Torun, Poland.
Domasłowski, W. (1966). Badania nad strukturalnym wzmacnianiem kamieni roztworami zywic epoksdowych. *Biblioteka Muzealnictawa i Ochrony Zabythków* B, XV, (English summary).
Domasłowski, W. (1975). Profilactic conservation of ancient stone objects. Collective work, published by N. Copernicus University, Toruń, Poland.
Fitzgerald, G. P. (1971). Algicides. The University of Wisconsin, Literature Review No. 2, pp. 1–50.
Gargani, G. (1968). *In* 'Biodeterioration of Materials. Proceedings of the 1st International Symposium' (A. H. Walters and J. S. Elphick, eds.), pp. 252. Elsevier, Amsterdam.
Henderson, M. E. K. and Duff, R. B. (1963). *Journal of Soil Science* **14**, 236–246.
Hueck van der Plas, E. (1966). *International Biodeterioration Bulletin* **2**(2), 69–120.
Hueck van der Plas, E. (1968). *International Biodeterioration Bulletin* **4**(1), 11–28.
Kauffmann, J. (1960). *Corrosion et Anticorrosion* **8**, 87–95.
Kieslinger, A. (1966). Conference on the Weathering of Stones, Brussels, 1–40.
Krumbein, W. E. (1968). *Zeitschrift für Allegmeine. Mikrobiologie de l'Institut* **8**, 107–117.
Krumbein, W. E. (1972). *Revue Écologie Biologique Solaie, IX* **3**, 283.
Krumbein, W. E. (1973). Deutsche Kunst-und Denkmalpflege. Deutsche Kunstferlag München, Berlin, 54–71.
Krumbein, W. E. and Pochon, J. (1964). *Annals de l'Institut Pasteur, Paris* **107**, 724–732.
Lefevre, M. (1974). *Studies in Conservation* **19**, 126–156.
Myers, G. E. and McCready, R. G. L. (1966). *Canadian Journal of Microbiology* **12**(3), 477–484.
Penkala, B. (1961). Ph.D. Thesis: Technical University of Warsaw, Poland.
Pochon, J. (1964). Conférence Sociéte d'Encouragment pour l'Industrie Nationale, 4 juin, 33–47.
Pochon, J. (1966). Conference on the Weathering of Stones, Brussels, 99–118.
Pochon, J. and Jaton, C. (1968). *In* Biodeterioration of Materials: Proceedings of the 1st International Biodeterioration Symposium, Southampton. (A. H. Walters and J. S. Elphick eds.), pp. 258–268. Elsevier, Amsterdam.
Pochon, J., Coppier, O. and Tchan, Y. T. (1951). *Chemie et Industrie* **65**, 496–500.
Provasoli, L. (1958). *Annual Review of Microbiology* **12**, 279–308.
Przedpełski, Z. (1957). Konserwacja kamienia w architekturze. Wydawnictwo Budownictwo i Architektura, Warszawa.
Round, F. E. (1966). 'The Biology of Algae' Edward Arnold, London.
Shields, L. M. and Durrel, L. W. (1964). *The Botanical Review* **1–3**, 92–128.
Sokoll, M. (1977). M.Sc. Thesis. N. Copernicus University, Toruń, Poland.
Straszewska, G. (1977). M.Sc. Thesis; N. Copernicus University, Toruń, Poland.
Strzelczyk, A. (1967). *Biblioteka Muzealnictwa i Ochrony Zabytków* B. XIX, 123–128.
Thiebaud, M. and Lajudie, J. (1963). *Annales de l'Institut Pasteur* **105**, 353–358.

Walsh, J. H. (1968). *International Biodeterioration Bulletin* **4**(1), 1–10.
Webley, D. M., Henderson, M. E. K. and Taylor, J. F. (1963). *Journal of Soil Science* **14**, 102–112.

4. Wool

J. LEWIS

*International Wool Secretariat, Ilkley, Yorkshire LS29 8PB, England**

I. Introduction 81
II. Fleece Damage Caused by Micro-Organisms 84
III. Bacterial Proteolytic Enzymes in Removal of Wool from Sheepskins . . . 89
    A. The Sweating Process 92
    B. Enzyme Depilation 96
    C. Recovery of Wool from Skin Pieces 97
IV. Microbial Damage of Processed Wool 98
    A. Conditions that Favour Development of Mildew on Wool 99
    B. Detection of Mildew Damage 101
    C. Testing of Textile Materials for Resistance to Microbial Attack . . . 103
    D. Chemical Control of Mildew 110
    E. Eradication of Mildew on Infected Goods 114
    F. Characteristics of Selected Treatments 114
V. Damage of Domestic Wool Textiles by Detergents Containing Proteolytic Enzymes 116
    A. Effects of Enzyme Detergents on Wool 116
    B. Treatments to Inhibit Degradation of Wool in Enzyme-Detergent Solutions 122
VI. Conclusion 127
VII. Acknowledgements 128
References. 128

I. INTRODUCTION

Keratin, the substance of wool and related materials such as hair and horn, is a complex of sulphur-containing proteins, stabilized by inter-molecular disulphide cross links, salt linkages and Van der Waals forces.

* Present address: Dinoval Chemicals Ltd., 54 Dundonald Road, London SW19 3PH.

It is generally believed that these crosslinks are largely responsible for the natural insolubility of keratins and for their relatively high resistance to digestion by proteolytic enzymes. When exposed to these enzymes, not more than 10% by weight of the wool fibre can be solubilized, even when the fibre is disintegrated by mechanical or chemical means, or has had its surface layer or cuticle removed (Alexander and Hudson, 1963).

The portion of the fibre that is digested is accounted for by the enzyme lability of intercellular material as well as nuclear remnants, cytoplasmic debris and the endocuticle (Dobb, 1963). A loss of this order associated with proteolysis, however, leads to severe disaggregation of the fibre which, when exposed to mechanical strain, will break and fibrillate readily, resulting in further deterioration. The remaining 90% of the wool, mainly comprising the keratin of the *ortho*- and *para*-cortex (Mercer, 1953), cannot be digested by tryptic enzymes unless the disulphide bonds have been ruptured (Geiger *et al.*, 1941) as, for example, by certain oxidative and reductive processes. Some chemically treated wools are therefore more susceptible to proteolysis (Lewis, 1972) and to microbiological damage (Lewis, 1973a, b) than untreated wool. Prolonged exposure of untreated wools to certain bacterial enzymes, however, has been reported to disrupt individual cortical cells to reveal an intracellular, fibrillar structure, interpreted to indicate digestion of the intracellular matrix (Molyneux, 1959).

In practice, wool suffers very little damage from contact with proteolytic enzymes and, in those cases where degradation is experienced, it is associated, generally, with one of the following causes: (i) fleece damage caused by micro-organisms; (ii) proteolytic activity in the removal of wool from skins; (iii) microbial damage of processed wool; (iv) damage of domestic wool textiles by detergents containing proteolytic enzymes.

In all cases, the proteolytic enzymes originate predominantly from micro-organisms and operate by hydrolysing those peptide bonds adjacent to the amino-acid side chains which satisfy their particular requirements of specificity. Most bacteria can degrade exogenous proteins using the products as sources of carbon, nitrogen and energy. Proteins must be broken down extracellularly by micro-organisms which use them as sources either of nitrogen or amino acids, because protein molecules are too large to pass through most microbial cell membranes.

Many bacteria produce proteases capable of hydrolysing proteins and these accumulate in the growth medium in various ways. In some cases, enzymes are detectable in the medium, simply as a result of the rupture of older cells and consequent release of intracellular proteases. With truly extracellular enzymes, however, large quantities of proteolytic enzymes are concentrated in the medium, irrespective of the age of the culture.

Extracellular proteases cleave the majority of proteins into peptides, after which the reaction is reported to cease (Lamanna et al., 1971). By then, however, it will have proceeded far enough to produce products whose sizes allow them to be assimilated by living cells. In older cultures, proteases appear and their presence is accompanied by peptidases which hydrolyse peptides to individual amino acids. The latter two types of enzyme are thought to be of intracellular origin and appear in the medium only when cells rupture. When proteases and peptidases are present, certain substrate proteins can be hydrolysed completely to provide amino acids for those organisms in the culture that are still alive. Degradation of protein fibres by microbial activity, therefore, may be summarized as follows.

Protein → proteose → peptone → polypeptide → amino acid → carbon dioxide + ammonia + water.

Many bacteria cannot be sustained by pure, complex proteins but are able to break these down, provided another source of nitrogen is present, in an assimilable form. Certain highly proteolytic organisms, however, including *Bacillus subtilis*, *Chromobacterium prodigiosum* and *Proteus vulgaris*, can digest protein in the absence of other sources of nitrogen (Rettger and Sperry, 1915; Rettger and Berman, 1918; Rettger et al., 1916). More recently, keratinolytic organisms such as *Candida albicans* (Kapica and Blank, 1957) and *Tritirachium album* Limber (Ebeling et al., 1974) have been reported which are capable of growing on native keratins as sole carbohydrate and nitrogen source.

With the majority of wool-infecting organisms, decomposition of polypeptides is increased by addition, to various media, of another carbon source such as carbohydrate (Raistrick, 1919). Similarly, the activity of saprophytic species is enhanced, considerably, by the presence of fats and fatty acids, attributable to production of easily assimilable nutrients from the degradation of the fats (Race, 1946).

In sheepskins, fleeces and wool textile materials, additional sources of nitrogen, carbohydrates and fats, such as animal excreta, soaps, oils, fatty acids or perspiration, are common and constitute different levels of impurity which are often more than adequate to initiate microbial growth. Consequently, the number of species of proteolytic bacteria capable of affecting raw or soiled wool is much larger than that of the truly keratinolytic organisms capable of digesting clean, uncontaminated wool. Metabolism of proteolytic bacteria depends on, and is regulated by, a complex system of enzymes and catalysts. The activity of these is influenced by a variety of environmental factors including moisture content, temperature, pH value and the oxidation–reduction potential of the system. The extent to which these factors influence growth of proteolytic bacteria and associated damage to wool has been described, in detail, by Race (1946).

II. FLEECE DAMAGE CAUSED BY MICRO-ORGANISMS

The presence of large numbers of micro-organisms in fleece wool has been reported (Prindle, 1935) and their numbers, in certain fleece types, fluctuated considerably with rainfall (Mulcock and Fraser, 1958). One feature of their presence is the production of various spectacular stained effects when the environment is conducive to development of the causative organism (Fraser and Mulcock, 1956). In most cases, little physical damage to the fibre takes place, and staining often results from accumulation of particulate pigments. Some of these can be removed, more or less completely, by aqueous or solvent scouring, but in other cases the staining is extremely resistant to removal (Henderson, 1968).

With certain infecting organisms, however, staining is accompanied by extensive degradation of the fibre, evident both macroscopically and microscopically. Such damage inevitably downgrades the fleece and causes a number of problems in subsequent processing and dyeing of the affected wool.

Typical of this fault is the damage known as pink rot (Waters, 1932). This is manifested by a discolouration of the wool ranging from bright pink to creamy-pink. Although the attendant fibre damage is often severe, it is generally localized and only a small proportion of the fleece is usually affected. Typical areas are located behind the foreleg or on the side, and in one of the natural openings in the fleece. Only 2.5–5.0 cm in the middle of the staple or lock of wool are affected. Pink rot is found among a variety of breeds of sheep, and its development seems to be initiated by prolonged wetness of the fleece. Fibres become rotten and fused into a stringy, pink-coloured mass (Fig. 1). The organism responsible is capable of digesting the material between the cortical cells of wool fibres (Fig. 2) and resembles *Bacillus subtilis* (Molyneux, 1959) an organism known to exhibit considerable proteolytic activity. Control of infection and degradation by this and related organisms consists primarily in selecting animals with a well-ventilated fleece (Hayman, 1953; Fraser, 1957).

Wool displaying a fault known as black fungus tip (Fig. 3) is character-ized by a tarry black discolouration of the staple end (Mulcock, 1959). This infection was identified only as recently as 1959 to be of microbial origin, and is found principally in a variety of New Zealand wools from

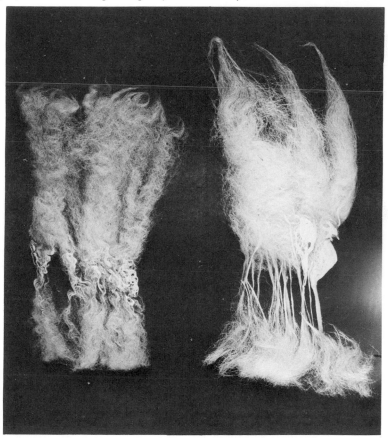

Fig. 1. Staples of wool affected by pink rot.

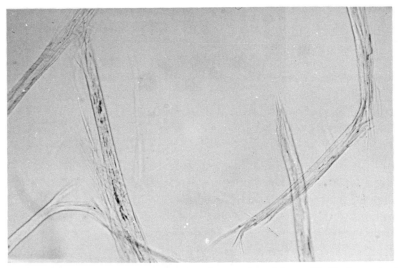

Fig. 2. Wool fibres damaged by the micro-organism causing pink rot. Fibrillation and release of cortical cells are evident.

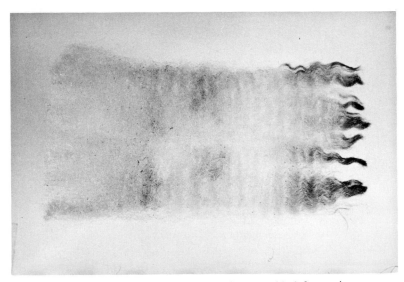

Fig. 3. A staple of wool exhibiting the phenomenon known as black fungus tip.

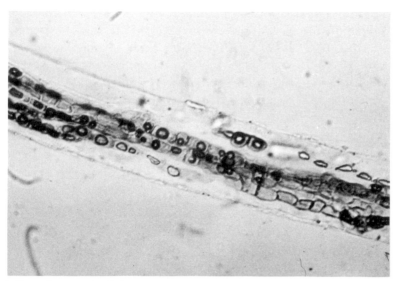

Fig. 4. Extensive growth of the causative organism, *Peyronellaea glomerata*, within the fibre. Magnification × 620.

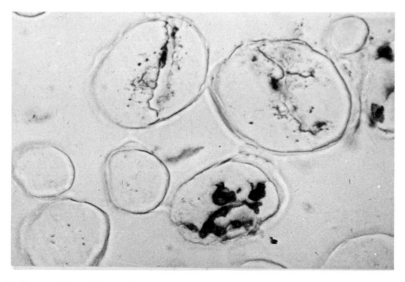

Fig. 5. Cross-sections of fibres affected by *Peyronellaea glomerata* revealing extensive development of the organism within the fibres. Magnification × 330.

sheep pastured in areas of both high and low rainfall (Mulcock, 1961). The attendant discolouration, caused by the presence of fungal mycelium growing in the tip of the fibres (Figs. 4 and 5), cannot be removed by scouring or solvent extraction and is probably the most permanent discolouration of wool known (Henderson, 1968). Microscopic examination of infected fibres has shown that black pigmented fungal hyphae are located within fibres and, where fruiting structures are formed, the cortical cell structure of the fibre becomes disrupted (Fig. 6). The fungus has been identified as *Peyronellaea glomerata* (Mulcock, 1961).

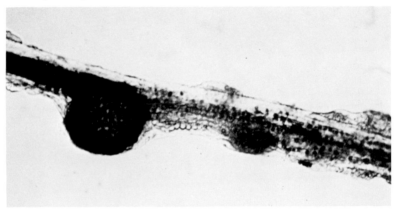

Fig. 6. Rupture of a wool fibre by a fruiting body of the fungus, *Peyronellaea glomerata*. Magnification × 330.

The effects of sunlight and weathering may damage the tips of staples considerably, and this damage appears to be a prerequisite to infection and deterioration by this organism. The basal portion of the staple is not attacked, even when incubated with a culture of this organism under optimum physiological conditions. Henderson (1968) suggested that this condition was a relatively unimportant economic factor in the past, since the infected portion of the fibre is so weak mechanically that it was lost in the older, more vigorous, methods of wool scouring. Increasing use of more gentle scouring procedures may exacerbate problems associated with a stained and degraded wool of this type. Unfortunately, no effective means for controlling the development of this infection can be recommended at present.

III. BACTERIAL PROTEOLYTIC ENZYMES IN
REMOVAL OF WOOL FROM SHEEPSKINS

Wool removed from the skins of slaughtered sheep and lambs (slipe wool) accounts for approximately 10% of the World's wool supply. In countries such as the United Kingdom, New Zealand and South America, where meat production is the primary product of sheep farming, the ratio of slipe wool to total wool production is higher than that of several other wool-producing countries. In most of these, establishments exist for dewoolling sheepskins, often as an integral part of large freezing complexes for dealing with animal carcases for the

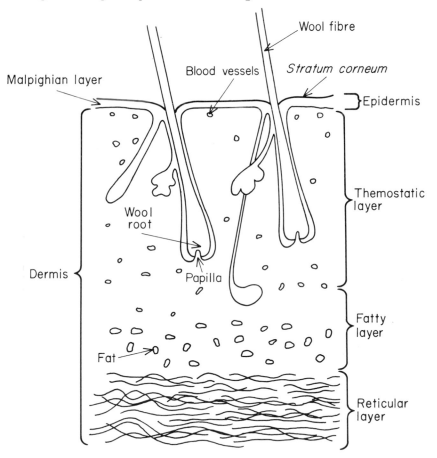

Fig. 7. Diagram of a vertical section through sheepskin. After Lennox (1945).

frozen-meat trade. In addition, skins are exported for treatment, or fellmongering, in other countries and, of these, France and the United Kingdom are the largest processors in Europe.

Fellmongering is an industry based on removal of wool from sheep-skins, and various bacterial and chemical processes are in commercial use (Carrie and Woodroffe, 1960). The two products of processing are wool fibre and the dewoolled skin, or pelt, used widely in the leather trade. It is uncommon to remove wool from the skin by shearing, since the maximum length of fibre is not obtained in this way and the pelt remains covered with fibre which renders it unsatisfactory for leather manufacture. Instead, processes are used that allow the wool to be loosened and pulled, more or less intact, from the pelt leaving it devoid of fibre. An understanding of fellmongering processes is assisted by consideration of the gross structure of the skin of the sheep (Anon, 1950a). The skin consists essentially of two main layers, a thin epidermis or outer layer and a much thicker layer known as the dermis. These layers are capable of subdivision as outlined in Figure 7.

The epidermis is formed by growth and division of the very active layer of cells in the basal or Malpighian layer. These cubical cells divide, become flattened and keratinized as they move towards the outer surface of the skin where they form the *stratum corneum*. The keratinized layer is resistant to bacterial attack whereas the Malpighian layer is particularly susceptible. The dermis consists of two main layers of approximately equal thickness separated by a thin fatty layer. The outer or thermostatic layer (known in leather manufacture as the grain layer) of the dermis contains the wool follicles (Figs 8 and 9) and the underlying or reticular layer consists of a reticulate structure of bundles of collagen fibres.

Removal of wool from sheepskins is effected either by bacterial digestion of the prekeratinous follicle bulb, known as sweating, or by application of a depilatory chemical or enzyme preparation on the flesh side of the skin, known as painting. Preliminary operations are common to both methods, and involve soaking the skins in pits of water to remove blood and loose dirt in the case of fresh skins and, in a longer process, to restore the moisture content of dried skins to a value close to that of fresh ones. Antiseptics are sometimes added to the liquor to retard growth of anaerobic bacteria, many of which attack collagen and damage the skin. After soaking, the skins are passed through a machine to remove solid contaminants, such as vegetable

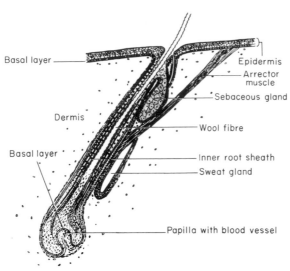

Fig. 8. Detailed structure of a wool follicle.

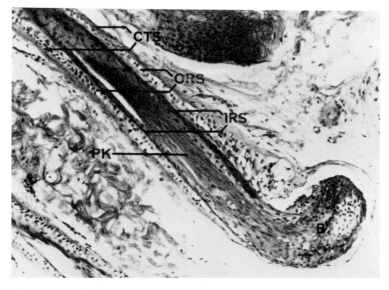

Fig. 9. Vertical section through sheepskin showing the structural features of a wool follicle involved in removal of wool by sweating processes. Abbreviations: CTS, connective tissue sheath; ORS, outer root sheath; IRS, inner root sheath; PK, prekeratinous zone; B, follicle bulb.

burrs and lumps of soil, and some soluble contaminants such as dried perspiration. Excess water is then removed prior to subsequent processing.

The choice of process for dewoolling skins depends mainly on the relative effect of the process on the quality of the wool and pelt produced. In general, where preservation of fibre length and quality is important, as is the case with merino skins, sweating is preferred, at the expense of the pelt, in this case a secondary consideration since it is thin and wrinkled. Sweating involves digestion of the prekeratinous region of the fibre only (which is very short), whereas the painting process digests or damages the fibre to at least (and often beyond) the surface level of the skin (Anon, 1951). With skins from crossbred animals, where a relatively better pelt is produced and where preservation of fibre length is not so important, painting processes are generally preferred.

A. The Sweating Process

This process is effected by bacteria present on the skin and involves digestion of the bulb and prekeratinous region of the fibre in the follicle,

Fig. 10. Sheepskins hanging in a sweating chamber.

and also of the Malpighian layer. Sweating is conducted in specially designed chambers, where the deburred and drained skins are hung over rods (flesh side out, Fig. 10) or from hooks (flesh sides together). The chambers are provided with ventilators in walls and ceilings with which to control the temperature and humidity of the atmosphere. According to these conditions and the type of skin, the process of sweating may take between two and eight days before the wool is readily removable. The progress of the process is monitored, subjectively, by assessing the ease of removal of wool from the shoulder region of skins, by the development of a glistening appearance on the flesh side (bacterial slime) and by the smell of ammonia.

The principal organisms responsible for loosening of the fibres are aerobic and vary depending on the source of the skins. With Australian skins, a strain of *Proteus vulgaris* predominates (Maxwell, 1945) while, on Canadian, English and American skins, the active organism is a species of *Pseudomonas* (Maxwell, 1950). These bacteria multiply in and digest the Malpighian layer and the prekeratinized fibre but do not attack the keratinized fibre or the epidermis (Ellis, 1945). On pulling to remove the wool fibres, a small amount of epidermis may remain attached to the roots of the wool, and these 'skin flakes' provide one means of differentiating between slipe wool and shorn wool, even in scoured samples.

During the period of active multiplication of the bacteria, a gradual migration of the organisms towards the wool roots takes place. It has been shown, however, that bacteria do not penetrate to the base of the root until about ten hours prior to completion of sweating at 25°C (Maxwell, 1945). During bacterial action, a considerable quantity of ammonia is evolved but, although this is a contributory factor, it is insufficient in itself to cause loosening of the fibre.

Lennox and Maxwell (1945) provide convincing evidence that complete loosening of wool is dependent on bacterial action by relating the ease of removal of wool from the skin to the rate of multiplication of the bacteria present, for different temperatures of sweating. Their results (Fig. 11) show very good agreement between the time of loosening of the wool (for which a physical test was devised) and the time to attain minimum bacterial count, defined as the minimum number of organisms always present when the wool is loosened completely. It was also shown that the temperature of the sweating process significantly affects the activity and rate of multiplication of bacteria present. Maximum activity increased rapidly from 10°C to 22°C, more slowly up to 30°C and, thereafter, up to 40°C, insignificantly. It is generally recognized, however, that damage to the pelt is minimized by fermenting for a longer time under relatively cool conditions (Anon, 1966a).

High humidity is also desirable in order to prevent evaporation of moisture from the skins and to provide an adequate environment for extracellular digestion of the substrate by bacteria. Various rudimentary techniques are used to control temperature and humidity depending on the local climate and, in the main, these involve running hot or cold water over the floor of the sweating chamber. Sophisticated installations utilize steam injection, refrigerated coils and humidifiers to control conditions precisely.

Typical advantages claimed for the sweating process used to recover wool from skins include the lack of use of noxious chemicals, of

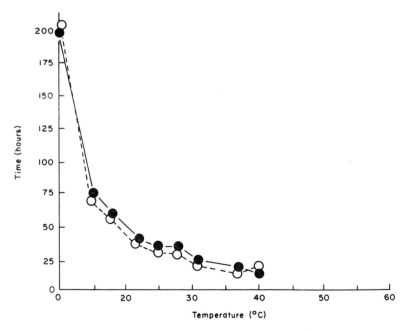

Fig. 11. Graphs showing the effect of temperature in sweating on the rate of wool loosening and on the rate of multiplication of skin bacteria. ● indicates time for complete loosening of wool; ○, time to attain standard bacterial count. After Lennox (1945).

significant loss of wool and of problems associated with disposal of a chemical effluent. Sweated wool has a characteristic tapered end to the root instead of the normal bulbous appearance of wool roots pulled from fresh skins. In shorn wool, of course, the latter remain embedded in the skin.

The precise sequence of events in the mechanism of the sweating process has not been elucidated completely. It is not known whether bacteria penetrate the skin and act *in situ* around the follicles, or whether enzymes diffuse to this region, through the reticular layer, from bacterial slime on the flesh side. A large number of bacterial species have been isolated from skin slime but the majority do not exhibit depilatory activity in pure culture. Maxwell (1945) showed that *P. vulgaris* had a marked depilatory potential, but little is known of the combination of enzymes secreted by this bacterium.

Soluble proteolytic enzymes produced by bacteria, and which are liberated on drying of sweated wool, have been examined in detail in a series of studies, with a view to obtaining better control of the sweating process. Separation of molecules with proteolytic activity in an eluate from gel chromatography has been carried out using isoelectric focussing (Delsol and Sergent, 1975). In this technique, a continuous current is established across an electrolyte system in which the pH value increases progressively from the anode to the cathode. The gradient of pH values is stabilized by the use of low molecular-weight amphoteric molecules called ampholines. During separation, proteins and other amphoteric molecules are repelled by the electrodes, and collect in a region where the local pH value is the same as that of their isoelectric points. Using this technique, three active fractions have been recovered, the isoelectric points of which are situated around pH 3.7, 4.3 and 6.8. This investigation showed that several types of proteolytic enzyme are involved in wool removal by sweating. Some success in elucidating the various types of peptide linkages broken by proteases produced by bacteria during the sweating process was reported in a subsequent paper (Delsol and Sergent, 1976). An effluent from compressed, processed wool was purified by filtration, gel chromatography and isoelectric focussing. Fractions were concentrated and assayed against a well-characterized protein, namely casein, the amino-acid composition of which is known.

Confirmation of specificity was sought by subjecting this to both acid and enzyme hydrolysis. Fractions obtained near the isoelectric

Wools containing impurities, such as vegetable oils, soaps, suint, sizes, hygroscopic and nitrogenous materials, are more liable to attack by mildew than is clean wool because these additives provide a readily digestible source of nutrients for the organisms. Microbial growths are often initiated in contaminants on the wool and subsequently spread to the wool fibres. Impurities may also support growth of organisms, sometimes pathogenic ones, which do not degrade the fibre itself but which are capable of staining it (Henderson, 1968). Contact of wool with jute, cotton and other cellulosic fibres, e.g. in packaging, in carpets or in blends, can also promote biodeterioration because of the enhanced liability of wool to become mildewed in the presence of these less-resistant fibres (David and Barr, 1972).

Wool stored under very adverse conditions will eventually become rotten. In the early stages of attack, the distal scale edges exhibit marked morphological changes (Fig. 12) and damaged fibre ends adopt a frayed appearance as cortical cells become disaggregated (Fig. 15,

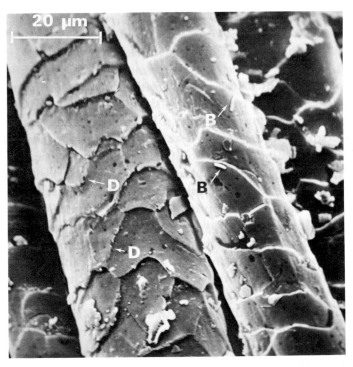

Fig. 12. Wool fibres exhibiting damage to the distal scale edges (D) characteristic of the early stages of attack by micro-organisms. Structures (B), possibly bacilli, are also present.

p. 105). Complete degradation takes place occasionally in the case of paper makers' felts and roller-lapping tops. Destruction of a wool textile material is effected by extracellular proteolytic enzymes which are secreted by micro-organisms onto the wool fibres. The function of these enzymes is to convert nondiffusible proteins into diffusible products which can be adsorbed and metabolized by micro-organisms. Growth of these is therefore not supported by the fibre *per se* but by its degradation products. The chances of attack are obviously higher with unsound wool or with wool that is damaged chemically or physically during processing, because the enzyme-resistant exocuticle may have been removed or disrupted. The more severe the processing damage, the more quickly will the fibres be affected by micro-organisms.

Fig. 13. A sample of fabric exhibiting streaking after piece dyeing. The paler thread was found, by microscopical examination, to contain mildew-damaged fibres.

B. Detection of Mildew Damage

Mildew damage to wool may not be identifiable easily until it has reached the stage where all, or a large part, of the infected substrate is rendered useless. The first indication of incipient mildew is the musty odour characteristic of mildewed materials in general. Macroscopic visible signs of mildew are the variously coloured growths and stains that develop on infected goods, but extensive growth is essential before these become visible to the naked eye. Infected wool which is subsequently dyed will invariably dye to a paler shade than undamaged fibres or yarns of the same batch (Fig. 13). This is accounted for by the mildewed areas exhibiting a resistant effect, preventing normal penetration of dyestuffs. Where the fibre is actually damanged, absorption of the dye may be rapid, but its fixation on boiling will be diminished and insufficiently-dyed areas will be produced. Redyeing to a darker shade using dyes of high molecular-weight may eliminate differential dyeing effects (Anon, 1950b).

Detection of fungal damage by examination of infected goods under ultraviolet radiation (Garner, 1967) is not satisfactory unless the damage has reached the stage where mildew attack is often obvious anyway. The procedure is, moreover, equivocal since certain auxiliaries used, or present in processing aids such as mineral oils, dyestuffs and optical brightening agents, also fluoresce in ultraviolet radiation. The most satisfactory method for identifying mildew infection and its consequences, therefore, relies on a microscopic examination of suspect fibres. The adhesive-tape technique (Lloyd, 1965) for microscopical examination of fibres makes it possible to detect the presence of fungi and bacteria on fibres in yarns and fabric without the necessity to tease out fibres, a process which, apart from introducing undesirable artefacts, inevitably destroys fragile fungal growths. With this procedure, a short length of cellulosic, adhesive tape, e.g. 'Sellotape', is pressed gently onto the surface of the suspect substrate and carefully peeled off. A 1 cm square of the tape is then excised and mounted, sticky side up, in a stain-mountant such as lactophenol–cotton blue. A cover slip placed over the tape prevents it from curling as it absorbs water. Some of the various forms of fungal and bacterial growths which may be revealed by this procedure are illustrated in Figure 14. Wool that has been wet processed after mildew attack is, however, unlikely to retain any surface growths and, in some cases therefore, it is necessary to tease out a few fibres for examination of the consequences of mildew damage in a similar mountant. Fungal deterioration may not be detectable in this case, but bacterial damage is revealed, even after considerable processing, by the presence of well-defined dark specks and streaks which tend to follow the cortical cell boundaries. Fibrillated fibres and frayed fibre ends are also evidence of mildew damage. Figure 15 shows wool fibres exhibiting progressively more severe bacterial damage.

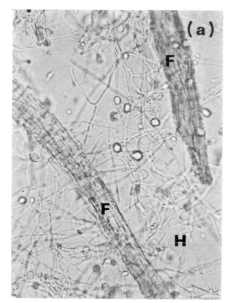
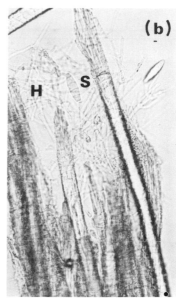

Fig. 14. Wool fibres (F) examined by the 'adhesive-tape technique' showing (a) fibrillation of infected fibres, and (b) mycelial growths comprising hyphae (H) and spores (S). The last vary in shape and size from 1 μm to 200 μm (Smith, 1971) and therefore identification of these on the basis of a microscopical examination is difficult. Magnification × 500.

C. Testing of Textile Materials for Resistance to Microbial Attack

Test methods for assessing the susceptibility of textile materials to micro-biological degradation, or for comparing the efficiencies of mildew-proofing products generally, involve a method for deliberately infecting the textile, combined with a physical method of testing, usually a tensile-strength test, for measuring the extent of the resultant de-gradation. Procedures for infecting and promoting growth of mildew on the test specimen are necessarily severe because of the accelerated nature of the test. Most methods involve exposure of materials under

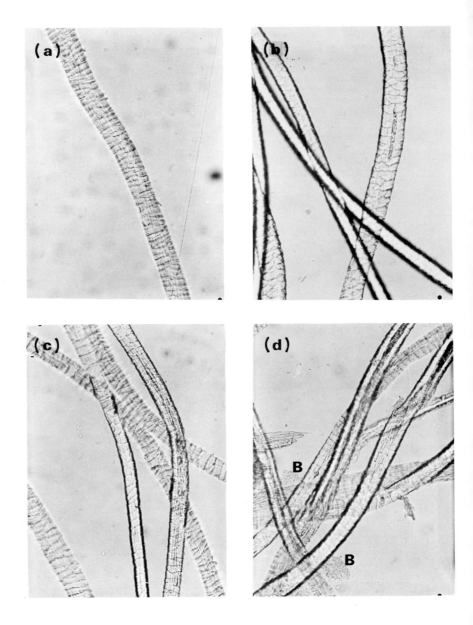

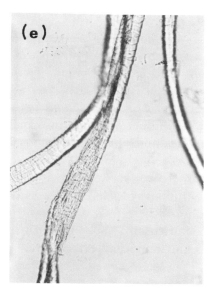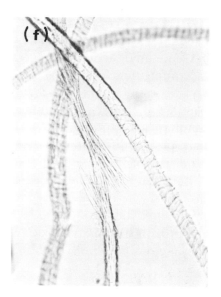

Fig. 15. Appearance of untreated, undamaged fibres prior to infection (a) and after incubation with an infecting medium for one to five weeks (b–f respectively). As the incubation period increased, longitudinal streaking characteristic of bacterial damage in wool was supplemented by fibrillation and fraying of damaged fibre ends. Bacterial colonies (B) are evident also. The magnification of all figures is × 300.

test to controlled physical, chemical and biological conditions which encourage rapid deterioration by single species, or by mixed cultures of fungi and/or bacteria. In this way, the test should provide an indication of the extent of microbial damage in a conveniently short time. Many different procedures for infecting textile substrates have been devised (Siu, 1951) in an attempt to standardize experiments. The difficulty lies principally in the tremendous variability of the organisms involved. The tests most commonly employed are as follows.

1. Pure-Culture Techniques

In such tests (Marsh, 1945), the sample of fabric is exposed to attack by a pure culture of a known micro-organism for specified periods of time (Fig. 16). In the case of wool fabrics, it is preferable to inoculate unsterilized samples with keratin-destroying organisms because the physical processes sometimes used for sterilization, such as autoclaving and dry-heat, may be injurious to the fibres (Burgess, 1924). Micro-

2. Mixed-Culture Techniques

bial deterioration of protein fibres involves the combined, simultaneous or sequential action of several species, and the use of a mixed inoculum, therefore, can be more severe and more representative of typical in-use conditions than the pure-culture procedure. The advantage of a mixed inoculum (Wellman and McCallan, 1945) is the simultaneous exposure of test samples to a number of organisms which may possess different tolerances to biocides and varied physiological requirements.

Both the single- and mixed-culture procedures have disadvantages, however, when they are used for testing mildew-proofed or inherently mildew-resistant materials. Notable among these is accumulation in the growth medium of toxic residues or waste products produced by the micro-organisms, which may lower the activity of the culture.

3. Perfusion Techniques

With this procedure (Eggins *et al.*, 1968), supplementary nutrients are supplied continuously to a porous substrate which is held in contact

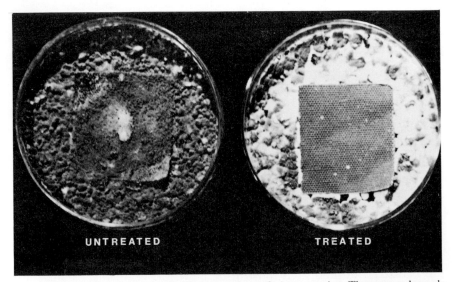

Fig. 16. Samples of fabric inoculated with a pure culture of micro-organism. The untreated sample on the left is overgrown with fungus. The treated sample on the right remains free from surface growth although extensive growth is present on the supporting medium.

with a single or mixed inoculum of micro-organisms. Expended growth medium, together with waste metabolic products, is at the same time removed. Some of the disadvantages of the former techniques are therefore obviated by continual supplementation of substrate with nutrients and by removal of inhibitory products by perfusion. In consequence, the environment is made more favourable to growth of bacteria and fungi, and thus the substrate more susceptible to deterioration. The apparatus (Fig. 17) consists essentially of a glass-fibre wick connected to both ends of a strip of textile material held in contact with an appropriate inoculum in a glass Petri-dish. The strip carrying the nutrient salts solution is enclosed by an autoclavable plastic sleeve to maintain sterility. The tail wick, carrying exhausted nutrient and metabolic wastes, is exposed to the air to encourage drying and to maintain a constant flow of nutrient through the assembly.

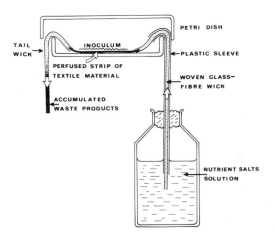

Fig. 17. Schematic diagram of perfusion apparatus. The whole assembly can be sterilized, prior to inoculation, by autoclaving. See text for details.

4. Soil-Burial Method

The most drastic and popular test method involves burial of strip-test samples in composted soil contained in trays, soil beds, or in the field, for specified periods of time (Batson *et al.*, 1944). After washing gently

Fig. 18. Soil-burial procedure. Note the insertion of test strips into a prepared soil bed by means of a wooden wedge.

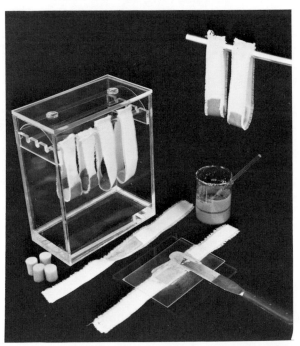

Fig. 19. The apparatus employed for evaluating mildew resistance by the 'soil-infection method'.

and conditioning the strips, the decrease in breaking strength, generally as a percentage, is measured as a criterion of microbiological resistance. In laboratory exposure, soil is often employed in trays of 7.5–15 cm depth which is maintained at 95% relative humidity and 32°C throughout the test. The depth to which samples are inserted by, for example, a wooden wedge, should be controlled to leave, say 5 cm of a 15 cm strip of cloth exposed above the level of the soil (Fig. 18). In some cases, it may be preferable to bury the strip in a horizontal position, 2.5 cm below the surface of the soil. Difficulties in standardizing soil-burial procedures involve the high degree of variability of results from different soils, in ensuring adequate uniformity, compactness, aeration and moisture content of the soil, and an equivalent degree of contact of the soil particles with the test strips.

Fig. 20. Test strips of worsted serge after evaluation by the 'soil infection method' for two weeks. The two strips in the top half of the print are untreated and clearly are heavily infected by microbial growths. The treated strips in the bottom half of the photograph, while stained slightly by the soil-compost medium, do not support any surface growth.

5. Soil-Infection Method

In this test (Lloyd, 1955), strips of cloth measuring 2.5 cm wide by 30 cm are infected by partially coating them with a soil-compost suspension thickened to a paste with Kieselguhr. They are incubated while suspended over water (Fig. 19) and are kept wet by means of a polyether-foam plug impregnated with water. Each strip is hung by its ends to form a closed loop with the infecting medium on the outside. After incubation at 26°C–28°C for the desired period of time (2–4 weeks should be adequate), the strips are rinsed gently to free them from soil (Fig. 20), dried, conditioned and tested. Advantages of this method are: (i) uniform contact of the soil medium with the specimens; (ii) standardized conditions of humidity; (iii) uniformity of moisture content and aeration between the strips; (iv) numerous tests may be performed concurrently because of the convenient size of the apparatus and the test specimens involved and (v) there are no difficulties in maintaining sterility or storing large volumes of composted soil.

D. Chemical Control of Mildew

Chemical treatments for complete inhibition of mildew formation on wool are, in the main, unnecessary provided that the wool goods are stored carefully in well ventilated surroundings in a cool, dry atmosphere. Fungal spores cannot germinate without moisture and therefore if the amount of water in the material can be kept below a certain critical level, fungal growth will be prevented. These apparently simple storage requirements are, however, sometimes difficult to satisfy. Particular instances are unfavourable conditions during wet processing, and storage and use of wool and wool goods in leaky warehouses, in tropical climates and during shipping, particularly if impermeable polythene containers are used.

Treatments with antiseptic substances are the most effective means of controlling growth of mildew and of eradicating incipient infection. Antiseptics are not necessarily fungicidal since they often merely inhibit growth of fungal spores which can be very resistant to toxic substances. The action of an antiseptic is limited by factors such as the availability of nutrients present with it, the moisture content of the fibre, the concentration of antiseptic and the duration of contact of fungi with wool. It has been noted that antiseptic substances can be specific to a substrate, and completely different efficiencies are possible on different materials (Smith, 1971).

Many commercial mildew-proofing products are nevertheless

effective for wool, and treatments are available for aqueous application by exhaustion, padding or spraying to stock (loose fibre), sliver (an assembly of fibres), yarn, fabric and garments. Processes for solvent application, for example in dry-cleaning machines, are also feasible. As far as wool goods are concerned, however, sufficient protection is required only to prevent infection during processing and to impart some resistance to goods stored or shipped under non-ideal conditions. Modern commercial products are based on dichlorophen (5,5-dichloro-2,2-dihydroxydiphenylmethane), quaternary ammonium compounds, (mainly n-alkyl dimethyl (ethylbenzyl) ammonium chlorides), chlorinated phenols and fatty esters of chlorinated phenols and organotin compounds.

The majority of these products will prevent formation of mildew on wool, provided that sufficient product is correctly applied. Three outstandingly effective products for wool are Sandocide liquid (Sandoz) and Prestofen GDC (GAF), both based on dichlorophen and Mystox B (Catomance), composed of pentachlorophenol as a water-dispersible protein complex. Application of approximately 3% product based on weight of wool (or approximately 1–1.5% solids on weight of wool) will confer exceptional rot resistance under the most severe conditions, as indicated from the results in Table 1 (Lewis, 1973b).

Table 1

Mildew-proofing effects of some selected applications

| Sample | Percentage reduction in tensile strength after incubation for | | | |
	two weeks	three weeks	four weeks	five weeks
Untreated	68.2	86.8	91.0	100
Sandocide liquid (3%)	5.3	4.6	0.9	4.8
Prestofen GDC (3%)	0	0	0	0
Mystox B (3%)	3.6	0	6.3	2.2

Treatment concentrations indicate product application on weight of wool.

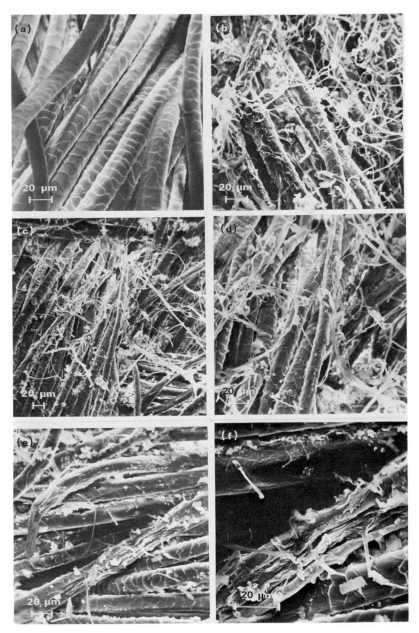

Fig. 21. Scanning electron microscope appearance of untreated, undamaged fibres prior to infection (a) and after incubation with an infecting medium for periods between two and five weeks (b–f respectively). Note the well developed fungal mycelia, the loss of scale structure and extensive fibrillation of mildew-damaged fibres.

An examination of infected, untreated and mildew-proofed fibres by scanning electron microscopy (Figs. 21 and 22) confirms the effectiveness of these treatments and illustrates clearly the formation of fungal mycelial mats and the extensive fibrillation of infected, untreated fibres.

Products based on salicylanilide (e.g. Shirlan), copper naphthenate, chromium-containing compounds and organomercurial compounds have largely fallen into disuse, either because of the discolouration which may be produced on treatment or on exposure to light, or because of problems with human and environmental toxicity. Volatile antiseptics, such as formaldehyde or ethylene oxide, are efficient for

Fig. 22. Scanning electron micrographs of wool fibres treated with selected mildew-proofing compounds (3% on the weight of wool) after incubation for four weeks with an infecting medium. (a) shows fibres treated with Mystox B, (b) Sandocide liquid and (c) and (d) with Prestofen GDC. In the last micrograph, soil particles adhering to fibres result from incomplete rinsing-off of the soil-compost medium.

gaseous sterilization of infected goods but, because of handling difficulties and because such treatments can offer temporary residual action
only, or none at all, they are not used for conferring subsequent mildew-
resistance to wool.

E. Eradication of Mildew on Infected Goods

The fairly good resistance of wool to attack by mildew means that it
is often possible to detect incipient damage, as evidenced by surface
growths and odour, before serious physical degradation is produced.
At this stage, it is, of course, essential to sterilize the goods to destroy
growth and de-activate proteolytic enzymes absorbed by the wool to
prevent deterioration and staining from proceeding to the extent that
the material becomes useless.

Mildew growth may be terminated, and residual resistance to further
attack conferred, by treating infected material with the products listed
above. Where a wet-processing stage is undesirable, solvent-soluble
application of products containing, for example, lauryl pentachlorophenol, or gaseous disinfectants may be used. If none of these procedures
is practicable, the infecting organisms may be destroyed by autoclaving
or drying at 100°C–150°C for 5 minutes but, in this case, residual
microbiocidal activity will not be imparted to the sterilized goods. It is
therefore important to ensure that re-infection is prevented by storage
in a dry atmosphere.

Stains produced by microbial degradation are difficult to remove.
They resist washing, alkaline and acid milling and bleaching with
hydrogen peroxide. Hot-reducing treatments with sodium hydrosulphite (sodium dithionite) are effective, however, on undyed goods
(Anon, 1958). Differential dyeing effects usually can be masked satisfactorily only by dyeing the goods black.

F. Characteristics of Selected Treatments

The ideal biocide for wool textiles should be inhibitory, in relatively
low concentrations, to a wide variety of micro-organisms, non-toxic to
humans and animals, biodegradable, colourless, non-yellowing and
odourless, applicable by a variety of finishing procedures, non-volatile,

and without deleterious effects on dyestuff, fibres and handle. It must also be cheap and possibly possess moth-proofing properties. Some of these aspects are discussed below.

1. Effectiveness

Products containing dichlorophen or pentachlorophenol have been found to be extremely effective mildew-proofing compounds. Their fastness to washing and dry-cleaning has not been examined because these properties are of relatively little importance for most likely end uses in the wool industry.

2. Toxicity

Concentrated products containing dichlorophen are highly alkaline and should be handled carefully. Dichlorophen is, however, one of the least toxic of the common biocides (acute oral LD_{50} values in mammals of 1,000 to 3,000 mg/kg body weight have been reported) and it is improbable that any toxic effects could result from contact with dichlorophen-treated materials.

Pentachlorophenol (with an oral toxicity LD_{50} value of 280 mg/kg in female rats) is physiologically harmful by skin absorption, and contact of undiluted products containing this chemical with eyes and skin must be avoided. Toxic effects would occur only after mishandling large quantities of products. Treated goods are most unlikely to produce any physiological or dermatological problems.

Both dichlorophen and pentachlorophenol are toxic to fish, and undiluted waste liquors containing high concentrations of these compounds should not be discharged in effluent to inland waterways.

3. Biodegradability

Tests on products containing dichlorophen and pentachlorophenol have shown that the bacterial activity of sewage effluent from an industrial area is completely inhibited by very low concentrations (approximately 1 p.p.m.) of these chemicals. The discharge of large quantities into sewage works would therefore be disastrous. Discharge of toxic effluent can be minimized by pad application (the recommended method for Prestofen GDC) but the exhaustion level of products containing dichlorophen and pentachlorophenol, applied by dyeing techniques, has been found to be insufficient to guarantee tolerable levels of residual chemicals in the waste liquor. Residual dichlorophen at 40 p.p.m. and residual pentachlorophenol at 35 p.p.m. was obtained in laboratory experiments. An effluent containing these concentrations would therefore require dilution and/or chemical pre-treatment before discharge.

4. Effects on Dyestuffs and Fibres

No deleterious effects on dyestuffs have been noted, and physical testing and scanning electron-microscope examinations have failed to reveal any fibre damage which may be attributed to treatment with dichlorophen or pentachlorophenol, at concentrations used for mildew-proofing.

5. Economics of Treatment

The cost of imparting effective mildew resistance to wool compares favourably with the costs involved for moth-proofing.

V. DAMAGE OF DOMESTIC WOOL TEXTILES BY DETERGENTS CONTAINING PROTEOLYTIC ENZYMES

A. Effects of Enzyme Detergents on Wool

Enzyme detergents contain proteolytic enzymes obtained as extra-cellular products of *Bacillus subtilis* (Friedman and Barkin, 1969) or *Bacillus licheniformis* (Anon, 1970). The enzymes are extracted by processes involving centrifugation, precipitation and filtration, and it is claimed (Anon, 1970) that the efficiency of these operations is such that the concentration of viable micro-organisms and spores in the final enzyme concentrate is exceedingly small. The enzymes are incorporated into detergent formulations in such a manner as to ensure maximum shelf life and reasonable stability to washing conditions for the enzyme in the product (Langguth and Mecey, 1969).

The enzyme preparation is stable under alkaline conditions even at fairly high temperatures. In domestic use, they catalyse hydrolysis of protein-containing substrates comprising common domestic stains and soils, such as perspiration, blood and egg yolk. They are claimed to be particularly effective in removing stains of this type but, in general, these 'biological detergents' do not show any significant advantage over conventional, non-enzymic detergents (Anon, 1969a).

Table 2

Wool samples shrink-resist treated by various procedures. For further details, see the text and Figure 23

A. 4% Dichloroisocyanuric acid-peroxide rinse continuous process (Anon, 1968).
B. 4% Dichloroisocyanuric acid-perioxide rinse omitted (Anon, 1969b)
C. Untreated wool (control for A and B)
D. 1.75% Permonosulphuric acid—batch process (B.P. 538, 429)
E. Untreated wool (control for D)
F. 3% Dichloroisocyanuric acid – 1% potassium permanganate (Murphey, 1969)
G. 2.5% Dichloroisocyanuric acid – 1.5% potassium permanganate (Murphey, 1969)
H. Untreated wool (control for F and G)

Fabrics made from protein fibres such as wool, silk and mohair have been found to be degraded when soaked in solutions of enzyme detergents and, since most manufacturers now recognize the dangers inherent in the use of these products, detergent packs are usually labelled with appropriate precautions. This was not the case in the early years of use of these products and, even now, problems still arise occasionally from misuse, apparent disregard of the instructions, or from ignorance of the fibre content of the article involved.

In a laundering procedure designed to remove protein-based stains, the major hydrolytic reaction that occurs in an enzyme-detergent solution is a proteolytic one, and this clearly represents a potential danger to wool fibres. Furthermore, additional constituents of the detergent formulation, notably builders such as sodium bicarbonate and sodium sesquicarbonate, and bleaches such as sodium perborate, are also capable of causing alkaline and oxidative damage to wool fibres.

Modifications to the cuticular structure of wool and, more importantly, the rupture of various chemical bonds in wool by oxidative and reductive reactions are some consequences of several common wet-finishing treatments applied to wool fibre, yarn, fabrics and garments. Typical of these are the processes for making wool resistant to shrinking during washing (Table 2). The possibility that these treatments might increase the enzyme lability of the fibre has been examined in detail using a commercial enzyme-containing detergent representative of an international selection of products (Lewis, 1972). In a preliminary experiment designed to evaluate the effect of enzyme-detergent solution on various shrink-proofed wools, it was shown that oxidative shrink-resist treatments (which operate by partial degradation of the surface of the fibre) promote further deterioration during soaking (Fig. 23).

Wool treated with 4% dichloroisocyanuric acid was used to optimize the conditions for proteolytic activity, and these experiments indicated maximum dissolution of the wool at pH 8.5, 45°C, and a detergent concentration of 12.5 g/l. In the shrink-proofing of wool with dichloro-

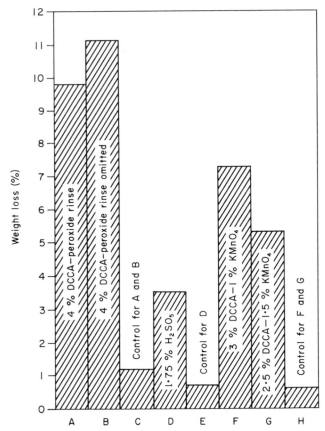

Fig. 23. Loss in weight of various wools, shrink-resist processed by oxidative techniques (for details, see Table 2) after soaking in solutions of enzyme detergent. Included also were untreated wool controls that had not received a shrink-resist treatment. Conditions of incubation were pH 8.5, 45°C, 12.5 g of detergent/l and a 24-hour soak. DCCA indicates dichloroisocyanuric acid.

isocyanuric acid, the major reaction is one in which disulphide bonds are broken and converted into sulphonic acid residues (Swanepoel and van Rooyen, 1969). The surface of the fibre is also modified and, after washing, shows fragmentation of the cuticle (Hepworth *et al.*, 1969). An increase in the level of chlorination of the fibre might be expected, therefore, to increase the rate and extent to which the treated wool is hydrolysed in enzyme-detergent solution. This was confirmed in an experiment, summarized in Figure 24, which shows a linear relationship between the degree of the chlorination and the weight loss resulting from soaking in enzyme-detergent solutions.

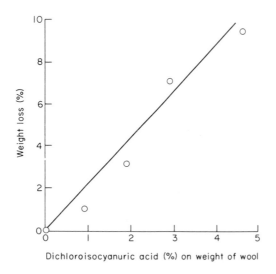

Fig. 24. The effect of chlorination on the weight loss of treated wool in enzyme-detergent solutions. Conditions of soaking as in Figure 23.

A comparison of the effects of repeated soaking in enzyme-detergent solutions on fabrics treated by two completely different shrink-proofing processes was prompted by these findings, since alternative less damaging shrink-proofing processes are now widely available. These involve coating the fibre surface with a microscopically thin layer of synthetic polymer. The overwhelming proportion of the World's snrink-proofed wool, for example, is processed by subjecting untreated wool to a low level of chlorination followed by treatment with Hercosett resin, a polyamide–epichlorhydrin prepolymer (Lewis, 1978). Experiments have been carried out to examine the effects of repeated soaking

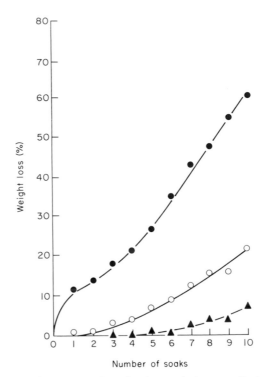

Fig. 25. Loss in weight of some treated wools when soaked repeatedly in enzyme-detergent solutions. Conditions of soaking were as in Figure 23. ● indicates changes in dichloroisocyanuric acid-treated wool, and ▲ changes in Hercosett-treated wool. Included also was an untreated wool control (○) that had not received a shrink-resist treatment.

in enzyme-detergent solutions on weight loss, bursting strength and microscopial appearance of untreated, dichloroisocyanuric acid-treated and Hercosett-treated fabrics (Lewis, 1972).

Figures 25 and 26 respectively show the loss in weight and the decrease in bursting strength of knitted samples made from similar untreated and treated wool. Samples of each set of fabrics obtained during the sequential incubation in enzyme-detergent solutions were then examined with a scanning electron microscope (Figs. 27, 28 and 29). This investigation proved conclusively that treatment of wool with Hercosett resin retards the weight loss that occurs when wool is soaked in solutions of enzyme-detergents. The resin coating was also shown to remain effective in preventing the fabric from felting, even after prolonged exposure to proteolytic enzymes. Untreated wool suffered a greater loss in weight and became weaker, whereas oxidative shrink-resist treatments with dichloroisocyanuric acid clearly rendered the fabrics highly susceptible to loss in weight and strength when soaked in these products. Scanning electron microscopy provided clear evidence of the diminished deleterious effects of proteolytic enzyme-detergents on resin-treated wool fibres.

It should be noted that the incubation period used in the reported experiments is likely to be much more severe, in terms of time and

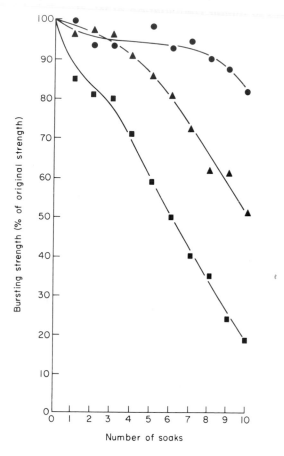

Fig. 26. Decrease in bursting strength of wools, soaked repeatedly in enzyme-detergent solutions. Conditions of soaking were as indicated in Figure 23. ■ indicates changes in dichloroisocyanuric acid-treated wool, and ● in Hercosett-treated wool. Included also was an untreated wool control (▲) that had not received a shrink-resist treatment.

temperature of soaking, than the conditions usually encountered in normal domestic laundering with these products. However, the combined effects of soak and wear, or wash and wear, when enzyme-detergents are employed, might be expected to result in premature fibre and fabric breakdown, if wool has been treated with oxidizing agents. Damage to untreated wool or resin-treated wool is unlikely under normal domestic conditions where presoaking is omitted.

B. Treatment to Inhibit Degradation of Wool in Enzyme-Detergent Solutions

In a subsequent publication, a search for methods of decreasing the

Fig. 27. Effects of enzyme-detergent solution on untreated wool. (a) Fibres prior to soaking exhibiting the typically sharp scale margins characteristic of untreated merino wool fibres. (b) After soaking for 3 periods of 24 hours. (c) After soaking for 6 periods of 24 hours. The mechanism by which fibre breakdown proceeds is clearly one in which release of scales from fibres exposes the less-resistant cortex which is then attacked by proteolytic enzymes leading to fibrillation of this structure into cortical cells.

rate and extent of enzymic hydrolysis is described (Lewis, 1974). Again, wool treated with dichloroisocyanuric acid was selected for detailed investigation because of its susceptibility to degradation by detergents containing proteolytic enzymes. Treated wool has been modified by additional treatment with formaldehyde, a chemical known to participate in a great number of reactions with wool (Reddie and Nicholls, 1971a). It was anticipated that modification in this way should hinder the hydrolytic effects of the enzyme by a blocking, cross-linking or

Fig. 28. Effects of enzyme-detergent solution on wool fibres previously shrink-resist-treated with 4% dichloroisocyanuric acid. The degradative effect of this shrink-resist treatment on wool fibres is evident, even prior to soaking in enzyme-detergent solution. (a) shows considerable fragmentation of the fibre surface and partial removal of scales. Also shown are fibres after soaking in enzyme-detergent solution for 3 periods of 24 hours (b) and after soaking for 6 periods of 24 hours (c). Extensive removal of the cuticle and fibrillation of the cortex have occurred. Protracted exposure to the effects of the detergent has produced a substantial diminution in fibre diameter. This is attributable to a loss of cortical cells resulting from digestion of the membranes which bind them together.

enzyme-poisoning mechanism, or by a combination of one or more of these, thereby minimizing degradation by proteolytic enzymes.

Reaction of wool with 0.2M formaldehyde at the boil effected a substantial decrease in enzyme lability (Fig. 30) even after a relatively short period of treatment (Lewis, 1974). The resistance of this treatment to leaching was examined by repeatedly soaking variously treated samples of wool in freshly prepared solutions of enzyme detergent. The data in Figure 31 indicated that the treatment was essentially durable to repeated soaking and that the modification of dichloroisocyanuric acid-treated wool with formaldehyde lowered the weight loss when soaked in enzyme-detergent solutions to the level that might be expected of

Fig. 29. Effects of enzyme-detergent solution on wool fibres shrink-proofed by the chlorine–Hercosett process. (a) shows fibres prior to soaking in enzyme-detergent solution, (b) fibres after soaking for 2 periods of 24 hours, and (c) after soaking for 6 periods of 24 hours. There is no evidence of fibre damage.

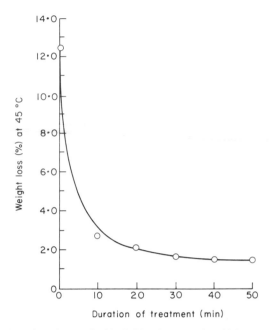

Fig. 30. Insolubilization of wool treated with dichloroisocyanuric acid by reaction with formaldehyde (0.2 M) at the boil for relatively short periods at pH 3.5 and 100°C. Assessment of weight loss was made after soaking for 24 hours with enzyme-detergent solution (12.5 g/l) at pH 8.5 and 45°C.

completely untreated wool. From a consumer's aspect, this is probably satisfactory, particularly when the severe conditions of the laboratory soaking procedure are considered. Treatment of previously untreated wool with formaldehyde clearly imparted a high degree of resistance to degradation by enzyme-detergent solutions.

The mechanism by which enzyme inhibition is imparted to wool by treatment with formaldehyde was also examined. During oxidative/reductive shrink-proofing operations, two major chemical reactions take place which are relevant. These are:

(1) Oxidation of disulphide bonds to form sulphonic acid groups (Swanepoel and Van Rooyen, 1969):

$$\text{Wool–S–S–Wool} \xrightarrow{\text{Cl}_2/\text{NaHSO}_3} 2 \text{ Wool–SO}_3\text{H}$$

(2) Cleavage of peptide bonds to form carboxylic and primary amine groups (Swanepoel et al., 1970):

$$\text{Wool–CONH–Wool} \xrightarrow{\text{Cl}_2/\text{NaHSO}_3} \begin{matrix} \text{Wool–COOH} \\ \text{Wool–NH}_2 \end{matrix}$$

Scission of disulphide bonds and the presence of free amino groups appear to constitute a prerequisite for severe attack of wool by proteolytic enzymes. The fact that proteolysis can be obviated by treatment with formaldehyde is thought to result from a modification of wool such that fixation of the enzyme on the protein anions is rendered impossible (Marsh, 1951). Reddie and Nicholls (1971a) showed that the extent of reaction of formaldehyde with amino groups of lysine was increased only slightly by a considerable increase in formaldehyde concentration. The formaldehyde content of wool increased markedly above pH 9 (Reddie and Nicholls, 1971b), but the extent of the reaction at the amino residues of wool remained relatively constant over wide

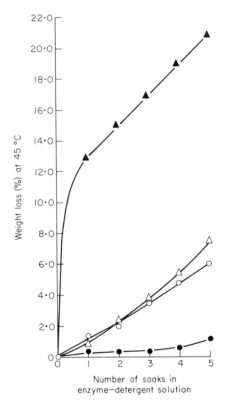

Fig. 31. Effect of repeated soaking of formaldehyde-treated (0.2 M) wools in enzyme-detergent solutions for periods of 24 hours. Treatment conditions were pH 3.5 and 20 minutes at 100°C. Weight loss in soaking in enzyme-detergent solution was assessed as in the caption to Figure 30. ▲ indicates the behaviour of untreated and △ of formaldehyde-treated wool, both of which had been shrink-resist treated with dichloroisocyanuric acid, while ○ is of untreated wool and ● is of similar wool treated with formaldehyde.

ranges of pH value. On the other hand, Reddie and Nicholls (1971a) reported that increased reaction temperature resulted in a significantly greater reaction at the amino groups.

The diminished hydrolysis of formaldehyde-treated wool in enzyme-detergent solutions correlates with these observations, in that inhibition of proteolysis showed little variation over ranges of formaldehyde concentration and pH value, but was highly dependent on the reaction temperature. By contrast, there is very little correlation between the results of this investigation and the results reported for reaction with the phenolic residues of tyrosine or the indole residues of tryptophan (Reddie and Nicholls, 1971a).

VI. CONCLUSION

Considering the diverse and severe pasturing conditions to which wool fibres are exposed during growth of the fleece, and in view of the great variety of micro-organisms present, surprisingly little microbial damage takes place at this stage. Of the minute quantity of wool involved, damaged fibre is almost completely eliminated by sorting or by the damaged, fragile fibres being removed by various mechanical processes involved in converting fibre into fabrics.

In the industrial processing of skins, such as in fellmongering, the susceptibility of the prekeratinous region of the fibre to digestion enables fibres to be removed from skins without adverse effect on the bulk of the fibre above skin level. That wool fibres are significantly more resistant to degradation than skin also allows the latter to be digested preferentially, in alternative, though sometimes crude, methods of removing fibre from skin pieces. More recently, however, sophisticated techniques have been developed which rely largely on optimizing conditions for bacterial decomposition of the skin.

Although proteolytic enzymes are capable of digesting a minor component only of the fibre, even after prolonged exposure, loss of the more enzyme-labile material lowers the physical strength of wool considerably. In domestic conditions, occasional problems arise through misuse of detergents containing proteases but, in the main, the incidence of damage of a proteolytic nature is very low indeed. Similarly, although mildew will grow eventually on wool stored under adverse conditions, its presence is generally manifested at an early stage and simple corrective action can be taken to prevent recurrence of the problem.

In conclusion, therefore, it is evident that wool exhibits a very much higher resistance to degradation by micro-organisms than many other natural commodities of animal or vegetable origin. On reflection, this is hardly surprising since animal fibres, such as wool and hair, have evolved primarily as protection from the elements which, depending on geographical location, can vary from one extreme of climate to the other.

VII. ACKNOWLEDGEMENTS

I am grateful to the Textile Physics Laboratory at Leeds University for scanning electron microscope facilities and, in particular, to Mr. T. Buckley for valued technical assistance. Acknowledgement is also due to Professor A. E. Henderson of Lincoln College, New Zealand for photographs and samples of infected fleece wools and to Dr. G. Delsol at the I.T.F., Sud, for the photograph of a 'sweating' chamber.

REFERENCES

Alexander, P. and Hudson, R. F. (1963). *In* 'Wool—Its Chemistry and Physics', Second edition, p. 296. Chapman and Hall, London.

Anon C.B.C.C. Screening Files N.A.S./N.E.C., Dow Chem. Co. Ltd.

Anon (1950a). *Wool Science Review* Part 1, **6**, 13.

Anon (1950b). *Wool Science Review* Part 2, **6**, 31.

Anon (1951). *Wool Science Review* **7**, 12.

Anon (1958). *Bulletin Trimestriel Centre Textile Chambre de Commerce, Roubaix* **40**, 5. Translation LT54—British Hat and Allied Feltmakers Research Association.

Anon (1966a). *Wool Science Review* **29**, 11.

Anon (1966b). *Wool Science Review* **30**, 27.

Anon (1968). *Wool Science Review* **34**, 1.

Anon (1969a). *Which?* September, p. 267.

Anon (1969b). Fisons Technical Brochure.

Anon (1970). *Soap, Cosmetics and Chemical Specialities* **46**, 111.

Batson, D. M., Teunisson, D. J. and Porges, N. (1944). *American Dyestuff Reporter* **33**, 423 and 449.

Burgess, R. (1924). *Journal of the Textile Institute* **15**, T573.

Carrie, M. S. and Woodroffe, F. W. (1960). *In* 'Fellmongers Handbook', p. 39. New Zealand Department of Scientific and Industrial Research, Information Series No. 25.

David, J. and Barr, A. R. M. (1972). *Textile Institute and Industry* **10**, 173.

Delsol, G. and Sergent, C. (1975). *Bulletin Scientifique de l'Institute Textile de France* **4**, No. 13, 1.

Delsol, G. and Sergent, C. (1976). *Bulletin Scientifique de l'Institute Textile de France* **5**, No. 19, 283.

Dobb, M. G. (1963). Ph.D. Thesis: University of Leeds, U.K.

Ebeling, W., Hennrich, N., Klockow, M., Metz, H., Orth, H. D. and Lang, H. (1974). *European Journal of Biochemistry* **47**, 91.

Eggins, H. O. W., Malik, K. A. and Sharp, R. F. (1968). *In* Proceedings of First International Biodeterioration Symposium. (A. H. Walters and J. J. Elphick, eds.). Elsevier, London.

Ellis, W. J. (1945). *Bulletin No. 184*, C.S.I.R. (Australia) p. 227.

Friedman, S. D. and Barkin, S. M. (1969). *Journal of the American Oil Chemists Society* **46**, 81.

Fraser, I. E. B. (1957). *Australian Journal of Agricultural Research* **8**, 281.

Fraser, I. E. B. and Mulcock, A. P. (1956). *Nature, London* **177**, 628.

Garner, W. (1967). *In* 'Textile Laboratory Manual', 3rd edn., volume 6, p. 153. Heywood, London.

Geiger, W. B., Patterson, W. I., Mizell, L. R. and Harris, M. (1941). *Journal of Research of the National Bureau of Standards* **27**, 459.

Green, G. H. (1955). *Journal of Applied Chemistry* **5**, 296.

Hayman, R. H. (1953). *Australian Journal of Agricultural Research* **4**, 430.

Henderson, A. E. (1968). 'Growing Better Wool', pp. 69–90. Reed, Wellington, New Zealand.

Hepworth, A., Sikorski, J., Tucker, D. J. and Whewell, C. S. (1969). *Journal of the Textile Institute* **60**, 513.

Kapica, L. and Blank, F. (1957). *Dermatologica* **115**, 81.

Lamanna, C., Mallette, M. F. and Zimmerman, L. N. (1971). 'Basic Bacteriology—Its Biological and Chemical Background', 4th edn., p. 893. Williams and Wikins Co., Baltimore.

Langguth, R. P. and Mecey, L. W. (1969). *Soap and Chemical Specialities* **45**, No. 9, 60.

Lennox, F. G. (1945). *Commonwealth Scientific and Industrial Research Bulletin, No. 184* (Australia) p. 9.

Lennox, F. G. and Maxwell, M. E. (1945). *Commonwealth Scientific and Industrial Research Bulletin, No. 184* (Australia) p. 143.

Lewis, J. (1972). *Textile Institute and Industry* **10**, 360.

Lewis, J. (1973a). *Wool Science Review* **46**, 17.

Lewis, J. (1973b). *Wool Science Review* **47**, 27.

Lewis, J. (1974). *Textile Research Journal* **44**, 707.

Lewis, J. (1978). *Wool Science Review* **55**, 23.

Lloyd, A. O. (1955). *Journal of the Textile Institute* **46**, T653.

Lloyd, A. O. (1965). *International Biodeterioration Bulletin* **11**, 10.

Marsh, J. T. (1951). 'Introduction to Textile Finishing', p. 508. Chapman and Hall, London.

Marsh, P. B. (1945). *Proceedings of the Conference on Biological Testing and Test Organisms*, Washington D.C., April 17. U.S. National Defence Research Communication.

Maxwell, M. E. (1945). *Commonwealth Scientific and Industrial Research Bulletin, No. 184* (Australia) p. 89.

Maxwell, M. E. (1950). *Australian Journal of Applied Science* **1**, 497.

Mercer, E. H. (1953). *Textile Research Journal* **23**, 388.

Molyneux, G. S. (1959). *Australian Journal of Biological Sciences* **12**, 274.

Mulcock, A. P. (1959). *Nature, London* **183**, 1281.

Mulcock, A. P. (1961). Ph.D. Thesis: Lincoln College, New Zealand.

Mulcock, A. P. and Fraser, I. E. B. (1958). *Australian Journal of Agricultural Research* **9**, 704.

Murphey, T. P. (1969). *Knitted Outerwear Times* **23**, 48.

Onions, W. J. (1962). *In* 'Wool—An Introduction to its Properties, Varieties, Uses and Production', p. 41. Benn, London.

Pressley, T. A. (1950). *Australian Journal of Applied Science* **1**, 484.

Prindle, B. (1935). *Textile Research Journal* **6**, 23.

Race, E. (1946). *Proceedings of the Symposium of the Society of Dyers and Colourists*, 1946, p. 67. Society of Dyers and Colourists, Bradford, England.

Raistrick, M. (1919). *Biochemical Journal* **13**, 446.

Reddie, R. N. and Nicholls, C. H. (1971a). *Textile Research Journal* **41**, 841.

Reddie, R. N. and Nicholls, C. H. (1971b). *Textile Research Journal* **41**, 303.

Rettger, L. F. and Sperry, J. A. (1915). *Journal of Biological Chemistry* **20**, 445.

Rettger, L. F. and Berman, N. (1918). *Journal of Bacteriology* **3**, 367.

Rettger, L. F., Berman, N. and Sturges, W. S. (1916). *Journal of Bacteriology* **1**, 15.

Shaposhnikov, N. V., Kozlova, E. I. and Azova, L. G. (1964). United States Clearinghouse Translation No. AD719550 1971. From *Vestnik Moskovskogo Universiteta* **2**, 58.

Siu, R. G. H. (1951). *In* 'Microbiol Decomposition of Cellulose' (R. G. H. Siu, ed.), p. 326. Waverley Press, Baltimore.

Smith, G. (1971). *In* 'An Introduction to Industrial Mycology' (G. Smith, ed.), 6th edn., p. 298. Arnold, London.

Swanepoel, O. A. and Van Rooyen, A. (1969). *South African Wool Textile Research Institute Technical Report No. 125.*

Swanepoel, O. A., Van der Merwe, J. P. and Grabherr, H. (1970). *Textilveredlung* **5**, 200.

Tristram, G. R. (1953). *In* 'The Proteins' (H. Neurath, ed.), volume 1, Part A., p. 216. Academic Press, New York, U.S.A.

Waters, R. (1932). *New Zealand Journal of Agriculture* **44**, 35.

Wellman, R. H. and McCallan, S. E. A. (1945). *OSRD Report 5683, Washington DC.* United States National Defence Research Communication.

Yates, J. R. (1968). *Journal of the American Leather Chemists Association* **63**, 474.

Yates, J. R. (1974). *Revue Technique des Industries du Cuir* **66**, 130.

5. Hides and Skins

BETTY M. HAINES

British Leather Manufacturers' Research Association, Moulton Park, Northampton NN3 1JD, England

I. Introduction	. . . .	. 131
II. Deterioration by Autolysis	. . . .	. 133
III. Deterioration by Bacteria	. . . .	. 136
IV. Curing	. . . .	. 141
A. Salting .	. . . .	. 141
B. Drying .	. . . .	. 143
C. Dry-Salting .	. . . .	. 144
D. Storage	. . . .	. 144
V. Alternative Methods of Cure	. . . .	. 145
A. Biocides	. . . .	. 145
B. Chilling	. . . .	. 146
C. Freezing	. . . .	. 146
VI. Deterioration by Moulds	. . . .	. 146
References .	. . . .	. 147

I. INTRODUCTION

Deterioration in the skin can begin immediately after death, when enzymes naturally present in the skin begin to break down skin proteins and fats by autolysis (Tancous, 1970). Once the skin is removed from the animal, the inner surface of the skin becomes contaminated, bacteria penetrate the skin, and their action rapidly overtakes that of autolysis. Until the skin is permanently preserved by tannage it is necessary to protect the raw skin from both forms of deterioration. It is customary to refer to the skins of larger animals such as cattle, as hides, but in this paper only the term skin will be used.

Animals, such as cattle, sheep, goats and pigs, produce the skins that are most commonly used in leather production. These animals are slaughtered and the skin removed at abattoirs where the main concern is for the meat, the skin being regarded as a by-product to be processed elsewhere. Between the abattoir and the tannery, skins may be moved in a variety of ways depending on the commercial conditions of the country concerned. Skins may be processed into leather immediately after flay with the minimum of delay, but a more general practice is for the skins to pass through one or more intermediate stages (Fig. 1). Tanneries specialize in the type of leather produced and so require the

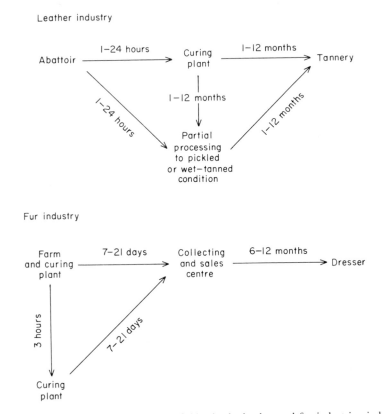

Fig. 1. Scheme illustrating the movement of skins in the leather and fur industries, indicating possible delay times.

supply of a particular type of skin. This makes it necessary to group skins according to type before transfer to the tannery.

There is considerable international trade in untanned skins involving World-wide movement of skins, a potentially biodegradable product. This marketing structure introduces delays, making it necessary to give some form of temporary preservation which will bring the skin into a non-putrescible condition to permit storage and transport, but which can be readily reversed at the start of leather processing. This preservation, referred to as curing, may be required to protect skins for periods of up to one year or more. A more recent practice is to transport skins in a partially processed condition. They are taken through the usual pretanning processes to either pickling, in which the skin, in the presence of a high concentration of salt is acidified to a pH value of about 1·5, or further to tannage with chromium salts. In either the pickled or the wet-tanned condition, the skin is protected against bacterial attack but is liable to mould growth, unless protected by fungicides.

In the fur industry, hair and skin is the prime product and consequently there is greater incentive to prevent deterioration, which can lead to loosing of hair. Skins are taken from a wide variety of fur-bearing animals and many are farm-reared, but the processing of the skin through to tannage is undertaken by specialized dressers. As in the leather industry, it is necessary to give the raw skin a temporary preservation or one to allow for skins to be accumulated, graded and distributed World-wide.

In both the leather and fur industries the skin is open to deterioration at three stages, namely: (a) between flay and completion of cure, the skin is most vulnerable to both bacterial and autolytic attack and it is at this stage that damage most frequently occurs; (b) in the cured state, if storage conditions are adverse; and (c) during the pretanning stages of leather processing if the skin is held at a neutral pH value for sufficient length of time to allow bacterial action to occur.

II. DETERIORATION BY AUTOLYSIS

Changes brought about by autolysis have been followed by holding skin in water, but under toluene to maintain aseptic conditions for a maximum time of nine months and at a temperature of 37°C (Tancous, 1970). Digestion of skin proteins was initiated by enzymes: one isolated

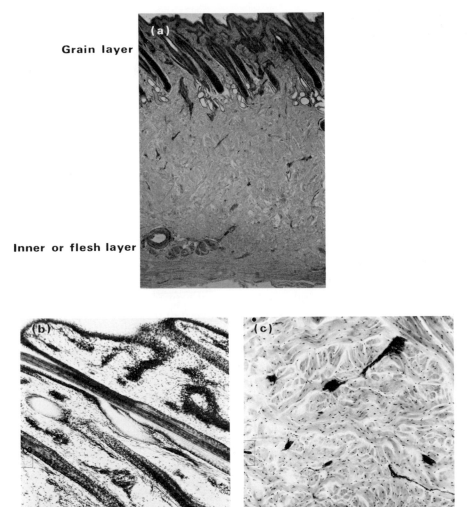

Grain layer

Inner or flesh layer

Fig. 2. Photomicrographs showing the structure of sections through newly flayed cattle skin. (a) Shows a thin section through a skin with the grain-layer region at the top and the inner or flesh layer at the bottom. Magnification × 15. Higher magnifications are also shown of the outer grain layer (b) and of the inner flesh layer (c); magnification of both photomicrographs × 40.

from the system gave a positive reaction for cathepsin and was found
to be an amidase. Subsequently, deterioration was increased by the
lyotropic action of fatty acids, mainly acetic and butyric acids activated
by cysteine. The fatty acids were produced by enzyme action on skin
fats and cysteine from breakdown of the epidermis. In the first 20 weeks,
as the fatty acid content increased the pH value fell to 5.0. Later, the
pH value rose due to alkaline breakdown products of the protein and,
above pH 8.0, the lyotropic action of the acids ceased. The histological
changes that occurred were a complete loss of cellular tissues, detachment of the epidermis, breakdown of fat cells and migration of the fat
through the skin and extreme fibrillar separation of the collagen fibre
bundles.

Leather produced from skin kept under these conditions was loose
and stretchy. When the skin to which salt had been added was kept
under the same aseptic conditions, similar autolytic changes occurred
(Tancous and Jayasimhalu, 1973), but only when the salt in solution
in the skin moisture was 10% or less. Inhibition of autolysis was

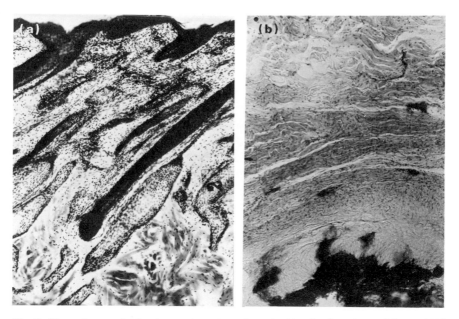

Fig. 3. Photomicrographs showing sections through cattle skin after fours hours delay at 30°C
after flay. Bacteria are beginning to penetrate into the skin, but cells in the corium and grain
layer (a) are still in good condition. There is, on the other hand, a dense accumulation of
bacteria on the flesh surface (b) shown by the dense black layer. Magnification of both photomicrographs, × 45.

achieved by addition to the salt of sodium pentachlorphenate, sodium trichlorphenate or alkylbenzyldimethyl ammonium chloride.

III. DETERIORATION BY BACTERIA

Skin deterioration brought about by bacteria proceeds far more rapidly than that caused by autolysis. The progress of bacterial attack, which can be followed microscopically, involves progressive breakdown of cellular tissues through the skin (Figs. 2–5).

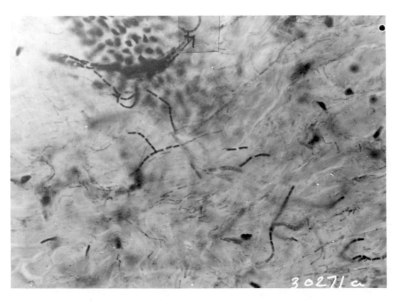

Fig. 4. Photomicrograph showing bacteria penetrating along the collagen fibre bundle of cattle skin. Only a few cells remain, and these are fibroblasts. Magnification × 760.

Although the outer surface of the skin is heavily contaminated by bacteria, the epidermis covering the outer surface acts as a barrier to bacteria for at least the first four to five days (Dempsey, 1969). Therefore, bacteria penetrate the skin from the inner or flesh surface and putrefactive changes proceed from the flesh surface inwards. At the time the skin is removed from the animal, the flesh surface is clean (Fig. 2), but it rapidly becomes contaminated by blood, dung and general dirt as it comes into contact with other skins. In the first few hours after flay, bacteria accumulate on the flesh surface (Fig. 3). They then penetrate into the skin, the cells along the fibres being destroyed *en route* (Fig. 4). The bacteria finally reach the outer layers of the skin, accumulating below the epidermis, and in the cellular tissues of the grain layer, i.e. the layer containing the hairs, sweat and sebaceous glands. Finally the cells in this region are destroyed, at which stage the epidermis and the hair-follicle walls become detached from the dermis, and the hairs become loose (Fig. 5). Aerobic bacteria initiate this process, but the main putrefactive changes are brought about by anaerobic bacteria (Tissier and Martin, 1902). These are the changes that are easily followed microscopically. What is more difficult to assess, is the degree to which the skin proteins are attacked, so affecting the final leather quality.

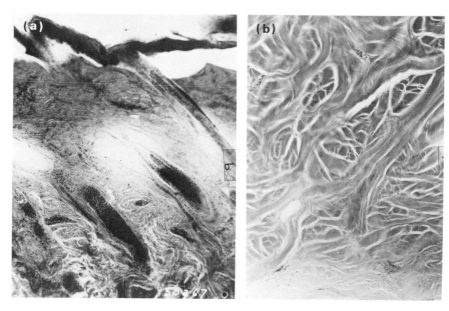

Fig. 5. Photomicrographs showing sections through cattle skin after 24 hours delay at 30°C after flay. (a) Shows a section through the grain layer, (b) a section through the inner flesh layer. Bacteria are seen to have penetrated through the full thickness of the skin. All cells have been destroyed, and the epidermis and hairs are detached from the dermis. Magnification of both photomicrographs, × 45.

Putrefactive bacteria are, in general, capable of breaking down cellular and interfibrillary non-collagenous proteins, but only certain types of bacteria such as *Clostridium histolyticum*, produce the enzyme

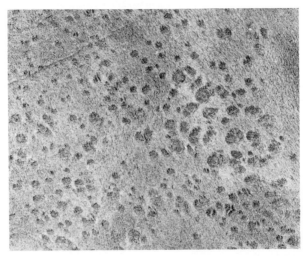

Fig. 6. Photomicrograph showing loss of tissues from the flesh surface of a sheepskin. Putrefaction has resulted in deep pitting of the suede surface. Magnification × 0.5.

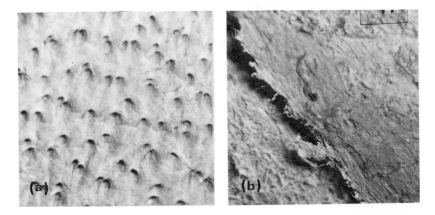

Fig. 7. Photomicrographs of the grain surface of leather from cattle skin. (a) Shows a normal grain surface with a hair follicle pattern; (b) shows an adjacent area of the same skin where putrefaction has destroyed grain tissue. Magnification for both photomicrographs, × 15.

collagenase that is capable of breaking down collagen, the leather-making raw material. As collagen is degraded, so skin tissue is lost from both the flesh (Fig. 6) and grain surfaces (Fig. 7) and eventually holes develop (Fig. 8). Such deterioration generally occurs at a later stage than the loosening of the hairs.

Proteolytic bacteria, though not producing collagenase, may be still capable of bringing about changes in the collagen fibre. The conformation of the collagen molecule makes it highly resistant to enzymes, but at the non-helical end of the molecule, the telopeptide, the amino-acid sequence is different from the main helical portion, and is more vulnerable to enzyme attack. The intra- and inter-molecular bonds are sited in the telopeptide region and, if these bonds are removed from the molecule as a result of enzyme action on the telopeptide, the collagen fibre will swell to a greater extent in acid and alkali, and will be less readily depleted again at neutral pH values. This can result in a loose textured leather.

Several methods have been developed to determine the degree of protein decomposition in the skin by estimation of breakdown products

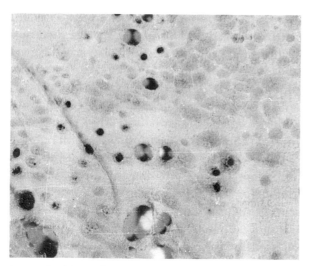

Fig. 8. Photomicrograph of the surface of a sheepskin leather, where loss of tissue due to putrefaction has resulted in holes being formed in the skin. Magnification ×0.5.

is achieved when the moisture content of the skin is lowered from 60% to a maximum of 48% and a minimum of 40%, and there is sufficient salt in solution in the skin moisture to reach at least 90% saturation (Benrud, 1968). In this condition the skin should remain in a well preserved condition for at least six months provided that the storage temperature and humidity are moderate.

Although sodium chloride is an excellent preservative, some bacteria can grow on salted skins. These are the halophilic bacteria which require for growth a minimum of 15% sodium chloride in their environment. These are both rods and coccal forms, classified as Halobacteriacae (Buchanan and Gibbons, 1974). The majority of halophilic bacteria are chromogenic, being coloured pink, red or violet, but some forms are colourless. It is the spread of these coloured bacteria over the flesh surface of the skin that has given rise to the term 'red or purple heat'.

Associated with these bacteria, but not classified in the same family, are other halophilic types which are able to grow over a wider range of salt concentrations from saturation down to 6%. There are also halo-tolerant types that can only grow at salt concentrations between 6 and 15%. Therefore, on the flesh surface of a red-heated skin, there is a mixed flora of halophilic and halotolerant bacteria. The specific environmental requirements of red-heat bacteria explain the pattern of incidence under commercial conditions. Red-heat bacteria are aerobic and, consequently, their growth is restricted to the surface of salted material. They can continue to grow when salt saturates the moisture present which is why it is not uncommon to find red heat on the flesh surface where solid salt is also present. For optimum growth they require a moisture content of 65% (Formisano, 1966) and a temperature of 33–40°C. This is in line with the general experience that red heat tends to develop when weather conditions are warm and humid. For optimum growth the bacteria require a pH value of 6.5–8.0 (Formisano, 1966). They will tolerate highly alkaline conditions but cease to grow at pH values below 6.0.

Halophilic bacteria are to be found naturally in sea water and salted lakes and, consequently, salt that is produced by evaporation is highly contaminated. Halophilic bacteria in themselves are unlikely to cause marked deterioration. Their growth is confined to the surfaces of the skin and any proteolytic action will also be superficial.

In 1966, Formisano isolated from red-heated cattle skins eleven different species that had gelatinolytic activity. Woods *et al.* (1973)

isolated two strains of halotolerant bacteria, *Achromobacter nocardia* and *Achromobacter iophagus*, that produced collagenase, but they also found the collagenase to be inactivated by salt in concentrations above 10%. This implies that, provided there is a sufficiently high salt content within and at the surface of the skin, there is little danger of the collagen itself being degraded. The danger increases when salted stock is held under unsuitable conditions of warmth and humidity. Excessive re-absorption of moisture by the salt will not only lower the salt concentration at the surface, but salt can also be lost by drainage away from the skin. The purple or violet forms of halophilic bacteria tend to be slower to develop, but they have the added disadvantage in that the violet colouration penetrates into the skin. The pigment has been shown to be soluble in the skin fat (Tancous, 1972). There have been no reports of the red colouration diffusing out of bacteria into the skin.

In practice, the incidence of red heat is usually taken as a warning that the skin may not be in the best condition and that it should be processed as soon as possible. Chemicals have been added to curing salt to prevent development of red heat, and those in current use are sodium metabisulphite, boric acid and naphthalene used at a concentration of 1–2% based on the salt weight.

B. Drying

The simplest method of curing is drying by evaporation and it is particularly suitable for countries where the climate is hot and dry. It is also the method employed to cure most fur skins. It is a satisfactory form of preservation provided that, during drying, there is free circulation of air over both surfaces of the skin, as the drying needs to proceed sufficiently rapidly to counteract putrefaction. Dried skins used by the leather industry come from tropical climates where the method of drying is to suspend the skin on frames kept in the shade or out of direct sunlight, so preventing the skins reaching too high a temperature. The practice of spreading skins on the ground to dry restricts evaporation to only one surface, and can result in considerable damage to the skin, not so much from putrefaction as from hydro-thermal shrinkage, a form of protein denaturation in which the fibrillan collagen assumes a rubber-like state. In these climates, ground temperatures can exceed 60°C (Sykes, 1954). Wet skins exposed to such

conditions during a drying period of one day or more will be readily
heat degraded. Further degradation of the damaged protein will then
occur as the skin passes through highly alkaline liquors in the early
stages of leather production.

In temperate climates, where fur skins are cured, it is necessary to
dry the skins in an enclosed area in which air is circulated at a controlled temperature and humidity. The skins are generally thin and
the time required to dry the skin completely is short compared with
the time that would be required for a sheepskin or cattle skin. It is
customary with some animals, such as the mink, to remove the skin
as a complete unopened skin, and to dry these skins fur out. To obtain
unimpeded air circulation over the flesh surface, the skin is stretched
on to grooved boards, and warm dry air is blown through each skin.
Problems arise when ill-fitting boards fail to separate the flesh surface
over the entire skin. There are then areas where the rate of drying is
slow, the warm conditions encourage bacterial action, and the
combination of the two result in hair becoming loose.

C. Dry-Salting

Dry-salting can combine some of the merits of wet-salting with some
of the merits of drying. The dry-salted skin is light in weight, it can
be kept indefinitely and yet it will soak back in water far more readily
than the air-dried skin.

Dry-salting is frequently used in countries where salt is readily available but where the hot climate would make storing of wet-salted skins
undesirable. It is also used for skins that are so thick that the skin would
be damaged by putrefaction or hydrolysis before completion of drying.
Dry-salting is carried out by wet-salting followed by air-drying the salted
skin. The precautions already referred to for prevention of damage
when curing by either salting or drying apply to this method.

D. Storage

Skins cured by any one of these three methods will remain in a satisfactory state of preservation providing storage conditions are not
adverse. Prolonged high humidity and heat are to be avoided.

V. ALTERNATIVE METHODS OF CURE

Salt is an excellent preservative for skin and, provided sufficient salt is applied and storage conditions are not adverse, the raw skin can be held for one to two years. The advantage of using salt is that no complex machinery or expensive plant is required for the curing procedure, and storage conditions for the salted skins are relatively undemanding. Consequently, a salt cure has permitted considerable latitude with regards to transport, time of purchase and time of use of the cured raw material. The disadvantage of salt lies in the relatively large quantity that is required, at least 8 kg of salt for every cattle skin. This in turn leads to an undesirable high salinity in the soak liquors entering the tannery effluent. The concentration of dissolved solids in trade effluent is causing concern to water authorities, particularly in countries with restricted water supplies. This has prompted a search for alternative methods of preserving skins over a limited period of time. Three methods are described below.

A. Biocides

Bacterial attack can be retarded by applying biocides to the raw skin, either by spraying the flesh surface of the skin or by immersing the skin in biocide. A wide range of biocides have been tested (Haines, 1975), but the general finding so far has been that a biocide application comparable in cost to the current cost of salt is likely to preserve a raw skin for no longer than six days at 26°C which is equivalent to warm summer conditions. Longer preservation times have been achieved, but at a cost several times that of salt, or with a biocide that would be environmentally unacceptable (Cooper and Galloway, 1974; Hughes, 1974), or the sealing of raw skins in air-tight containers (Bailey and Hopkins, 1975). In the United Kingdom, a biocide has been used commercially. The flesh surface of skins was sprayed with an aqueous solution of a polymeric biguanide which preserved skins for three days (Haines, 1973).

B. Chilling

Bacterial activity is highly dependent on temperature, and trials have shown that the rapid chilling of skins to below $-1°C$, i.e. just above the freezing point of the skin, and storage at $-1°C$ allows the skin to remain in a good state of preservation for a maximum of three weeks (Haines, 1975). Since 1975, this method of temporary preservation has been in limited commercial use in the United Kingdom at costs comparable to that of salting (Mark, 1975). A three-week holding time is still a short time interval that requires close co-operation between hide market and tanner, and which imposes a change from the established trade pattern of sale, distribution and organization. It offers little latitude for delays such as holidays, machinery failures or variations in slaughter rates. The alternative is freezing and cold storage.

C. Freezing

Trials have been conducted where skins have been chilled to $0°C$ immediately after flay, and folded, hair out, into a block of convenient size for handling and frozen at $-30°C$ (Lhuede and Scroggie, 1971) or $-10°C$ (Haines, 1977).

Skins that were held for three months at $-10°C$ were thawed by immersion for 24 hours in water at room temperature, and then processed into leather. The leather produced was slightly softer than leather processed from salted skins, but no adverse features were found (Haines, 1977). As yet, freezing is not in use commercially, but the costing of the procedure has been discussed by Ferguson and Mack (1977).

VI. DETERIORATION BY MOULDS

Moulds growing on untanned skins or on damp leather start from surface infection, but will invade the full thickness of the skin. The most common moulds, belonging to the genome *Aspergillus* and *Penicillium*, give rise to black and green surface discolourations. Neither of these groups appears to cause much noticeable surface damage other than discolouration. There are, however, other moulds, though of rarer occurrence, that can cause some pitting of the skin surface. One is *Fumago vagans*.

Generally, mould growth through a raw skin or pickled skin is not usually accompanied by marked degradation of the collagen such as occurs with bacteria. Their effect is more subtle, but it can lead to a wide variety of defects in the final leather. For example, moulds break down fat, either the natural fat of the skin or fat applied during the leather processing. Fatty acids produced can migrate to the leather surface forming a fine white bloom. Fatty acids can react with chromium salts used in tannage to form chrome soaps, which then cause uneven dyeing. Breakdown of fats can lead to a variation in the absorbancy of the leather surface, so that surface coatings applied are unevenly absorbed and the final surface gloss is uneven. Moulds will grow readily on leather tanned with vegetable tans, as the relatively large amount of concentrated tans and non-tans provide nutrient for moulds. Moulds grow in vegetable tan liquors, producing an enzyme tannase that attacks the tans. Moulds will also grow on leather tanned with chromium salts, particularly where glucose has been used in production of the chromium tanning salts. Skins that are marketed in the pickled condition are treated with a fungicide. Skins are also treated with fungicide at the end of tannage to protect against mould growth either during relatively short delays in the tannery or longer term storage and transport of tanned material.

REFERENCES

Bailey, D. G. and Hopkins, W. J. (1975). *Journal of American Leather Chemists Association* **70**, 372.

Benrud, N. C. (1968). *Journal of the American Leather Chemists Association* **63**, 417.

Buchanan, R. E. and Gibbons, N. E. (eds.) (1974). *Bergey's Manual of Determinative Bacteriology*, 8th ed. Williams and Wilkins, Baltimore.

Cooper, D. R. and Galloway, A. C. (1974). *Journal Society of Leather Technologists and Chemists* **58**, 120.

Dempsey, M. (1969). *Journal Society of Leather Trades Chemists* **53**, 32.

Ferguson, R. and Mack, D. (1977). *Raw Hides and Skin Seminar, British Leather Manufacturers Research Association*, p. 93.

Formisano, M. (1966). *United States Agricultural Research Service Project* No. URE 15, (60), 5.

Haines, B. M. (1973). *Journal of Society of Leather Technologists and Chemists* **57**, 84.

Haines, B. M. (1975). *Raw Hides and Skin Seminar, British Leather Manufacturers Research Association*, p. 50.

Haines, B. M. (1977). *Raw Hides and Skin Seminar, British Leather Manufacturers Research Association*, p. 76.

Hughes, I. R. (1974). *Journal of Society of Leather Technologists and Chemists* **58**, 100.

Koppenhoefer, R. M. and Somer, G. L. (1939). *Journal of American Leather Chemists Association* **34**, 34.

Kritzinger, C. C. and van Zyl, J. H. M. (1954). *Journal of American Leather Chemists Association* **49**, 207.

Lhuede, E. P. and Scroggie, J. (1971). *Australian Refrigeration, Air Conditioning and Heating Journal* **25**, 22.

Mark, M. (1975). *Raw Hides and Skin Seminar, British Leather Manufacturers Research Association*, p. 71.

Schmitt, R. R. and Deasy, C. (1963). *Journal of the American Leather Chemists Association* **58**, 577.

Stuart, L. S. (1940). *Journal of American Leather Chemists Association* **35**, 554.

Sykes, R. L. (1954). *Journal of Society Leather Trades Chemists* **38**, 304.

Tancous, J. (1970). *Journal of American Leather Chemists Association* **65**, 176.

Tancous, J. (1972). *Journal of American Leather Chemists Association* **67**, 344.

Tancous, J. and Jayasimhalu, K. (1973). *Journal of the American Leather Chemists Association* **68**, 132.

Thomson, J. A., Woods, D. R. and Welton, R. L. (1971). *Unpublished Results Leather Industries Research Institute, South Africa, Research Bulletin*, No. 573.

Tissier, H. and Martin, K. (1902). *Annals de l'Institut Pasteur, Paris* **16**, 865.

United States Military Specification for Wet-Salted Hides (1970). MIL-H-43250B (GL).

Woods, D. R., Welton, R. L. and Thomson, J. A. (1972). *Journal of the American Leather Chemists Association* **67**, 217.

Woods, D. R., Rawlings, D. E. and Cooper, D. R. (1973). *Unpublished Results Leather Industries Research Institute, South Africa, Research Bulletin*, No. 643.

6. Metals

J. D. A. MILLER

Corrosion and Protection Centre, University of Manchester Institute of Science and Technology, Manchester, England

I. Introduction	150
A. Basic Principles of Electrochemical Corrosion	150
B. Classification of Microbial Corrosion Processes	153
II. Corrosion by Concentration Cell Formation	155
A. Mechanism	155
B. Microbial Growth and Concentration Cell Corrosion in Industrial Situations	156
C. Control of Microbial Growth in Cooling Water Systems	159
III. Sulphuric Acid Corrosion	162
A. Acid Production by Thiobacilli	162
B. Corrosion by Thiobacilli	163
C. Detection of Causative Organisms	164
IV. Aviation Fuel Tank Corrosion	165
A. Historical	165
B. The Fungus *Cladosporium resinae*, and Other Contaminants	167
C. Mechanism of Corrosion of Integral Fuel Tanks of Subsonic Aircraft	171
D. The Possibility of Fungal Growth in Supersonic Aircraft	173
E. Prevention and Control	174
V. Corrosion by the Sulphate-Reducing Bacteria	179
A. Historical	179
B. The Sulphate-Reducing Bacteria	181
C. Nature and Extent of Corrosion by the Sulphate-Reducing Bacteria	183
D. Mechanism of Iron Corrosion by the Sulphate-Reducing Bacteria	187
E. Corrosion of Other Metals by the Sulphate-Reducing Bacteria	194
F. Prevention and Control	195
G. Detection of Sulphate-Reducing Bacteria	198
References	199

I. INTRODUCTION

A. Basic Principles of Electrochemical Corrosion

Corrosion in aqueous or damp environments—the natural process whereby metals tend to a greater or lesser degree to revert to the chemically combined state—is too well known to need description, except to say that it manifests itself as a weight loss after any adherent corrosion product has been removed. This metal loss may either be uniform (and therefore conveniently measured as simply a weight loss) or take the form of pitting or other localized attack (when it is of vital importance to know the depth of the deepest parts of the attack in order to assess damage). Loss of metal, then, is often accompanied by formation of insoluble corrosion products which may be closely adherent—and hence sometimes protective, eventually stifling the process —or loose and bulky. The corrosion mechanisms known or postulated to be at work in microbial attack on metals are all relatively simple ones well known to electrochemists.

Two interrelated, but quite distinct, processes occur on the surfaces of metals in wet or damp situations to bring about the overall reaction we know as corrosion.

1. The Anodic Reaction

If a metal is immersed in water or an aqueous solution, it passes into solution as cations leaving electrons behind. Clearly, this separation of electrical charge (electrons in the metal, cations in solution) cannot continue indefinitely and, in the case of the more noble metals, a dynamic equilibrium is set up between metal and hydrated cation that is reflected in an equilibrium potential characteristic of the two. Equation (1) shows a typical dynamic equilibrium for metal M.

$$M \rightleftharpoons M^{z+}_{(aq).} + ze \tag{1}$$

By contrast, the (base) engineering metals, for a reason explained below, do not exhibit this dynamic equilibrium but suffer corrosion.

If some means exists for removing electrons left behind in the metal when a cation forms, the equilibrium is not established but there is a nett dissolution; that is, the anode reaction (electron release) is favoured. Thus, if a base metal in a suitable aqueous system is put in electronic contact with a more noble metal in the same system (with which, therefore, it is also in ionic contact, completing the circuit), electrons will flow from the base metal (the anode) to the less negatively charged noble metal (the cathode) where they will be consumed in a cathodic reaction. It will be shown below that electrons can equally be consumed on the surface of the base metal itself, that is, some parts of the surface may act as cathodes.

2. Cathodic Reactions

Electrons arriving at the cathode have several possible fates. For ferrous materials in natural waters including neutral salt solutions, accumulation of electrons at the cathode is prevented in aerated conditions principally by the oxygen reduction reaction:

$$O_2 + 4e + 2H_2O \rightarrow 4OH^- \tag{2}$$

An electron-consumption reaction is termed a cathode reaction.

In the absence of oxygen, the principal cathode reaction is:

$$H_3^+ + e \rightarrow \tfrac{1}{2}H_2 + H_2O \tag{3}$$
$$\text{Hydrated proton}$$
$$\text{from solution}$$

Equation (3) represents an overall reaction proceeding *via* formation of atomic hydrogen which is adsorbed on the metal surface. Normally this then becomes gaseous (molecular) hydrogen by one of the reactions (4) or (5) below, though under some conditions it can be absorbed by the metal instead. The combination reaction, atomic to molecular hydrogen, is slow and is often the rate-limiting step in the cathodic reaction in corrosion. It can be catalysed, for example by platinum.

$$2H_{(ads)} \rightarrow H_2 \tag{4}$$
$$H_{(ads)} + H^+ + e \rightarrow H_2 \tag{5}$$

In aerated acidic conditions, protons (hydrogen ions) are more readily available than oxygen and the cathodic reaction shown in equation (3) becomes the dominant means of consuming electrons released in the anodic reaction. In neutral solutions, protons are in short supply and, in the absence of dissolved oxygen, the cathodic process becomes sluggish. Since it is a matter of experience that, when corrosion occurs, there is no build-up of electrical charge, it follows that the rate of electron release (anodic reaction) and electron consumption are in balance. Therefore when the cathodic reaction is sluggish, the corrosion rate is low. Thus, although the essential feature of the corrosion reaction, metal loss, is an anodic process, it can be seen from this account that it is frequently under cathodic control.

Clearly, any agent that accelerates reactions (4) or (5) will accelerate the cathodic, and hence the anodic, reaction. These would include protons, catalysts, and agents capable of removing molecular hydrogen. A cathode in which the electron-discharge reaction is limited by the restrictions imposed by the hydrogen evolution reaction, equation (3), is frequently said to be polarized; though the situation is easily reversible by conditions favouring reactions (4) and (5) or by ingress of oxygen to facilitate the oxygen reduction reaction, whereas other familiar types of cathodic polarization (such as obstruction by corrosion product) are not.

As already stated, iron (or steel) and other base metals used in construction corrode 'spontaneously' in a wet environment without the aid of a more noble metal in electronic and ionic contact with them. In the case of iron, for example, the presence of millscale, irregularities of crystalline structure and other chance variations in the physical or chemical nature of the surface will result in the metal surface becoming differentiated into often minute anodic and cathodic areas. Anodic sites of this sort are frequently annihilated by dissolution so that the next anodic reaction must occur at another site; thus, when there is no constraint on the choice of site of the anodic reaction, uniform corrosion will result since the metal surface will be dissolved away in near-random fashion. However, if some local inhomogeneity in the surface localizes the anodic site, localized corrosion will occur. For example, if the surface is bathed in a solution the concentration of which varies from one point to another (resulting in concentration cells), or if one area of the metal surface is occluded by some means and hence is less oxygenated than another (resulting in differential aeration cells, the anodes of which are the less well oxygenated areas), sustained and therefore severe localized attack can occur. A differential aeration cell is, of course, merely a common kind of concentration cell. Other metals, including ones more noble than iron, are subject to these influences on their surfaces.

B. Classification of Microbial Corrosion Processes

Microbial corrosion may perhaps be defined as metal loss caused, or accelerated, by microbial action at one or both of the two sites controlling electrochemical corrosion, namely the anode and the cathode. Some writers draw a distinction between direct and indirect mechanisms of microbial corrosion, without making it clear how they define these terms. The distinction would appear to me to be that, if a microbe interlinks an electrode process with its own metabolism, it is a direct effect; otherwise it is an indirect effect. Thus, removal of cathodic hydrogen by sulphate-reducing bacteria for oxidation inside the organisms to yield energy, or removal from an alloy of certain metals required by living organisms as trace elements, would be direct effects (this latter process is highly unlikely, since only extremely minute amounts of these metals are required and almost all natural and industrial environments must contain sufficient of such metals in solution; but the mechanism has been suggested to explain microbial corrosion of aircraft alloys). Conversely, the differential aeration cells set up by coherent microbial growth that happens to occur on a wet metal surface, or the corrosion caused by sulphide that arises as an end-product of the anaerobic respiration of sulphate-reducing bacteria growing nearby, would constitute indirect mechanisms. Similarly, direct attack on (biodegradation of) a protective organic coating by microbes, exposing bare metal to some form of electrochemical corrosion, is an indirect microbial corrosion mechanism; though it is outside the scope of this chapter.

An attempt at a more detailed breakdown of microbial corrosion mechanisms than a simple direct/indirect one is presented below; the possibility is acknowledged that more than one of these mechanisms may operate in a given situation.

1. Absorption of oxygen and other nutrients by microbial growths adhering to damp or wet metal surfaces, causing concentration cells to be set up between the interior of the growth and the immediate environment. In such an electrochemical cell, the oxygen-depleted zone beneath the coherent microbial growth is anodic with respect to the exposed metal bordering the microbial material, and anodic dissolution occurs. Microbial growth can occur in almost all industrial situations in which water is present and temperatures are not severe throughout, since such an enor-

mous range of substances can serve as nutrients for at least a few types of microbe. In general, only closed systems and those in which the temperature is too high throughout to allow microbial growth to occur are likely to remain free from corrosion by this mechanism.

2. Utilization of various substances that can result in alteration in the chemical or physical nature of the environment in deleterious ways. For example, certain additives in lubricating oils may be degraded, reducing the protective effect of the oil against corrosion and wear; and emulsifiers in cooling and cutting emulsions may be degraded, resulting in breaking of the emulsion and loss of its special properties. This subject is dealt with in Chapter 9 (p. 259), and will not be enlarged on here.

3. Liberation from the organisms of certain end-products of growth. This can be conveniently subdivided as follows.

   (a) Production of substances with surfactant properties, leading to emulsification of water present in lubricating oil systems. This problem occurs particularly in the lubricating oil of marine diesel engines, in which the circulating systems are so large and complex that the exclusion of dirt, microbes and water is virtually impossible. The presence, due to emulsification, of water droplets in the circulating oil can lead to corrosion of bearings and journals. Such emulsification by microbial action may seem anomalous in view of the breaking of emulsions referred to in mechanism (2) above; however, different organisms are involved, growing in different conditions. This topic is also dealt with in Chapter 9 (p. 259).

   (b) Production of sulphuric acid by certain species of autotrophs belonging to the genus *Thiobacillus*.

   (c) Production of carboxylic acids as metabolic end-products or by leakage of tricarboxylic acid-cycle intermediates. These acids, though weaker than sulphuric acid, are nevertheless corrosive particularly to nonferrous metals.

   (d) Production of sulphide ions by the sulphate-reducing bacteria. These ions are known to be aggressive to metals. Other proposed mechanisms of corrosion by the sulphate-reducing bacteria are described below.

4. Interference with the cathodic process in the absence of oxygen

by the sulphate-reducing bacteria, either directly by removal of hydrogen or indirectly by the formation of iron sulphides.

Some of these mechanisms will now be described in detail.

II. CORROSION BY CONCENTRATION CELL FORMATION

A. Mechanism

Under this heading falls all microbial corrosion that is due to neither production of aggressive substances or surfactants that affect lubricating oils, nor to uptake or breakdown of substances whose removal alters the infected substance deleteriously. It depends merely on the uptake of normal nutrients including oxygen by an adherent microbial growth, thus setting up concentration gradients, and is one type of an extremely well known and common form of corrosion of which crevice attack and corrosion beneath occluded parts of metals are examples.

All types of microbes that can grow in a coherent colony, or mat, or mass of 'slime' on damp or immersed metal are potentially harmful from this standpoint, and it would be unhelpful to list all of the species of bacteria, yeasts, filamentous fungi and algae that have been found in such mixed growths.

Figure 1 shows diagrammatically the mechanism of anode formation by a differential aeration cell set up under a microbial colony. In such

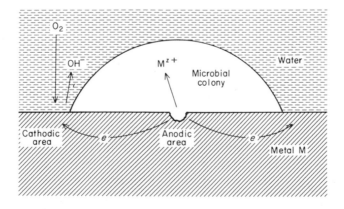

Fig. 1. Diagrammatic representation of a coherent mass of microbial growth on a metal surface, showing the setting up of a differential aeration cell.

cases, active growth of the organisms keeps the oxygen concentration near to zero in the centre; but, once the electrochemical cell is established and the colony increases in bulk, even the death of organisms in the interior of the colony need not extinguish the cell since there is now a substantial mechanical barrier to the ingress of oxygen (moreover, the biochemical oxygen demand of the mass of autolysed organisms will assist removal of oxygen that diffuses inwards). This type of microbial corrosion is thus the only one that does not depend on active growth, with the possible exception of iron sulphide corrosion initiated by the sulphate-reducing bacteria (see Section V, p. 179).

A serious additional problem arises if the anaerobic conditions at the centre of the colony initiate growth of any sulphate-reducing bacteria that became entrapped in the colony in the early stages of growth.

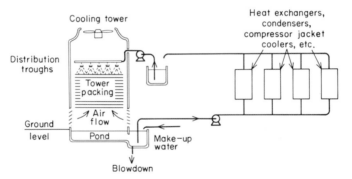

Fig. 2. Diagram of an open recirculating industrial cooling water system with an induced-draught cooling tower.

B. Microbial Growth and Concentration Cell Corrosion in Industrial Situations

This type of corrosion is rife in industrial cooling-water systems, especially of the open recirculating type. A typical system of this kind is shown diagrammatically in Figure 2. The principle is the cooling of the water, after it has passed through heat exchangers, condensers etc., by evaporation of a proportion of it in the cooling tower. In the tower the water is sprinkled, usually from perforated troughs, over slats of wood (or in some modern towers a plastic infill), and the cooled water collects in the 'pond' at the base of the tower ready for re-circulation. Many cooling towers are of the 'forced draught' type, with fans at the base forcing air up through the water droplets and across the water films on the slats (tower packing). The wide top of the tower is open to sunlight. The alternative arrangement is 'induced draught', in which a single large fan above the troughs draws air up through the falling water. This arrangement excludes some, but not much, light from the top of the tower; moreover, the open bottom of such a tower admits light.

The loss by evaporation is replaced by make-up water. Unless this make-up water is perfectly pure, there is a concentration effect causing a build-up of electrolytes in the water to an unacceptable level from the standpoint of corrosion; and this concentration of dissolved solids makes the recirculating water resemble an algal culture medium. Because of this undesirable concentration effect, water is deliberately bled off such a system ('blowdown') at a calculated rate, and replaced by more make-up water. The rate of blowdown is kept as low as possible so as not to waste water and chemicals that are put into it for various purposes.

The availability of mineral nutrients, coupled with sunlight, a good growth temperature in many parts of the tower, and efficient aeration, favours heavy growths of algae. Such algal material (usually mixtures of green and blue-green algae and diatoms) eventually breaks off under its own weight and drops into the pond, finds its way into dark parts of the system, dies and decomposes, liberating a complex mixture of organic nutrients for heterotrophic organisms to utilize. Photosynthetic organisms are, as always, the primary producers, creating conditions in which other forms of life can flourish.

Even without algal growth, though, an open recirculating system can easily accumulate enough organic detritus, from dead leaves to oil, to enable heterotrophs eventually to gain a foothold in the system. The resulting mixed microbial growth, usually referred to as 'slime' (and sometimes, largely erroneously, as 'algae') by maintenance engineers, tends to occur wherever the microbes can anchor to minute irregulari-ties on the metal surface, or there is a sudden drop in water velocity in the system, such as where a pipe feeds into a heat exchanger, and settling out can occur. Rates of heat transfer in heat exchangers so fouled are much decreased. The slimy nature of such microbial growth is due to the presence of various microbes—pseudomonads and others—that produce capsules or sheaths of mucilaginous material, usually polysaccharide in nature, enabling the organisms to cohere and frequently entrapping other, non-slime-forming organisms. Kelly (1965) identified some of the commonest bacterial constituents of microbial slime as *Bacillus subtilis*, *B. cereus*, and species of *Flavobacterium*, *Aerobacter* and *Pseudomonas*. Algal growths contained predominantly species of *Chroococcus*, *Oscillatoria*, *Chlorococcus*, *Ulothrix*, *Scenedesmus* and *Navicula*.

The algae themselves are troublesome for other reasons than that they introduce organic matter into the system. Metal assembly bolts fastening the cooling tower packing material can be corroded by their presence by the same mechanism as corrosion under slime. Moreover, algae present another problem not connected with corrosion. A heavy growth tends to block the outlet holes in the troughs at the top of the

Fig. 3. Section of a cast-iron pipe showing extensive tuberculation from Miller and King (1975).

tower (see Fig. 2), the function of which is to ensure that the water is sprinkled evenly over the infill and descends as a film. The result is that water descends unevenly, and may even spill over the top of troughs and descend as a cascade, making the efficiency of the tower drop to a small fraction of the theoretical value (see the graph in Purkiss, 1972).

Another microbiological problem that frequently occurs in iron pipe-work in industrial cooling systems as well as in water mains and a variety of other situations is that caused by the iron bacteria. Two types are involved, namely members of the filamentous genus *Sphaerotilus* (Chlamydobacteriaceae) and related forms such as species of *Crenothrix* and *Leptothrix* (which may both be forms of *Sphaerotilus*) and the stalked unicellular bacterium *Gallionella* (Caulobacteraceae). Both types are chemolithotrophic autotrophs, obtaining energy from oxidation of ferrous to ferric ions which results in extensive local deposition of ferric hydroxide. The originally delicate filaments of *Sphaerotilus* sp. become coarse and bulky by this process, and form easily visible tufts (in this state it is sometimes referred to as the 'sewage fungus'). Inside iron pipes, this and *Gallionella* sp. eventually form large tubercles which enable pitting corrosion to occur underneath and can greatly restrict water flow. The biological origin of tubercles in pipes has sometimes been disputed; yet it is said that chlorination of the water helps to control tuberculation, confirming its microbial nature. An advanced stage of tuberculation of a pipe is shown in Figure 3.

C. Control of Microbial Growth in Cooling Water Systems

One approach to the problem of microbial corrosion in industrial cooling-water systems is to use corrosion-resistant materials for con-struction wherever possible. Castleberry (1969) has briefly surveyed the corrosion-resistant metals used for fittings in cooling towers. Various resistant materials such as stainless steel and graphite are employed for manufacturing heat exchangers. However, the problem is not merely one of preventing corrosion. As already stated, microbial growth seriously lowers the efficiency of heat exchangers and cooling towers, and so must be suppressed. Microbial control is therefore used instead of, or as well as, attempts at corrosion-proofing the system.

Much has been written about the use of biocides in industrial cooling-water systems. Numerous substances, mostly organic, have been tested, and programmes of treatment described that were found to be successful in a specified system. The term 'biocide', one generally disliked by microbiologists, nevertheless is rather useful in an industrial context despite, or perhaps because of, its imprecision. It is used to denote any chemical that has been tested in an industrial situation, probably on an empirical basis, for effective killing and/or prevention of undesirable microbial growth and found satisfactory. It would thus appear to have the same meaning as the old term 'disinfectant'; and whether a biocide kills or merely functions as a biostat is usually not important. There is in fact a term 'biostat', used by more cautious writers, just as some use the more specific term 'microbicide' instead of biocide. Few biocides can remove a really heavy, well-established growth (and, in any case, this may be undesirable for large-scale deposition of dead microbes, perhaps in another and inaccessible part of the system, means that the treatment will create as many problems as it has solved); and they are generally used for *preventing* growth, or preventing a recurrence of growth, in a system that has been mechanically or chemically cleaned as far as possible.

In this connexion, it became clear that slimy growths, even when quite small, were unharmed by many biocides and could continue to grow and spread. A recent, and rather hideous, coinage is 'slimicide', used to denote a biocide (or mixture of biocides) that is effective in breaking up slimy microbial growths—a valuable property in the context of treating cooling-water systems.

The properties desirable in a biocide are as follows. (1) It should be safe to handle in concentrated form (not injurious to skin; not spontaneously flammable or explosive). (2) It should be non-toxic to higher forms of life at the concentration used. (3) It should be ultimately biodegradable, or decompose, so that it does not persist when discharged into rivers or sewage systems and infringe antipollution regulations. A high dilution coefficient also helps in this respect. (4) It should have a broad spectrum of activity against microbes; that is, it should control all troublesome species likely to be encountered. (5) It should be able to penetrate and break up microbial slime. (6) It should have an acceptable cost when used at an effective concentration. (7) It should be non-reactive towards dissolved organic substances. (8) It should be compatible with other additives (corrosion inhibitors and antinucleating agents). (9) It should be compatible with materials of construction;

that is, non-injurious to metal and to wood if used in the cooling tower. (10) It should have low volatility and odour. (11) It should be non-foaming at an effective concentration. (12) It should be stable in dilute solution, surviving repeated heating and exposure to light, during its residence time in the system that is to be protected.

In open circulating cooling systems, biocides may be used in two ways, firstly as a shock treatment to follow mechanical cleaning of a badly fouled system in order to kill or remove any inaccessible microbial growths, and secondly as a lower level, or residual, concentration, maintained for a long period or permanently, to prevent recurrence of growth. These treatments may be combined in an 'intermittent feeding' system whereby at intervals a 'slug' of biocide is introduced, and the dose repeated when the biocide concentration has fallen to 25% of the initial concentration (that reached when the 'slug' has thoroughly mixed with the recirculating water). This time interval can be calculated from knowledge of the capacity of the system and the rate of loss by blowdown (see p. 157) and windage (loss of droplets in the air current through the cooling tower, as distinct from loss by evaporation which does not involve loss of a nonvolatile biocide).

This topic and the selection of a biocide are dealt with in detail in the Betz Handbook of Industrial Water Conditioning (1976), and further information on biocides is given by Miller (1971) and Purkiss (1971, 1972). Modern biocides include chlorine (or hypochlorite), quaternary ammonium compounds, chlorinated phenols, complex phenolic compounds, fatty amines and diamines (which are more effective at penetrating slime when combined with a quaternary ammonium compound; see Tehle, 1966) and various proprietary formulations containing two or more biocidal substances and relying on a synergistic effect between them.

It is not normally economically feasible to use biocides in once-through systems, in which water is discharged after performing its cooling function instead of being passed through a cooling tower and recycled.

Treatment for water-carrying pipes suffering from internal tuberculation involving iron bacteria is difficult. Chlorination, though said to prevent tubercle formation, will not remove tubercles once they have formed. The only successful treatment appears to be to grind the tubercles out with rotary grinding heads on long flexible rods, and coat the inside of the pipe with bituminous paint applied by a special rotary brushing device.

Not much appears to be known yet about internal microbial corrosion of submarine crude-oil pipelines. These systems could easily become infected with oil-utilizing microbes at the assembly stage. It is known, from the accumulation of hydrogen sulphide in basically 'sweet' crudes, that sulphate-reducing bacteria are present and active in it, thus presenting the possibility of a hazard arising from mixed growth causing differential aeration corrosion and sulphide corrosion. The frequent passage of 'pigs' along such pipelines should help to prevent extensive microbial growth from occurring.

III. SULPHURIC ACID CORROSION

A. Acid Production by Thiobacilli

There are several species in the genus *Thiobacillus* (Pseudomonadales), of which all but one are strict autotrophs. Some confusion exists as to the delineation of the species in biochemical terms, but they all share the ability to oxidize either sulphur or various of its more reduced compounds to obtain energy for the fixation of carbon dioxide. A related form, frequently called *Ferrobacillus ferrooxidans* but now regarded by most authorities as belonging to the genus *Thiobacillus*, can also oxidize ferrous to ferric iron for carbon dioxide fixation. Thiobacilli are motile, usually short, Gram-negative rods; they are strict aerobes, except for one species which is a facultative anaerobe.

Purkiss (1971) stated that the following interlinked reactions are performed by mixed cultures of thiobacilli acting on elemental sulphur or sulphide:

$$2H_2S + 2O_2 \rightarrow H_2S_2O_3 + H_2O \tag{6}$$

$$5S_2O_3^{2-} + 4O_2 + H_2O \rightarrow 5SO_4^{2-} + H_2SO_4 + 4S^0 \tag{7}$$

$$2S^0 + 3O_2 + 2H_2O \rightarrow 2H_2SO_4 \tag{8}$$

Probably only *T. thiooxidans* (including *T. concretivorus*, which may be slightly different) and *T. ferrooxidans* can produce hazardous quantities of sulphuric acid. The former is said to remain active at pH 0.7, corresponding to more than 5% sulphuric acid (Kempner, 1966). These species only grow in conditions that are already acidic (pH about 5).

Thiobacillus ferrooxidans is important economically because of its ability to leach metal ores, i.e. to solubilize them so that the metal can be recovered without recourse to smelting. One such reaction of this species that is of interest is its action on pyrite (Le Roux *et al.*, 1974). The overall reaction can be written:

$$4FeS_2 + 15O_2 + 2H_2O \rightarrow 2Fe_2(SO_4)_3 + 2H_2SO_4 \tag{9}$$

In this reaction, the organism clearly oxidizes the sulphur as well as the ferrous iron in the substrate, showing its affinity with the other thiobacilli.

Thiobacilli are fairly commonly encountered in Nature. They probably obtain their energy from simple sulphur-containing breakdown products of sulphide minerals, or on sulphide produced in an adjacent locality (or in the same locality at a previous time, when conditions were anaerobic) by sulphate-reducing bacteria. Normally they are not found in large numbers but, if a local high concentration of a suitable substrate occurs, their activity may be great and result in a zone of pH 2 or lower. Such conditions can arise in sewage, in which sulphide concentration may be high owing to the action of putrefactive clostridia unless aeration is efficient. Suitable conditions can also arise locally in 'made-up' ground, into which industrial waste material sometimes finds its way. For this reason, sulphuric acid corrosion of underground pipelines is sporadic and unpredictable.

B. Corrosion by Thiobacilli

A classic case of mineral-acid corrosion in the soil was documented by the then National Chemical Laboratory at Teddington in Great Britain (Report; Chemistry Research 1956, 1957). Severe corrosion of steel gas pipes and cast-iron water mains occurred in a newly developed residential site in two and a half years. The peaty soil surrounding the pipes had a pH as low as 2.0 and contained sulphuric acid; it was not made-up ground and there had been no previous industrial activity in the district. At lower levels, the soil contained large numbers of sulphate-reducing bacteria and the sulphides produced diffused upwards to the zone containing the pipes and provided an energy source for thiobacilli. The sulphate-reducing bacteria were supported in turn by sulphate-bearing springs. Draining and liming the soil was recommended. The former helps aeration and hence discourages growth of sulphate-reducing bacteria, and the latter renders conditions unsuitable for growth of acid-producing thiobacilli.

Metal sewage pipes can be attacked in the splash zone by thiobacilli; so, incidentally, can concrete sewage pipes, with disastrous results. The use of internal acid-resistant protective coatings, or the introduction of as much turbulence in the liquid flow as possible, by the use of for example weirs to improve aeration and thus discourage sulphide pro-duction, are the usual remedies.

Corrosion by acid-producing thiobacilli is easily recognizable merely by measurement of pH values in the local soil. Alternatively, the presence of the causative organism may be confirmed by using the bacteriological growth test given in Section C (this page).

Pyrite oxidation by *T. ferrooxidans* produces its troublesome effects principally in mines. Water seepage is often a problem in mines, and the presence of pyrite and of *T. ferrooxidans* can result in the pH value of the mine water dropping to 2–3, rendering it highly corrosive to pumping machinery and pipes. Means of combating the problem are the use of lime and the selection of acid-resistant metals such as stainless steel for the pumping equipment. Again, the problem can readily be recognized by the pH value of the water and, in addition, by a brown deposit of basic ferric sulphate.

C. Detection of Causative Organisms

The acid-producer *T. thiooxidans* can be grown in the following medium (Starkey, 1935): $(NH_4)_2SO_4$, 0.4 g; KH_2PO_4, 4 g; $CaCl_2$, 0.25 g; $MgSO_4$, 0.5 g; water to 1 litre. The medium is dispensed into conical flasks (say 100 ml to 250-ml flasks), about 1 g of elemental sulphur is sprinkled on the surface, the flasks are plugged with cotton wool and sterilized by steaming for one hour on each of three successive days. The pH value should be approximately 5. About 1 ml or 1 g of the material to be tested is inoculated into 100 ml of medium and the flasks are incubated at 30°C. The pH value should be checked daily; a rapid drop to about 2.5 indicates growth of *T. thiooxidans*. This can be con-firmed by microscopic examination for the bacteria, which are especi-ally abundant on and around the sulphur particles.

IV. AVIATION FUEL TANK CORROSION

Various types of microbes, including fungi, excrete organic acids (tri-carboxylic acid-cycle acids and others) during active aerobic growth or under conditions of nitrogen starvation. In addition, organic acids are a common end-product of fermentative microbial metabolism. Within large masses of mixed microbial growth the local concentration of these acids may cause serious corrosion problems, particularly to certain non-ferrous metals. Nevertheless, the harmful effect of organic acid producers is probably more often due to damage to protective coatings and electrical insulation materials, exposing the underlying metal to attack by many agencies both biological and non-biological. Such deterioration of protective and insulating coatings is particularly rife in the warm, humid conditions of the tropics. Use of inert polymeric coatings and insulation, and of effective fungicides, appears to have greatly decreased the magnitude of this problem in recent years. One microbial problem due to organic acid production which has been largely, but not completely, overcome in recent years is the interesting one of corrosion of jet-fuel tanks, and particularly of aircraft integral fuel tanks.

A. Historical

Before the development of the gas turbine engine for propulsion of aircraft, no microbiological problem connected with fuel had been encountered. The fuel (petrol, for piston engines) had been carried in tanks constructed for the purpose. The turbojet, however, was found to run more economically and safely on kerosene-type fuels which, unlike petrol, contain a preponderance of straight-chain paraffins. More-over, about the time of the rapid development in aircraft design neces-sitated by the greater speeds achievable by the jet engines, it was realized that the wings offered a convenient storage space for the fuel. The box structure within was easily compartmentalized and sealed against leaks, and pumps and piping were provided to enable fuel to be transferred to maintain trim during flight.

At this point the problem appeared, and the technical literature of the late 1950s and 1960s abounds in case histories of fuel filter blockages, pump failures and wing-tank corrosion which, it was realized, were associated with 'sludge' on the tank bottoms. Examination of this sludge showed that it was composed of bacteria, yeasts and filamentous fungi (Prince, 1961). The problem was clearly a World-wide one. Both commercial and military aircraft were affected, especially those based in, or visiting, the tropics. The principal microbial constituent of the sludge on a dry weight basis was also the same the World over: a fungus, variously called *Hormodendrum resinae* Lindau, *Cladosporium resinae* (Lindau) de Vries and *Amorphotheca resinae* Parbery, and commonly known in the aircraft industry as the 'kerosene fungus' (see, for example, Hazzard, 1963). Other fungi are commonly present, though, as well as bacteria and yeasts as already mentioned, and the roles and inter-actions of the various organisms involved in sludge formation have not yet been satisfactorily investigated. Churchill (1963) compiled the fol-lowing list of microbes reported to be present in significant amounts in aircraft tank sludge: *Pseudomonas aeruginosa, Aerobacter aerogenes, Sphaero-tilus natans,* and species of *Clostridium, Bacillus, Desulfovibrio, Micrococcus, Spicaria, Fusarium, Aspergillus, Penicillium* and *Cladosporium.*

Integral fuel tanks of aircraft frequently contain water, for various reasons. Firstly, water is slightly miscible with kerosene. The water content in parts per million is numerically approximately the same as the fuel temperature in °F; hence, water is precipitated by the chilling that occurs in the stratosphere. Secondly, as the fuel level in the wing tanks falls during flight it is replaced by air, which is often humid. Thirdly, water as a separate phase is present further back in the system (that is, in fuel storage tanks on airfields) for similar reasons and because of occasional ingress of rainwater, and can become entrained in the fuel when pumped into bowsers and thence into aircraft. Aircraft are provided with drainage points in the lower wing surface for removal of water, but as pointed out by Elphick and Hunter (1968) water puddles accumulate readily on wing tank bottoms in modern aircraft that lack a significant wing dihedral angle facilitating drainage.

The water bottoms in aircraft were often reported to be 'dirty', containing mineral salts and rust particles (from storage tanks— kerosene has a pronounced penetrating action and readily dislodges rust and scale) as well as microbes, and the corrosion associated with the sludge was at first generally attributed to the inanimate matter; the micro-organisms being dismissed as consequential, not causal. Efforts were made to improve 'housekeeping' and exclude this dirt, but, of course, it was impossible to exclude water and dissolved nutrients (even with the use of coalescer/filter units when transferring fuel) from the aircraft and, for that matter, bacteria and fungal spores. Eventually it was realized that corrosion of the aluminium alloy under microbial deposits—characterized particularly by pitting, exfoliation and inter-granular attack—was caused in some way by microbes.

B. The Fungus *Cladosporium resinae*, and Other Contaminants

The brownish sludge, known to maintenance workers ever since kero-sene was introduced as an aircraft fuel, takes its colour from its major component *Cladosporium resinae*. This fungus is able to grow on kerosene and on various straight-chain paraffins, as sole carbon-and-energy source. It has been called the 'creosote fungus' since it is commonly found growing on creosoted timber; it also grows on the resin of conifers, from which it was first isolated, and in the soil (Parbery, 1968). Its spores can frequently be demonstrated in the atmosphere.

It was originally believed to be an imperfect fungus, and ascus formation appears not to occur in mycelium in fuel tanks or in labora-tory cultures originally isolated from contaminated fuel. Parbery (1969) found a perfect form which he named *Amorphotheca resinae*, though the universally encountered imperfect form is still referred to as *C. resinae*. Figure 4 is a photomicrograph of the sporulating imperfect form, and Figure 5 shows a typical growth of *C. resinae* in a wing tank. The fungus is variable in colour; shades from pinkish-brown through grey-brown to olivaceous green have been noted. It is also morphologically variable, especially with regard to the conidiophores, and the species has been subdivided so as to take this variation into account (Sheridan, 1972).

The optimum temperature for growth is stated to be 30–35°C; growth is very slow at 10°C and is negligible at 40°C. The mycelium is clearly able to remain viable during many hours of exposure to subzero tem-peratures; the fuel during a long haul at subsonic speeds may reach temperatures as low as −40°C. On landing, the residual fuel will take a considerable time to warm up to ambient temperature; indeed, ice may persist in the integral tanks of aircraft used on transatlantic flights for up to 24 hours after landing (Elphick, 1971). This implies that, in commercial aircraft operated in temperate zones and with as quick a turn-round time as possible, there will only be very short periods in

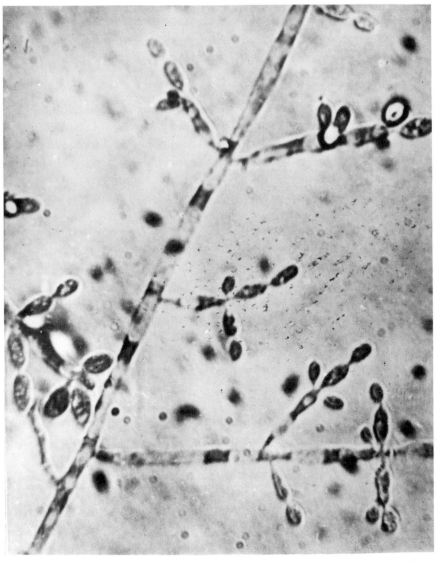

Fig. 4. Photomicrograph of *Cladosporium resinae*. Magnification ×466. From Miller and King (1975).

which growth can occur. Moreover, it explains the fact that, in general, such aircraft suffer markedly less from this problem than do private and military aircraft which tend to be grounded for much longer periods. There is an element of unpredictability, though: growth may occur in one of several identical aircraft operating on the same schedule.

In such cases it is customary to attribute the trouble to a heavily contaminated batch of fuel, and this may well be true in some cases. Fuel spillages on soil are known to lead to a proliferation of hydro-carbon-utilizing microbes, and it is a short step for *C. resinae* from there into the storage tank. Heavy growths have occurred from time to time in airfield storage tanks. Hill *et al.* (1967) showed that spores from such growth easily enter the oil phase and can retain their viability there for considerable periods (weeks or months, extrapolating from their curves) until they meet conditions suitable for germination.

Growth of *C. resinae* in a diphasic system (kerosene overlying water)

Fig. 5. *Cladosporium resinae* (dark patches) growing in an aircraft integral fuel tank. From Miller and King (1975).

occurs first at the interface. Discrete drops of water on a tank bottom tend to become anchored by the mycelium, hindering drainage through the outlets provided. Water is one of the end-products of growth, and mycelial growth spreads, uniting with other growing colonies and eventually forming an extensive mat. Disturbance on emptying and refilling the tank causes extensive pick-up of spores by the fuel.

The question of nutrient supply is a rather interesting one. Apart from the obvious carbon source, it is at first sight surprising that there should be a sufficient supply of the elements essential to life. It can be assumed that these arrive in the aircraft wing tank (*a*) as non-hydrocarbon substances always present to some degree in fuel, (*b*) as, at least in some cases, fuel additives, (*c*) as impurities ('dirt') picked up from inadequately cleaned storage tanks and pipework, and through tank covers inadvertently left open, and (*d*) as airborne particulate matter passing into tanks as fuel levels drop. Solid material that settles in a tank has a fair chance of contacting water, and water-soluble nutrient substances will partition into the water phase. Most potential nutrients will be utilized by one or more components of the mixed microbial flora, and the synergistic effects usual with mixed cultures probably operate.

The above account shows that there are several difficulties inherent in making laboratory studies of aircraft microbial corrosion. The microbiologist's instinctive tendency is usually to isolate in pure culture the organism that he considers to be responsible for the effect in order to achieve reproducibility. In fact, it may be impossible to perpetuate in the laboratory the mixed microflora obtained from a fuel tank without inducing a marked change in its composition. Hedrick *et al.* (1964a) experimented with a mixed culture consisting of *Pseudomonas aeruginosa*, *C. resinae*, *Aspergillus niger* and *Desulfovibrio* sp. After a long-term experiment (one year) the only two significant survivors were *Ps. aeruginosa* and *C. resinae*. Such changes will in part be an effect of temperature. It is extremely difficult to simulate the temperature cycle of the fuel in an aircraft wing, and hence an arbitrary temperature is decided on that will necessarily result in selection for such species as grow most rapidly at that temperature.

Another reason for the change in composition of the microflora during experimentation is selectivity imposed by the chosen growth medium. Here, as with the choice of temperature, the criterion is usually that the results should arise within a 'reasonable' period of time; hence, conditions in the water bottom are departed from quite markedly. Bushnell-Haas medium (see Parbery and Thistlethwaite, 1973) is frequently used, overlaid with fuel as sole carbon source; although Scott and Forsyth (1976) found that the medium of Klausmeier *et al.* (1963) gave better growth of organisms from aircraft tanks, especially of mesophilic fungi, than did Bushnell-Haas medium.

C. Mechanism of Corrosion of Integral Fuel Tanks in Subsonic Aircraft

One possible cause of corrosion under the microbial mats is formation of oxygen concentration cells, and this has never been disproved. Churchill (1963) appears to have been the first to suggest that aluminium alloy corrosion might be due to organic acids produced by mixed growths, amongst other causes.

Hedrick *et al.* (1964a), in the one-year experiment, found only a very slow rate of attack of aircraft alloys. In an attempt, presumably, to avoid the complication of a fairly concentrated artificial medium, they grew their mixed culture (referred to in Section I.B) in sterile deionized water and JP-4 fuel at a ratio of 4:1. Losses due to evaporation were made up at intervals by addition of water and fuel. In these conditions it is surprising that any significant growth occurred. The water:fuel ratio was orders of magnitude different from that in an aircraft so that, in the latter situation, a much greater supply of the other essential nutrients than carbon would be available even if only pure water entered the aircraft.

Several workers have concluded that corrosion of aluminium alloy by mixed microbial growths is due to bacterial action. Pseudomonads are probably of very common occurrence in the fungal mats, and Blanchard and Goucher (1967) believed that corrosive organic substances liberated into the aqueous phase by pseudomonads were responsible for the corrosion. Iverson (1967) attributed the attack to sulphate-reducing bacteria, which have sometimes been isolated from tubercles on the bottoms of aircraft integral tanks. He regarded the role of *C. resinae* also present as merely providing a matrix for growth of the sulphate reducers, giving local anaerobic conditions and a supply of nutrients. Tiller and Booth (1968) partially confirmed Iverson's findings by demonstrating attack on pure aluminium by sulphate-reducing bacteria.

Hedrick (1970) made corrosion studies of four aluminium alloys by mixed cultures of *C. resinae* and *Pseudomonas aeruginosa*. Pitting, surface exfoliation and intergranular attack were noted, and attributed to removal of zinc and magnesium from the alloy by the action of extracellular enzymes (see also Hedrick *et al.*, 1967). On the other hand, Miller *et al.* (1964) considered that galvanic cells produced by oxygen concentration gradients beneath fungal mats, which they stated may have an e.m.f. of up to 60 mV, were sufficient to cause rapid pitting of the alloy.

Substantial evidence has recently been obtained that *C. resinae* causes serious corrosion of aircraft aluminium alloys by organic acid production. McKenzie *et al.* (1977), and Akbar (1979), showed that pure cultures of the mould growing on Bushnell-Haas medium supplemented with kerosene caused pitting, intergranular corrosion and surface exfoliation of alloy specimens, followed by weight loss determinations and electron microscopy. Electron probe microanalysis showed that typical corroded regions consisted of a zone rich in copper and iron, surrounded by an area depleted in zinc, magnesium and the aluminium comprising the matrix of the alloy. It was considered that continued corrosion around such copper- and iron-rich segregates could eventually lead to their detachment from the alloy surface. Citric, isocitric, *cis*-aconitic and α-oxoglutaric acids were shown to accumulate during growth. These acids were tested individually against aircraft alloys 2024-T3 and 7075-T6; the former alloy corroded most rapidly in α-oxoglutaric acid and the latter in citric acid, and in both cases corrosion rates were comparable with those in cultures of the fungus. Akbar (1979) concluded that the results obtained with weak solutions of these organic acids showed that there was no need to invoke oxygen concentration cells to account for corrosion of the two alloys under mats of *C. resinae*. The findings of de Mele *et al.* (1979) lent support to the views of McKenzie *et al.* (1977). Shifts in pitting potential of aircraft alloys caused by *C. resinae* were correlated with production of acids, whereas cultures of *Ps. aeruginosa* produced no shift in pitting potential.

Although there is now a strong case for corrosion being caused by carboxylic acid production by *C. resinae*, contributory factors involving other organisms cannot be ruled out at this stage. Whatever the mechanism or mechanisms, the resulting pitting corrosion, besides sometimes leading to perforation of a wing and loss of fuel, acts as foci for initiation of stress corrosion. Figures 6 and 7 show corrosion of aircraft alloy.

D. The Possibility of Fungal Growth in Supersonic Aircraft

With the coming of supersonic jet aircraft it was realized (Hill, 1969) that the experience gained on subsonic types with regard to microbial growth might not be applicable to the different conditions. At supersonic speeds, friction due to the atmosphere results in warming of the fuel, in sharp contrast to chilling of the fuel in subsonic stratospheric

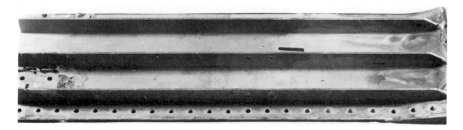

Fig. 6. Pitting corrosion inside an aircraft integral fuel tank exposed by removal of a mat of *Cladosporium resinae*.

Fig. 7. Electron micrograph showing exfoliation and pitting corrosion of an aircraft integral fuel tank. Magnification × 1,000.

flight. At speeds of about Mach 2 the temperature induced by kinetic heating can reach 120°C, at which there would be little tendency for free water to remain in the fuel and little likelihood of any organism surviving (Scott and Hill, 1971). On the other hand, such high speeds are normally maintained only for one or two hours' flight time and, if the fuel is not sterilized in this way, fuel tank temperatures will be favourable for growth for much longer periods than occur in subsonic aircraft, as pointed out by Elphick (1971). More information about the fuel conditions in supersonic transports is given by Thomas and Hill (1976a).

Fuel samples taken from 78 aircraft by Scott and Forsyth (1976) were membrane-filtered and the mixture of microbes retained by the filter was cultured at 30°, 37° and 45°C. Some 43 of the samples grew *C. resinae* when incubated at 30°C; only two did at 45°C, although at this temperature growth of the fungus *Aspergillus fumigatus* arose from 38 of the samples. In addition, *Bacillus cereus* occurred fairly frequently in the cultures incubated at 45°C. Thomas and Hill (1976a), working with a hydrocarbon-utilizing albino strain of *Aspergillus fumigatus* apparently not isolated from aircraft kerosene, concluded that it had the potential to grow in such situations at up to 50°C. In a later paper (Thomas and Hill, 1976b), this strain was reported to cause corrosion (characterized, unfortunately, only as weight loss and determination of aluminium in solution) of test strips of an aircraft aluminium alloy. In three weeks, the pH value of cultures fell from an initial value of 6.75 to 3.5, and it was inferred that organic acids were produced during growth.

E. Prevention and Control

The following are the main approaches to the problem of fungal corrosion in aircraft integral fuel tanks. (1) Application of a protective coating inside the tank. (2) Prevention of any fungal mycelium or spores from entering. (3) Exclusion of water. (4) Use of a biocide.

1. Application of Protective Coatings

If it is accepted as inevitable that microbes and water will enter the wing tanks, protection of the metal by a coating may be resorted to. The penalty for protecting the metal and allowing the fungus to grow, however, could be blockage of filters and pipes by detached pieces of mycelium. This method, then, can only be used as an adjunct to methods described below; although it is now in almost universal use. It is extremely difficult to apply an absolutely intact protective coating to the inside of an aircraft wing after fabrication: the working spaces, especially at the wing-tip end, are inadequate. The modern technique, therefore, is to apply one protective coating to wing-planks and related structures before assembly, and another afterwards.

In general, a protective coating for fuel tanks has to satisfy the following conditions: (a) it must not be affected by fuel, anti-icing additives, water or microbial products; (b) it must have good adhesion, and flexibility at low temperatures; (c) it should be easily applied by spray, brush or slushing to give a continuous film; and (d) it must be compatible with caulking and sealing compounds.

Because of the weight penalty of thick coatings, a film thickness for each coat of 25 μm is generally aimed for. Buna-N synthetic rubber and etch (wash) primers are often employed at this thickness. Other common protective coatings are epoxy resins and polyurethanes, used at somewhat greater film thicknesses (Scott, 1971). Experience has shown that very thin films give inadequate protection: they are all permeable to water, and fungal penetration can occur. Williams (1972) obtained evidence that Buna-N is metabolized, as well as physically penetrated, by *C. resinae*.

Polysulphide and other sealants, used at up to ten times the thickness quoted above, provide good protection. Scott (1971) stated that their use is worth consideration in critical areas where contamination is most likely (or has been found by experience to be most troublesome) and in places where access for cleaning or repair is particularly difficult. He described a typical coating procedure for British-built aircraft. Firstly, a coating of epoxy primer incorporating strontium chromate, 25 μm in thickness, is applied by spray to fuel tank components, including plumbing, before assembly. Critical areas inboard are further protected by brushed-on polysulphide sealant. After assembly, a slushed coating of Buna-N is given, again about 25 μm thick. This system has been found to give good protection against corrosion. Strontium chromate is a readily soluble biocide which is 'trapped' in the above coating procedure by the overlying Buna-N, but acts on any organisms that penetrate the topcoat. Strontium chromate is mentioned again in Section IV.E.4 (p. 178).

2 and 3. Prevention of Entry of Microbes and Water

These two aspects can be considered together, since many of the common-sense precautions apply to ingress of water, dirt and microbes. What enters the aircraft depends on what is in the bowser, which in turn depends on what is in the airfield storage tanks and in the road tankers that deliver the fuel to the airfield. Petroleum refiners state that their product is dry when taken away by the customer; but it is useless, for example, having thoroughly clean, dry road tankers if the flexible delivery pipes are carried in such a position that they can pick up water and dirt from the road. Application of good housekeeping and elimination of carelessness all along the line can do much to minimize ingress of water, dirt and microbes. In general, the major airports of the world now maintain a high level of fuel-quality control. Smaller, less frequently used airports, especially in tropical regions, might be expected not to achieve such high standards.

Despite all good housekeeping precautions, though, condensation is certain to occur in all fuel-storage tanks in humid climates; unless replacement air is made to pass through filters, microbes will enter too. Above-ground storage tanks are preferable to underground ones from the point of view of draining off accumulated water, and tilted cylindrical tanks or conical-based tanks allow better drainage. There is some uncertainty as to whether jet-fuel straight from the refinery, together with pure water, will support microbial growth in sufficient quantity to cause significant corrosion of aluminium alloy; but, either way, exclusion of impure water will certainly help to control growth. Scott and Forsyth (1976), in the investigation already mentioned in connexion with supersonic aircraft, obtained microbial growth from 72 of the 78 fuel samples taken, when incubated at 30°C using two different growth media, though considerable differences were found in heaviness of contamination and numbers of species present. Visual inspections of the insides of the integral tanks were not made, but there is no reason to suppose that many, or indeed any, of the aircraft contained visible signs of growth.

Filter/coalescers are commonly used for removing water and larger particulate matter from fuel. Micron-size filters could in principle be used, or possibly other forms of sterilization, when refuelling an aircraft, but such procedures would increase costs substantially (not least by the extra time that refuelling would take). Neither would it prevent direct infection of the aircraft from the atmosphere, unless the incoming air during flight was also filter-sterilized and the aircraft wing tanks—surfaces and the contained air—somehow made initially sterile. The use of various techniques for killing or removing microbes present in hydrocarbon fuels has been discussed by Hedrick *et al.* (1964b).

4. *Use of Biocides*

Biocide treatment is a universal remedy for fuel-tank contamination. The aim is to get the biocide to the aqueous phase in the tank, and there are three techniques. The first, and very common, technique is to use a fuel-soluble biocide that has a high partition coefficient with respect to water so that low concentrations in the fuel will build up into higher, toxic concentrations in any water that is present. The second is to package a water-soluble/fuel-insoluble biocide in small cartridges that can be placed in contact with known areas of water deposition. The third is to incorporate a water-soluble biocide in a protective coating; this has already been mentioned above (p. 175).

The first substance to be widely used in the first way was ethylene glycol monomethyl ether (also referred to as EGME, methyl cellosolve and fuel system icing inhibitor or FSII). As the last name and initials imply, this substance was first used as an antifreeze for fuel in very high-flying military aircraft; it was then found to inhibit fungal growth in aircraft in which it was used. Under severe operating conditions, it is only effective if added at every refuelling and, since a fairly high concentration (0.1–0.15%) is necessary in the fuel to suppress growth, it is an expensive treatment. Its use is therefore confined for the most part to military aircraft, which generally remain grounded for longer periods and whose fuels are more prone to microbial growth. The advantage of FSII is that, since it contains only the 'innocuous' elements carbon, hydrogen and oxygen, its use in fuel caused no alarm amongst aircraft engine manufacturers.

Many boron-containing compounds are effective fungicides, and certain organoborinanes have been found to have suitable properties for use in aircraft fuel. The proprietary mixture of organoboron compounds commonly used is effective at preventing initiation of fungal growth at about 135 p.p.m., at which concentration it is now permitted by the aircraft engine manufacturers for intermittent use. Its use was originally sanctioned only as a fill-and-drain procedure while an aircraft was grounded, because of the possible effects of combustion products on turbine blades.

A 'shock' treatment of about twice the above concentration is more effective, and a typical modern procedure, for preventive treatment when an aircraft is grounded for a reasonably long period, is to one-third fill the tanks with fuel containing 270 p.p.m. organoboron to cover the areas most likely to hold drainable water and to be infected, to leave as long as possible and then to dilute the biocide by refuelling to capacity before take-off.

A trial of another biocide of unspecified chemical composition has been reported by Thomas and Hill (1977).

The importance of contact time for both FSII and organoborinanes has been stressed by Hill (1970). Even so, these biocides have limitations. No amount of exposure will remove a well-established growth and, in these circumstances, the aircraft must be grounded and cleaned out, as must a storage tank in a similar condition. Spraying or rinsing with a cheap biocide such as methanol is then often carried out to kill inaccessible growth (Scott, 1971) after mechanical cleaning.

Amongst water-soluble/fuel-insoluble substances, chromates have been probably the most used. The United States Air Force has found 2% potassium dichromate in water bottoms to be effective in controlling microbial growth (Churchill, 1963).

Because of the objections to the continuous use of biocides in fuel, there has been a good deal of interest in combining the functions of protective coating and growth inhibitor by incorporating water-soluble biocides in the film (e.g. the system incorporating strontium chromate, referred to earlier). The limitation of such a system is the thinness of the coating, which renders it unsuited to supporting a sustained leaching of biocide at an effective concentration. Rubidge (1975) has cast doubt on the efficacy of strontium chromate as a fungicide in aircraft fuel tanks. An informative account of the use of biocides in aircraft has been given by Elphick (1971).

To sum up, the only feasible approach appears to me to be to minimize the entry of water, dirt and microbes by simple, common-sense precautions, to protect by coatings and by periodic shock treatment with a biocide, and to resort to cleaning out if necessary. Frequent visual inspection and removal of water bottoms, of both aircraft-fuel tanks and storage tanks, should be made an essential part of any maintenance programme, with particular attention being paid to aircraft that visit high-risk areas. Such measures are widely carried out now, with the result that the incidence of heavy microbial growths and wing corrosion has fallen dramatically over the last few years.

V. CORROSION BY THE SULPHATE-REDUCING BACTERIA

As indicated in the Introduction, corrosion by the sulphate-reducing bacteria appears to have two or more distinct mechanisms. These will be examined together in the present section. It is probably true to say that the role of these bacteria in the underground corrosion of pipes was the first of their undesirable activities to be recognized, and to be studied in detail; it is also probably their most important activity economically. It will therefore be considered first, and will figure largely throughout this section.

A. Historical

The anaerobic dissimilatory sulphate-reducing bacteria have been known to bacteriologists since before the turn of the Century, having first been described by Beijerinck (1895). It was clear that they performed a type of anaerobic respiration in reducing sulphate to sulphide; nevertheless, little more was discovered about them in the earlier days, owing no doubt to the difficulty of isolating them and of perpetuating them in any but the most crude cultures. Notable exceptions included the work of Baars (1930) and a study of Stephenson and Stickland (1931) in which they demonstrated a stoicheiometric uptake of gaseous hydrogen by non-growing bacteria supplied with a small quantity of sulphate according to the equation:

$$SO_4^{2-} + 4H_2 \rightarrow S^{2-} + 4H_2O \tag{10}$$

It was also suspected before the turn of the Century that microbes might play a part in corrosion of metals immersed in water. In 1910, Gaines suggested that both iron and sulphur bacteria were responsible in part for corrosion of buried ferrous metals, and Bengough and May (1924) considered that hydrogen sulphide produced by sulphate-reducing bacteria may play an important part in this type of corrosion.

Ten years later, von Wolzogen Kühr and van der Vlugt (1934)

noticed that severe pipe corrosion associated with sulphide occurred in wet soil in the polder districts of Holland which are low-lying areas of clayey soil, rich in organic matter, approximately neutral in pH value and often waterlogged, and in which therefore minimal corrosion of iron would be expected from electrochemical considerations. Iron sulphide occurred not only as an adherent corrosion product on the pipes themselves but also in the soil in the vicinity of the corroding pipes. From its presence, they suspected, and duly confirmed, the presence of sulphate-reducing bacteria.

Von Wolzogen Kühr and van der Vlugt (1934) then made the first attempt to provide a detailed explanation of underground corrosion by the sulphate-reducing bacteria in electrochemical terms. Clearly the normal cathodic electron acceptor, oxygen, was absent; but if a depolarizing agent was present that could remove adsorbed hydrogen from cathodic iron, this would explain the phenomenon. Bearing in mind the work of Stephenson and Stickland (1931) on uptake (oxidation) of molecular hydrogen, the Dutch workers proposed that the bacteria could similarly remove atomic hydrogen from polarized cathodic areas of iron, oxidizing it to protons and electrons and utilizing the reducing power so obtained for reduction of sulphate to sulphide. The overall mechanism they postulated is as below.

Anodic reaction	$4Fe \rightarrow 4Fe^{2+} + 8e$	(11)
Electrolytic dissociation of water	$8H_2O \rightarrow 8H^+ + 8OH^-$	(12)
Cathodic reaction	$8H^+ + 8e \rightarrow 8H$	(13)
Cathodic depolarization by bacteria (compare equation 10)	$SO_4^{2-} + 8H \rightarrow S^{2-} + 4H_2O$	(14)
Corrosion product	$Fe^{2+} + S^{2-} \rightarrow FeS$	(15)
Corrosion product	$3Fe^{2+} + 6OH^- \rightarrow 3Fe(OH)_2$	(16)
Overall reaction	$4Fe + SO_4^{2-} + 4H_2O \rightarrow$ $3Fe(OH)_2 + FeS + 2OH^-$	(17)

These equations predict that the molar ratio of iron corroded to iron sulphide produced is 4:1, but in fact wide deviations have been noted (values ranging from 0.9 to 48). Postgate (1979) states that the ratio of hydroxide to sulphide is influenced by the organic content of the environment. In a high content (such as in laboratory simulations of underground corrosion) the corrosion product will be entirely sulphide, since the hydroxide reacts with free hydrogen sulphide.

B. The Sulphate-Reducing Bacteria

Sulphate reduction is of widespread occurrence amongst microbes. The unique feature of the sulphate-reducing bacteria is that they perform a *dissimilatory* reduction ('sulphate respiration'), a form of anaerobic respiration superficially similar to nitrate reduction and carbon dioxide reduction. To accomplish this, they require an environment of low redox potential (-100 mV (NHE) or even more negative), rather than the mere exclusion of air, and this is usually achieved in the laboratory by incorporation into the medium of a reducing agent such as sodium sulphide, cysteine hydrochloride, ascorbic acid or sodium thioglycollate at a concentration of the order of millimolar. Under such conditions, a small inoculum will normally grow provided oxygen is excluded. Failure to recognize this important requirement for a low initial E_h value was undoubtedly responsible for much of the difficulty encountered by the earlier workers in obtaining and perepetuating pure cultures, and hence for the lack of progress in this field. One could also mention the conflicting reports on their physiology and metabolism during this early period of research, probably due to the same cause and to the presence of various unnoticed contaminants.

It follows that, once a sulphate-reducing bacterium has managed to produce a small amount of sulphide, the local conditions will then be more favourable. Growth thus becomes a self-reinforcing process, and can result in a population 'explosion' which has sometimes been noticed in industrial situations. Initiation of growth may occur in any 'microhabitat' from which oxygen is excluded, perhaps by vigorous growth of a mass of aerobic organisms. The zone favourable for growth will then increase in volume as sulphide diffuses outwards. Eventual inhibition of the bacteria by their own sulphide depends on such factors as the possibility of escape of hydrogen sulphide and of the removal of sulphide ions in an insoluble form by any heavy metal ions present. Anodic dissolution of an iron pipe probably plays a part in removing from solution sulphide ions that would inhibit growth; and it is believed by some workers that the slight depression of the E_h value resulting from ferrous ions produced by anodic dissolution can provide the trigger mechanism for a burst of growth.

Much valuable information on the science (and art) of the cultivation of sulphate-reducing bacteria is given by Postgate (1979).

Sulphate reducers are abundant throughout the World. They have been found in all the continents, including Antarctica, in soils of almost all kinds and in natural waters from fresh to saturated brine. They have been found at depths up to 71 m in clay, and 500 m in the sea (Willingham and Quinby, 1971). It is rare to take a soil sample in the United Kingdom from which they cannot be isolated. Dry, well aerated soils may contain hundreds or even thousands of viable organisms/g (Miller and Tiller, 1971). Counts of 10^5/g of soil and greater should be regarded as potentially dangerous. Lakes, rivers and the sea contain viable organisms in numbers usually related to the degree of pollution (or inversely related to the degree of oxygenation) of the water (Booth *et al.*, 1963, 1965b, 1967a). They are usually abundant in estuarine mud.

Two apparently unrelated genera of sulphate-reducing bacteria, both Gram-negative, are now universally recognized: *Desulfovibrio*, consisting mainly of curved rods, sometimes spirilla, occasionally straight, with single polar flagellum or lophotrichous, and containing the readily demonstrable pigments cytochrome c_3 and desulfoviridin (bisulphite reductase); and *Desulfotomaculum*, a sporulating genus, mainly straight rods, peritrichous, lacking the above two pigments but containing a *b*-type cytochrome. Both genera show considerable pleomorphism. Some recent isolates have been allotted to new genera, but the status of these is at present uncertain. A number of generic names from the earlier literature (*Spirillum*, *Microspora*, *Vibrio* and *Sporovibrio*) are now recognized as invalid, as is the name *Clostridium nigrificans* for *Desulfotomaculum nigrificans*. Most cases of anaerobic bacterial corrosion appear to be attributable to species of *Desulfovibrio*.

Early reports of autotrophy in *Desulfovibrio* spp. were mistaken and were probably due to the unwitting use of impure cultures and to traces of organic substances in 'autotrophic' medium. *Desulfovibrio* spp. grow well in the laboratory on various mono- and dicarboxylic acids such as lactate, pyruvate and malate, and on complex mixtures of carbon sources, though not on glucose, in the presence of sulphate under strictly anaerobic conditions. It is interesting to note that at least some strains are now believed to be mixotrophic, and able to use certain simple organic oxidations, or hydrogen oxidation, to reduce sulphate and yield energy for assimilation of certain other carbon compounds. Badzoing *et al.* (1978) found that 30% of the cell carbon of a strain of *Desulfovibrio* arose from carbon dioxide when grown in acetate and carbonate under molecular hydrogen. This carbon dioxide fixation being in considerable excess of normal (heterotrophic) bacterial carbon dioxide fixation, it may have helped to give rise to the early belief in autotrophy. Some strains of *Desulfovibrio* can grow by dismutation of, for example, pyruvate (Postgate, 1952) or fumarate (Miller and Wakerley, 1966) in sulphate-free medium; this provides a useful tool for studying the corrosion activities of such strains in the absence of sulphide, and is referred to again on p. 190.

The ability of at least certain strains to utilize molecular hydrogen during mixotrophic growth is interesting in connexion with the corrosion theories discussed later in this chapter. So is the question as to whether the hydrogenase of *Desulfovibrio* spp. catalyses the oxidation of gaseous hydrogen by the following two reactions, as is required by the theory of von Wolzogen Kühr and van der Vlugt (1934):

$$H_2 \rightleftharpoons 2H \rightleftharpoons 2H^+ + 2e \qquad (18)$$

Hydrogenase activity towards molecular hydrogen can easily be demonstrated in bacteria suspended in non-nutrient buffer at about neutrality under molecular hydrogen in a respirometer, by adding either sulphate or an oxidation-reduction dye of a suitable E_0' value such as benzyl viologen. Most, but not all, sulphate-reducing bacteria contain hydrogenase; *Desulfotomaculum orientis* does not, and others contain it but appear to be unable to couple hydrogen oxidation to sulphate reduction, at least in the conditions of a manometric experiment. An example is *Desulfovibrio salexigens*, strain California 43:63, which shows very little or no hydrogenase activity towards sulphate but extremely vigorous activity towards benzyl viologen.

The nutrition, hydrogen metabolism and other biochemical aspects of the sulphate-reducing bacteria have been reviewed by Le Gall and Postgate (1973) and Postgate (1979). Modern views on taxonomy are given in the latter work. Figures 8 and 9 show some typical sulphate-reducing bacteria.

C. Nature and Extent of Corrosion by Sulphate-Reducing Bacteria

Pipelines and other metal objects buried in the ground, and structures erected in estuaries, frequently show sulphide corrosion. Booth (1964)

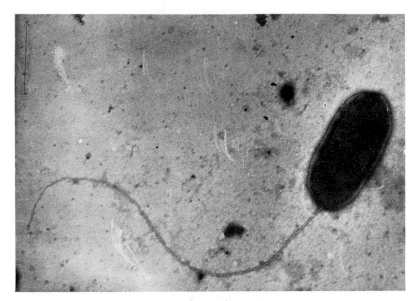

Fig. 8. Electron micrograph of *Desulfovibrio vulgaris* from an actively growing culture in low-iron medium. From Miller and King (1975). Magnification × 20,000.

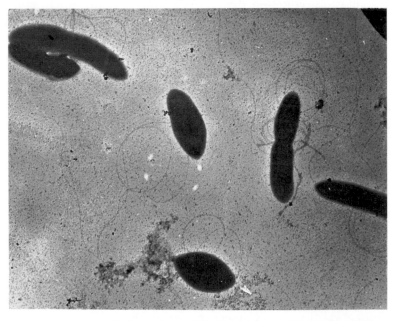

Fig. 9. Electron micrograph of the thermophilic, peritrichous sulphate-reducing bacterium *Desulfotomaculum nigrificans*. From Miller and King (1975). Magnification × 10,000.

estimated that at least 50% of failures of underground pipes in the
United Kingdom are due to bacterial corrosion. Pipeline corrosion is
most severe in wet clay or clay loam of about neutral pH value.
Progressive pitting corrosion is common: cast-iron pipes of wall thickness 6.3 mm have occasionally become perforated within a year of installation under these conditions, and perforation within four years is quite
common. Such corroded objects, if examined immediately after removal
from the soil, have a black corrosion product frequently smelling of
hydrogen sulphide (or which liberates hydrogen sulphide on treatment
with acid: a useful test to distinguish the iron sulphide corrosion pro-

Fig. 10. Piece of steel pipe cleaned to show bacterial corrosion. From Miller and King (1975).

Fig. 11. Piece of steel pipe turned (at right-hand end) to remove corrosion products and reveal
deep pitting. From Miller and King (1975).

duct indicative of bacterial action from black magnetite). The sulphide-containing corrosion product is often loose and, when lifted, reveals pits lined with bright metal (the anodic areas). A type of corrosion peculiar to cast iron is *graphitization*, in which the iron has been largely leached out of parts of the pipe leaving a matrix composed principally of graphite and soft enough to be penetrated by a sharp instrument or to give way under pressure. Typical examples of pipes that have corroded in anaerobic soil are shown in Figures 10 and 11. Sulphate-reducing bacteria can usually be detected in the corrosion product in such instances, and in the soil in the vicinity of the pipe in very much larger numbers than the 'background' count for that district.

A curious anomaly is that steel piles are reported not to be susceptible to corrosion by the sulphate-reducing bacteria to any marked degree (Romanoff, 1962). This point is referred to again, in Section IV.D, p. 192.

Local anaerobic conditions favouring growth of sulphate reducers can readily arise under heavy mixed microbial growths in various industrial situations, such as open recirculating cooling water systems (described by Purkiss, 1971) and in the paper-making industry (see, e.g., Soimajärvi *et al.*, 1978), thus exacerbating an already existing microbial corrosion problem (see Section II, p. 156). Kobrin (1976) described the corrosion by sulphate reducers of heat exchangers cooled by once-through river water, and Temperley (1965) reported catastrophic bacterial corrosion of condensers by salt-water strains of sulphate reducers in a once-through system on the Persian Gulf. In the latter instance, cast-iron condenser water boxes were completely graphitized in as little as 1000 hours of service.

Ships berthed in estuaries in which they rest on the bottom mud at low tide have been reported to suffer from this type of corrosion (Patterson, 1951). A laboratory study of corrosion by marine sulphate reducers isolated from depths of up to 500 m (Willingham and Quinby, 1971) showed severe attack on iron, especially at elevated hydrostatic pressures (200 atm; 20 MPa), and a study of the corrosion of structural steels in sea water by sulphate reducers has recently been made by Bultman *et al.* (1977). This work has interesting implications for offshore oil and gas rigs.

There is no incontrovertible evidence that any sulphate-reducing bacterium can utilize hydrocarbons as carbon-and-energy source for growth, yet these organisms cause considerable problems in the oil industry. Their growth in the presence of oil most probably depends on assimilable organic substances brought in by the aqueous phase if present, or on excretion of organic substances by aerobes such as pseudomonads that can utilize hydrocarbons and simultaneously provide an anaerobic microhabitat for sulphate reducers. There is evidently sufficient sulphate or other biologically reducible sulphur compounds in crude and other oils to support some growth, as well as the other nutrients already referred to in connexion with fungal corrosion of aircraft integral fuel tanks. Indeed, sulphate reducers have been implicated in corrosion of these tanks (Churchill, 1963), the mats of fungal growth evidently providing favourable conditions.

Roberts (1969) attributed leakages in long-term gas oil-storage tanks to the activities of sulphate reducers. Corrosion by them of oil-well equipment, such as pumps, well casings, holding tanks and pipelines, when the secondary recovery (waterflood) technique is employed is well documented (see e.g., Scott, 1965), and Farquhar (1974) has reviewed the role of sulphate reducers in oil-field corrosion.

Postgate (1960) reported cases of corrosion at high temperatures due to the activities of the thermophilic species *Desulfotomaculum nigrificans*, but there is little other documentation of corrosion by members of the genus *Desulfotomaculum*. A recent official survey of the extent and magnitude of microbial corrosion in British industry, in which sulphate-reducing bacteria figured largely, has been briefly reported by Wakerley (1979).

D. Mechanism of Iron Corrosion by the Sulphate-Reducing Bacteria

This subject has been discussed in detail by Davis (1967), Miller and Tiller (1971) and Iverson (1972). There is as yet no general agreement on the mechanism: various conflicting theories have been put forward and it may be that several of the proposed mechanisms operate concurrently. Most studies on the mechanism of anaerobic microbial corrosion have apparently been done with buried pipelines in mind; thus, mesophilic strains have been used, in sodium chloride-free medium at approximately neutral pH values.

Corrosion of an iron (or steel) pipe in a neutral anaerobic environment could occur if (1) the surface of the metal is organized into small anodic and cathodic areas, or (2) the whole surface in the anaerobic environment is anodic with respect to another part of the same structure in an oxygenated situation, resulting in a differential aeration cell.

Mechanism (1) is a well-known general corrosion phenomenon. As stated in Section I.A (p. 152) chance variations in the chemical and physical nature of the surface of iron, acquired during manufacture or during handling prior to installation, frequently result in the setting up of corrosion cells of sometimes minute size. Such cells could operate vigorously within a neutral anaerobic environment provided that either (a) the cathodic areas of the metal are depolarized by some means, or (b) an alternative local cathode is available which is less easily polarized than iron (in which case areas of the iron that were cathodic may become anodic with respect to the 'new' cathode), and is in electronic and ionic contact with the iron. Most investigations have assumed that the mechanism falls into category (a).

Mechanism (2), a 'long-line' corrosion effect having to act over distances of probably many metres and depending on good electronic and ionic conductivity between the anodic and cathodic zones, would only manifest itself to the pipeline engineer if the anodic area was relatively small and anodic dissolution resulted in perforation. Uniform weight loss over a large area need represent no hazard. For, no doubt, historical reasons (the perforated pipelines investigated by von Wolzogen Kühr and van der Vlugt (1934) were evidently in extensive areas of uniformly anaerobic soil), mechanism (2) has received little attention.

These possibilities have been investigated, along with others, by many workers since the Dutch investigators' theory was published. Some of the principal studies will now be mentioned in roughly chronological order.

An early reaction to the Dutch workers' theory was that bacteria could not cause such large-scale effects, while other workers set out to prove that the effect was one of anodic stimulation by biogenic sulphide rather than cathodic depolarization (Wanklyn and Spruit, 1952; Wormwell and Farrer, 1952; Hoar and Farrer, 1961). On the other hand, early attempts to confirm the Dutch workers' theory experimentally by demonstrating cathodic depolarization by sulphate-reducing bacteria were inconclusive.

Eventually, Horváth and Solti (1959) were able to show that cathodic depolarization was brought about, although, unfortunately, by an inadequately characterized organism. Independent studies by Booth and his coworkers at about the same time reached the same conclusion from observations of the effect of pure batch cultures of sulphate reducers on the polarization curves of mild steel. Cathodic depolarization occurred, though only when the bacteria (*Desulfovibrio vulgaris*, strain Hildenborough) were actively growing. *Desulfotomaculum orientis* (hydrogenase-negative) was found not to cause depolarization (Booth and Tiller, 1960). However, both organisms caused gradual formation, during about a week, of a partially protective film of iron sulphide on the mild-steel specimens. *Desulfovibrio salexigens* strain California 43:63, which as already stated shows hydrogenase activity only towards redox dyes in the respirometer, gave vigorous cathodic depolarization in these short-term growth experiments (Booth and Tiller, 1962). These authors concluded that the ability of a sulphate-reducing bacterium to utilize hydrogen, for whatever purpose, is the criterion for anaerobic microbial corrosion of ferrous metals, and that the earlier equation of corrosion with sulphate reduction was unwarranted. The Dutch workers' theory seemed to have been firmly verified.

Later, aware that short-term experiments with batch cultures gave very much lower corrosion rates than those often obtained in the field, and that the gradual formation of a protective film demanded investigation, Booth *et al.* (1965a) employed semicontinuous and continuous cultures. The result was that the close relationship between hydrogenase activity and cathodic depolarization became blurred. The thin iron sulphide films formed in early stages of semicontinuous culture kept corrosion rates relatively low but, after some 20–30 weeks of exposure to bacterial action, these films fractured and parts came detached. The films never re-formed, and thereafter the corrosion rate increased up to six-fold although it was still less than rates encountered in natural conditions. In these experiments, the direct correlation between corrosion and hydrogenase activity disappeared: the hydrogenase-negative *Desulfotomaculum orientis* also caused corrosion though to a lesser extent than the hydrogenase-positive strains used.

In continuous culture experiments, Booth *et al.* (1967a) found that, in a culture medium containing a high concentration of ferrous ions (0.5% ferrous ammonium sulphate), all strains whether hydrogenase-positive or -negative appeared to influence corrosion rates to the same degree. Corrosion/time curves were linear, and corrosion rates were high: values of the order of 220 mdd (mg dm^{-2} day^{-1}), comparable with rates found in clay soils, were obtained. Under such conditions, no real film formation ever occurred on the metal; instead, a completely unprotective, loose, bulky black mass of corrosion product formed around the steel specimen (which, it should be noted in connexion with more recent studies, would be in electronic and ionic contact with the metal). This corrosion product consisted of an equimolar mixture of iron sulphide and siderite (ferrous carbonate).

At about this time, it was discovered that vigorous depolarization of cathodes was caused by iron sulphide itself, prepared chemically by treating ferrous chloride with sodium sulphide, in a bacteria-free system (Booth *et al.*, 1968b). This seemed to explain a previous observation (Booth *et al.*, 1967b) that, in semicontinuous culture in a medium containing high concentrations of iron in which a bulky non-protective mass of corrosion product had formed, the vigorous depolarizing activity associated with the iron-rich medium continued for a long time after change-over to a medium containing a low concentration of iron. It was becoming clear that iron sulphide itself plays an important part in anaerobic corrosion.

A re-appraisal of the role of hydrogenase in this type of corrosion was now necessary. Booth *et al.* (1968b), using the sulphate-free dismutation medium of Miller and Wakerley (1966), showed that growing cultures of hydrogenase-positive strains did indeed cause cathodic depolarization, although in this system the corrosion rates for mild steel were very low.

In this connexion, it is interesting to note that hydrogenase-positive organisms other than sulphate-reducing bacteria have been tested for ability to contribute towards anaerobic corrosion by causing depolarization. Findings have been inconsistent. Booth *et al.* (1968b) concluded that neither hydrogen bacteria nor methane bacteria participate significantly in this process, whereas Mara and Williams (1971a, b) stated that various hydrogenase-positive bacteria that respire anaerobically using nitrate as a terminal electron acceptor effected cathodic depolarization and caused weight loss of mild-steel specimens. These findings were disputed by Ashton *et al.* (1973), both on the grounds of the non-operation of a depolarizing mechanism and the alternative explanation of attack by organic acids.

King and Miller (1971) inclined to the view that some direct depolarization of steel is carried out by the bacteria, but that depolarization or some other process brought about by iron sulphide is probably more important quantitatively. It is worth commenting here that I am unclear how an intact bacterium can remove hydrogen atoms adsorbed on a metal surface, as distinct from absorbing molecular hydrogen dissolved in water, and have been unable (unpublished results) to detect depolarizing activity in bacteria-free culture filtrates except for a very small activity in filtrates of old cultures in the stationary or decline phase of growth. There are several iron sulphides, and it will be necessary to consider them briefly.

It is generally agreed that the first iron sulphide formed on iron in the presence of sulphate-reducing bacteria is a thin film, protective to a greater or lesser degree, of mackinawite, a sulphur-deficient sulphide usually represented as $FeS_{(1-x)}$. As bacterial action proceeds, the film thickens as the sulphur content increases until eventually the film spalls. It is often considered that by this stage it is greigite (Fe_3S_4) (see, e.g., Mara and Williams, 1972), although King et al. (1976), who also showed that the rate of breakdown of protective iron sulphide films and the subsequent rate of corrosion are both proportional to the iron concentration in the growth medium, did not fully confirm this. They found greigite to be present even at very early stages; breakdown of the film was apparently due to the later transformation of some of the mackinawite to non-protective smythite (another form of Fe_3S_4) and pyrrhotite $(Fe_{(1-x)}S)$. This process, and the properties of the iron sulphides, are described in detail by Smith and Miller (1975).

The inability of any kind of iron sulphide film to re-form after breakdown of the original one strongly suggests a change in the electrochemical behaviour of the now exposed metal. Mara and Williams (1972) considered that the adjacent iron sulphides act as a cathode depolarizer, absorbing hydrogen in proportion to the cationic defects in the iron sulphides. However, King et al. (1973) found the iron sulphides to be more corrosive than this mechanism predicts. Besides being corrosive to iron in various degrees, all are cathodic to iron (Smith and Miller, 1975; Smith, 1980). It thus appears that bulky masses of iron sulphides, or lifted films that were once protective, can function as cathodes towards iron; and, although they have been shown not to be permanent cathodes in bacteria-free systems, in the presence of bacteria the reaction continues. The role of the bacteria could be either to 'regenerate' (or depolarize) the iron sulphide, enabling it to remain cathodic, to produce more iron sulphide by their growth reaction, or even to bring fresh iron sulphide surfaces constantly into contact with the steel by their movement. Figure 12 shows that bacteria grown in iron-rich medium are frequently coated with iron sulphide particles.

Quite different views from the above have been expressed by some investigators, such as Costello (1974, 1975) who considered that all cathodic depolarizing activity in cultures of sulphate-reducing bacteria is attributable to the cathodic activity of dissolved hydrogen sulphide produced by the organisms. Schaschl and Marsh (1963) maintained that bacterial action is at most a minor agency in anaerobic corrosion of steel in soil, and attributed it mainly to long cell action arising from differential aeration, zones of a pipe in anaerobic conditions being anodic to zones in drier and better aerated soil. Such anodic and cathodic zones, it was claimed, could be 30.5 m or more apart in soils of high conductivity.

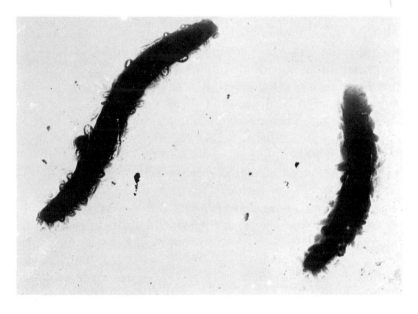

Fig. 12. Electron micrograph of washed sulphate-reducing bacteria (*Desulfovibrio vulgaris*) grown in high-iron medium. Note the particles of iron sulphide adhering to the cell walls. From Miller and King (1975). Magnification × 12,000.

It has already been stated that driven steel piles rarely suffer noticeably from bacterial corrosion. Rogers and Johnston (1971), who subscribe to the view that long-line currents play a major role in anaerobic corrosion, considered that for pipelines in backfilled beds differential compaction alone will usually ensure that alternate sections experience different degrees of aeration. A short, highly localized anaerobic section sandwiched between two fairly large aerated sections could corrode vigorously. They believe that a driven pile passes through insufficient aerated soil (bestowing cathodic properties) to cause significant corrosion at the lower, anaerobic, level.

In support of this theory, they described in a later paper (Johnston and Rogers, 1975) a study on an experimental pile in Australia. They found no corrosion at the anaerobic level, and rightly concluded that no long-line currents could have arisen and that neither was there local cell (bacterial) corrosion within the anaerobic zone. The latter, however, is not surprising; even the highest of the bacterial counts they obtained (10^3 bacteria/g of soil) is just within the range generally regarded as safe. Eadie (1977), also working in Australia, withdrew a pile after four years in the ground and found very little corrosion; such as was present was almost entirely in the upper, aerated, zone. Counts of sulphate reducers in the anaerobic zone were up to 10^5/g of soil.

The doubt about bacterial corrosion of driven piles, then, is unresolved. Limited studies have been made, in rather restricted areas of the Earth; but if it is true that they are 'immune' it can only be because the long-line current effect does not operate (as Johnston and Rogers, 1975, suggest) or that, for some reason, bacterial action cannot occur so readily in these conditions as in the much more disturbed surroundings of a backfilled trench around a pipeline. It should be borne in mind that, in other areas than those hitherto investigated, localized microbial corrosion may well occur; but, in contrast to pipe corrosion in which one perforation means failure of the pipeline, no noticeable impairment of the load-bearing capacity of the pile need result.

It appears that anaerobic corrosion of pipelines (and of other metallic structures in analogous environments) is frequently caused or exacerbated by sulphate-reducing bacteria, as stated by Booth (1964), although many workers now think that it is an indirect action either due to production of soluble sulphide by the bacteria or to iron sulphides arising from it rather than to a direct bacterial uptake of atomic or molecular hydrogen from cathodic areas for metabolic purposes. Iron sulphides are known to cause corrosion of iron; and sulphide ions in the soil and iron sulphide amongst corrosion products must, at least in the majority of cases, be biogenic, constituting strong circumstantial evidence for 'indirect' bacterial corrosion. Very large counts of sulphate reducers are obtained in the vicinity of corrosion where iron sulphides are present;

and corrosion rates obtained in continuous bacterial cultures have approached corrosion rates in anaerobic soils. This all points to sulphate-reducing bacteria playing an important role.

The most aggressive soils are clay loams; soils with an E_h value more negative than $+400$ mV with respect to the normal hydrogen electrode, corrected to pH 7, tend to be aggressive ($+430$ mV in the case of predominantly clay soils), as do soils of resistivity less than 2,000 ohm cm. Borderline cases can be resolved by water content. If this is greater than 20% the soil tends to be aggressive (Booth et al., 1967c, d). The ferrous ion concentration in the soil may be important: of two otherwise similar soils, one, with a mean ferrous ion content over a year of 333 μg/g soil, was aggressive, whereas the other, with a mean content of 59 μg/g, was non-aggressive (Miller and Tiller, 1971).

It may well be that several of the suggested mechanisms of microbial soil corrosion, as well as the non-biological long-line current effect, usually contribute to the overall phenomenon of anaerobic corrosion of steel. Laboratory experiments still have not simulated natural conditions in a very satisfactory way, and much more investigation is necessary.

E. Corrosion of Other Metals by the Sulphate-Reducing Bacteria

Very little work appears to have been done on metals other than iron and steel. Gilbert (1947) found severe corrosion of copper pipes in a moist clay soil; sulphate reducers were detected in large numbers and the corrosion product contained sulphide. Booth et al. (1967e) buried plates of aluminium, copper, zinc and lead, as well as mild steel, at twelve sites whose aggressiveness had been predicted on the basis of redox potential, resistivity and water content determinations. Predictions based on these criteria were found to be valid for mild steel at all but two of the sites, in which previously unnoticed local conditions caused more corrosion than was anticipated. Corrosion of the zinc and lead plates was also much as predicted, although it was not very clear whether the corrosion of these two nonferrous metals was initiated, or assisted, by microbial action. Neither copper nor aluminium behaved as predicted.

The effect of semicontinuous cultures of sulphate-reducing bacteria of a few months' duration on specimens of a copper–nickel alloy was

studied by Wan (1979). Weight loss was negligible over this period
and an apparently protective sulphide film formed; however, some
pitting was found under it. Various authors have dealt with the possible
role of sulphate reducers in causing corrosion of aircraft aluminium
alloys, and Hedrick (1970) has surveyed the work on this topic.

F. Prevention and Control

Prevention of corrosion by sulphate-reducing bacteria would ideally
entail preventing them from growing. Microbiological sterility, that is
the complete absence of living microbes, can only be achieved in a
defined system possessing microbe-proof boundaries, and such a system
almost never occurs in practice. We therefore have to be content with
controlling microbial growth (and hence corrosion) to an acceptable
level. As might be expected, oxygen is the cheapest and most effective
inhibitor of sulphate reducers and, despite its being the most efficient
cathodic electron acceptor, experience shows that an overall benefit
arises from efficient oxygenation.

The following scheme, summarized from Miller and Tiller (1971),
is convenient for classifying methods of preventing or retarding underground corrosion; it can be applied to aqueous situations where
appropriate.

1. Using non-corrodible materials.
2. Using corrodible materials.
    2.1. In a non-aggressive environment.
    2.2. In an aggressive environment:
        2.2.1. By giving the installation a non-aggressive surround:
            2.2.1.1. By arranging that the environment contains
                      nothing that will accept electrons or cations.
            2.2.1.2. By using a biocide in the environment to prevent
                      the growth of sulphate-reducing bacteria.
        2.2.2. By making the metal sufficiently negative with respect to
                 its environment to prevent cations escaping into it.
        2.2.3. By arranging a barrier, impervious to cations and/or
                 electrons, between the metal and its environment.

Some of these methods will be briefly discussed.

1. Use of Non-Corrodible Materials (Method 1)

The problem could be avoided altogether if chemically inert materials of adequate strength were available at a cost competitive with that of mild steel. Small-diameter polymeric pipes are now frequently used for water distribution; large-diameter cement and asbestos-cement pipes are satisfactory in unpressurized systems provided there is no possibility of attack by sulphur-oxidizing bacteria (sulphuric acid alters concrete to a soft chalky material). Stainless steel has occasionally been used for pipelines despite its expense. Load-bearing structural steel still has few serious rivals.

2. Construction of a Non-Aggressive Surround (Method 2.2.1.1)

Theoretically this would entail excluding from the environment of the metal any electron acceptors (the principal ones being oxygen and hydrogen ions) and water (to prevent cations forming from metal). This is clearly impossible to achieve, and in practice we aim for alkaline conditions (despite the greater tolerance of sulphate reducers to alkaline than to acid conditions) and good drainage (more in an attempt to achieve oxygenation than to suppress cation formation).

A non-aggressive surround for a pipeline consists ideally of chalk, though chalk–sand mixtures can be used and alkalinity may be dispensed with altogether by the use of gravel, rubble or clay-free sand. This back-filling must be done carefully to avoid damage to any protective coating on the pipe. Pipeline gradients should be controlled and drains or soak-aways provided at low points.

3. Use of Biocides (Method 2.2.1.2)

It would not be expected that a biocide, applied locally in an open system such as around a pipeline before backfilling, would remain in the vicinity for more than a short fraction of the desired life-span of the pipeline, unless the biocide was present in largish particles and was only very sparingly soluble yet effective at that concentration. Purkiss (1971) describes the use of such a biocide for pipelines.

In relatively closed or well-defined systems, such as industrial cooling-water systems and the interior of oil-well equipment, the use of biocides for controlling sulphate-reducing bacteria is much more feasible. The

National Association of Corrosion Engineers' publication (1972) lists five inorganic and 40 organic biocides for use in oil-field equipment, and further information is given by, for example, Saleh *et al.* (1964) and Postgate (1979, in Appendix 2).

4. Cathodic Protection (Method 2.2.2)

The principle of cathodic protection is to make the metal-to-environment potential so negative that cations cannot escape into solution from the metal. The structure is therefore made cathodic with respect to another nearby electrode installed for the purpose. There are two methods. Either a sacrificial anode of a baser metal than the structure to be protected is used, connected to the structure by a cable, or a large anode or ground-bed is made, often of graphite, fairly remote from the structure in order to protect a greater length of the structure, and made anodic by applying an e.m.f. between it and the structure. This technique was developed to provide protection against non-biological corrosion, and the accepted potential to give steel is -0.85 V with respect to the pipe-line engineer's copper/saturated copper sulphate electrode. Cathodic protection can also protect against corrosion by sulphate-reducing bacteria, though the structure must be at a more negative potential than this value. Booth and Tiller (1968) considered that -0.96 V ($Cu/CuSO_4$ electrode) is necessary.

5. Protective Coatings (Method 2.2.3)

Miller and Tiller (1971) summarized the qualities desirable in a pipe coating. It should be coherent, adherent, completely nonporous, mechanically resistant to the hazards encountered during delivery, laying and backfilling, and chemically resistant to prolonged contact with all of the environments liable to be encountered. Such an ideal coating, not surprisingly, does not exist. Therefore, particularly with large-diameter (and hence expensive) pipelines, it is customary also to apply impressed-current cathodic protection. Provided the coating is reasonably intact, the cost of the electrical energy consumed will be small.

An informative account of the use and effectiveness of the 'traditional' protective coatings was given by Logan and Parker (1965), and for recent developments the reader is referred to the Proceedings of the

2nd Conference on the Internal and External Protection of Pipelines (1977).

G. Detection of Sulphate-Reducing Bacteria

The most commonly used medium for detecting sulphate-reducing bacteria has probably been that of Baars (1930). This medium, as modified by Postgate (1966, 1969), has the following formulation: KH_2PO_4, 0.5 g; NH_4Cl, 1 g; $CaSO_4$, 1 g; $MgSO_4 \cdot 7H_2O$, 2 g; sodium lactate, 3.5 g; ascorbic acid, 1 g; thioglycollic acid, 1 g (but see note below); $FeSO_4 \cdot 7H_2O$, 0.5 g; tap water to 1 litre. For marine strains, add 2.5% NaCl or use seawater instead of tap water. Adjust pH value to 7.0–7.5. The medium always contains a precipitate and, when dispensing the medium into vessels, care should be taken to keep this precipitate uniformly suspended since its presence appears to aid growth.

The medium should be sterilized immediately after being made up by autoclaving at 121°C for 15 minutes, and should be used as soon thereafter as possible owing to the instability of the reducing agents. It is interesting to note that Postgate (1979), when quoting his modification of Baars (1930) medium and his own solid medium 'E', gives concentrations of 0.1 g/l for ascorbic acid and thioglycollic acid instead of 1 g/l of each as in his original descriptions; and that Grossman and Postgate (1953) stated that 1 g of thioglycollic acid/l was inhibitory to *Desulfovibrio* sp. I suggest using 0.1–0.5 g/l of each of these reducing agents.

Small screw-capped glass bottles of around 10 ml capacity are sterilized, and samples (about 2 g maximum) of soil, corrosion product, or water placed in them. Sterile medium is then poured in to fill the bottles completely, and the caps are replaced so as not to trap any air. For examination for *Desulfovibrio* spp., the bottles are incubated usually at 30°C (marine strains from temperate zones may make better growth at a lower incubation temperature), and for the thermophile *Desulfotomaculum nigrificans* at 55°C. In either case, blackening of the medium, which usually occurs within 2–3 days, indicates growth of sulphate reducers. Blackening is due to the formation of ferrous sulphide in this iron-rich medium.

REFERENCES

Akbar, S. A. (1979). Ph.D. Thesis: University of Manchester.
Ashton, S. A., Miller, J. D. A. and King, R. A. (1973). *British Corrosion Journal* **8**, 185.
Baars, J. K. (1930). Doctoral dissertation, Technical University of Delft.
Badzoing, W., Thauer, R. K. and Zeikus, J. G. (1978). *Archives of Microbiology* **116**, 41.
Beijerinck, M. W. (1895). *Zentralblatt Bakteriologie und Protistenkunde* **1** (abt. 2), 1.
Bengough, G. D. and May, R. (1924). *Journal of the Institute of Metals* **32**, 81.
Betz Handbook of Industrial Water Conditioning, 7th edition (1976). Betz Laboratories Inc., Trevose, Pennsylvania 19047.
Blanchard, G. C. and Goucher, C. R. (1967). *Electrochemical Technology* **5**, 79.
Booth, G. H. (1964). *Journal of Applied Bacteriology* **27**, 174.
Booth, G. H. and Tiller, A. K. (1960). *Transactions of the Faraday Society* **56**, 1689.
Booth, G. H. and Tiller, A. K. (1962). *Transactions of the Faraday Society* **58**, 2510.
Booth, G. H. and Tiller, A. K. (1968). *Corrosion Science* **8**, 583.
Booth, G. H., Cooper, A. W. and Tiller, A. K. (1963). *Journal of Applied Chemistry* **13**, 211.
Booth, G. H., Shinn, P. M. and Wakerley, D. S. (1965a). *Comptes Rendus du Congres International de la Corrosion Marine et des Salissures, Cannes, 1964*, p. 363.
Booth, G. H., Cooper, A. W. and Tiller, A. K. (1965b). *Journal of Applied Chemistry* **15**, 250.
Booth, G. H., Cooper, A. W. and Cooper, P. M. (1967a). *Chemistry and Industry* 2084.
Booth, G. H., Robb, J. A. and Wakerley, D. S. (1967b). *Proceedings of the Third International Congress of Metallic Corrosion, Moscow*, vol. 2, section 5, p. 542.
Booth, G. H., Cooper, A. W., Cooper, P. M. and Wakerley, D. S. (1967c). *British Corrosion Journal* **2**, 104.
Booth, G. H., Cooper, A. W. and Cooper, P. M. (1967d). *British Corrosion Journal* **2**, 109.
Booth, G. H., Cooper, A. W. and Tiller, A. K. (1967e). *British Corrosion Journal* **2**, 116.
Booth, G. H., Cooper, A. W. and Tiller, A. K. (1967f). *British Corrosion Journal* **2**, 222.
Booth, G. H., Elford, L. and Wakerley, D. S. (1968a). *British Corrosion Journal* **3**, 68.
Booth, G. H., Elford, L. and Wakerley, D. S. (1968b). *British Corrosion Journal* **3**, 242.
Bultman, J. D., Southwell, C. R. and Hummer, C. W. (1977). *Reviews on Coatings and Corrosion* **2**, 187.
Castleberry, J. R. (1969). *Materials Protection* **8** (3), 67.
Churchill, A. V. (1963). *Materials Protection* **2** (6), 19.
Costello, J. A. (1974). *South African Journal of Science* **70**, 202.
Costello, J. A. (1975). Ph.D. Thesis: University of Cape Town.
Davis, J. B. (1967). 'Petroleum Microbiology', Chapter 9. Elsevier, London.
De Mele, M. F. L., Salvarezza, R. C. and Videla, H. A. (1979). *International Biodeterioration Bulletin* **15**, 39.
Eadie, G. R. (1977). *BHP Technical Bulletin* (Broken Hill Pty. Co. Ltd., Melbourne, Australia) **21**, 13.

200   J. D. A. MILLER

Elphick, J. J. (1971). *In* 'Microbial Aspects of Metallurgy' (J. D. A. Miller, ed.), pp. 157–172. Medical and Technical Publishing Co., Lancaster.

Elphick, J. J. and Hunter, S. K. P. (1968). *In* 'Biodeterioration of Materials' (A. H. Walters and J. J. Elphick, eds.), pp. 364–370. Elsevier, Amsterdam.

Farquhar, G. B. (1974). *'Corrosion/74': National Association of Corrosion Engineers Annual Conference, March 1974*, paper 1. National Association of Corrosion Engineers, Houston, Texas.

Gaines, R. H. (1910). *Industrial and Engineering Chemistry* **2**, 128.

Gilbert, P. T. (1947). *Journal of the Institute of Metals* **73**, 139.

Grossman, J. P. and Postgate, J. R. (1953). *Proceedings of the Society of Applied Bacteriology* **16**, 1.

Hazzard, G. F. (1963). 'Fungal growths in aviation fuel systems. Part 4: Fungi in aviation fuel systems in Australia and the Far East'. Report 252, Defence Standards Laboratories, Canberra, Australia.

Hedrick, H. G. (1970). *Materials Protection* **9** (1), 27.

Hedrick, H. G., Miller, C. E., Halkias, J. E., Hildebrand, J. F. and Gilmartin, J. N. (1964a). *Developments in Industrial Microbiology* **6**, 117.

Hedrick, H. G., Carroll, M. T. and Owen, H. P. (1964b). *Developments in Industrial Microbiology* **6**, 133.

Hedrick, H. G., Crum, M. G., Reynolds, R. J. and Culver, S. C. (1967). *Electrochemical Technology* **5**, 75.

Hill, E. C. (1969). *Proceedings of the Third Conference on the Global Impact of Applied Microbiology, Bombay*, p. 65. World Health Organization.

Hill, E. C. (1970). *Journal of the Institute of Petroleum* **56**, 138.

Hill, E. C., Evans, D. A. and Davies, I. (1967). *Journal of the Institute of Petroleum* **53**, 280.

Hoar, T. P. and Farrer, T. W. (1961). *Corrosion Science* **1**, 49.

Horváth, J. and Solti, M. (1959). *Werkstoffe und Korrosion* **10**, 624.

Iverson, W. P. (1967). *Electrochemical Technology* **5**, 77.

Iverson, W. P. (1972). *In* 'Advances in Corrosion Science and Technology' (M. G. Fontana and R. W. Staehle, eds.), pp. 1–42. Plenum Press, New York and London.

Johnston, R. R. M. and Rogers, P. J. (1975). *Sixth International Congress on Metallic Corrosion, Sydney, Australia*, paper 11–1.

Kelly, B. J. (1965). *Materials Protection* **4** (7), 62.

Kempner, E. S. (1966). *Journal of Bacteriology* **92**, 1842.

King, R. A. and Miller, J. D. A. (1971). *Nature, London* **233**, 491.

King, R. A., Miller, J. D. A. and Smith, J. S. (1973). *British Corrosion Journal* **8**, 137.

King, R. A., Dittmer, C. K. and Miller, J. D. A. (1976). *British Corrosion Journal* **11**, 105.

Klausmeier, R. E., Kanzig, J. L., Jones, W. A. and Hammersley, V. L. (1963). *Developments in Industrial Microbiology* **4**, 306.

Kobrin, G. (1976). *Materials Performance* **15**(7), 38.

Le Gall, J. and Postgate, J. R. (1973). *Advances in Microbial Physiology* **10**, 81.

Le Roux, N. W., North, A. A. and Wilson, J. C. (1974). *Tenth International Mineral Processing Congress, London*, p. 1051. Institution of Mining and Metallurgy, London.

Logan, A. G. and Parker, W. D. (1965). *Journal of the Institute of Gas Engineers* **5**, 351.

Mara, D. D. and Williams, D. J. A. (1971a). *Chemistry and Industry* 566.

Mara, D. D. and Williams, D. J. A. (1971b). *Corrosion Science* **11**, 895.

Mara, D. D. and Williams, D. J. A. (1972). *In* 'Biodeterioration of Materials' (A. H. Walters and E. H. Hueck-Van der Plas, eds.), vol. 2, pp. 103–113. Applied Science Publishers Ltd., London.

McKenzie, P., Akbar, S. A. and Miller, J. D. A. (1977). *Microbiology Group Symposium 'Microbial Corrosion affecting the Oil Industry'*, Sunbury, *1976*, p. 37. Institute of Petroleum, London.

Miller, J. D. A. (1971). *In* 'Microbial Aspects of Metallurgy' (J. D. A. Miller, ed.), pp. 1–33. Medical and Technical Publishing Co. Ltd., Lancaster.

Miller, J. D. A. and King, R. A. (1975). *In* 'Microbial Aspects of the Deterioration of Materials' (R. J. Gilbert and D. W. Lovelock, eds.), pp. 83–103. Academic Press, London and New York.

Miller, J. D. A. and Tiller, A. K. (1971). *In* 'Microbial Aspects of Metallurgy' (J. D. A. Miller, ed.), pp. 61–105. Medical and Technical Publishing Co. Ltd., Lancaster.

Miller, J. D. A. and Wakerley, D. S. (1966). *Journal of General Microbiology* **43**, 101.

Miller, R. N., Herron, W. C., Krigrens, A. G., Cameron, J. L. and Terry, B. M. (1964). *Materials Protection* **3**(9), 60.

National Association of Corrosion Engineers (1972). *TPC Publication number 3: 'The Role of Bacteria in the Corrosion of Oil Field Equipment'*. National Association of Corrosion Engineers, Houston, Texas.

Parbery, D. G. (1968). *In* 'Biodeterioration of Materials' (A. H. Walters and J. J. Elphick, eds.), pp. 371–380. Elsevier, Amsterdam.

Parbery, D. G. (1969). *Australian Journal of Botany* **17**, 331.

Parbery, D. G. and Thistlethwaite, P. J. (1973). *International Biodeterioration Bulletin* **9**, 11.

Patterson, W. S. (1951). *Transactions of the North East Coast Institution of Engineers and Shipbuilders* **68**, 93.

Postgate, J. R. (1952). *Research* **5**, 189.

Postgate, J. R. (1960). *Progress in Industrial Microbiology* **2**, 47.

Postgate, J. R. (1966). *Laboratory Practice* **15**, 1239.

Postgate, J. R. (1969). *Laboratory Practice* **18**, 286.

Postgate, J. R. (1979). 'The Sulphate-reducing Bacteria'. Cambridge University Press, Cambridge.

Prince, A. E. (1961). *Developments in Industrial Microbiology* **2**, 197.

Proceedings of the Second International Conference on the Internal and External Protection of Pipes (1977). *passim*. BHRA Fluid Engineering, Cranfield.

Purkiss, B. E. (1971). *In* 'Microbial Aspects of Metallurgy' (J. D. A. Miller, ed.), pp. 107–128. Medical and Technical Publishing Co. Ltd., Lancaster.

Purkiss, B. E. (1972). 'Biotechnology of Industrial Water Conservation'. M & B Monograph CE/8, Mills and Boon Ltd., London.

Report: 'Chemistry Research 1956' (1957). Her Majesty's Stationery Office, London.

Roberts, G. A. H. (1969). *British Corrosion Journal* **4**, 318.

Rogers, P. J. and Johnston, R. R. M. (1971). *BHP Technical Bulletin* (Broken Hill Pty. Co. Ltd., Melbourne, Australia) **15**, 2.

Romanoff, M. (1962). *Monograph number 58: 'Corrosion of Steel Pilings in Soils'*. National Bureau of Standards, Unites States Department of Commerce, Washington, D.C.

Rubidge, T. (1975). *International Biodeterioration Bulletin* **11**, 133.

Saleh, A. M., Macpherson, R. and Miller, J. D. A. (1964). *Journal of Applied Bacteriology* **27**, 281.

Schaschl, E. and Marsh, G. A. (1963). *Materials Protection* **2**(11), 8.

Scott, J. A. (1971). *Proceedings of the Society of Automotive Engineers National Air Transportation Meeting, Atlanta, Georgia, May 1971*, p. 1. Society of Automotive Engineers, New York.

Scott, J. A. and Forsyth, T. J. (1976). *International Biodeterioration Bulletin* **12**, 1.

Scott, J. A. and Hill, E. C. (1971). *Institute of Petroleum Symposium: 'Microbiology 1971'*, p. 25. Institute of Petroleum, London.

Scott, W. R. (1965). *Materials Protection* **4**(2), 57.

Sheridan, J. E. (1972). *International Biodeterioration Bulletin* **8**, 65.

Smith, J. S. (1980). Ph.D. Thesis: University of Manchester.

Smith, J. S. and Miller, J. D. A. (1975). *British Corrosion Journal* **10**, 136.

Soimajärvi, J., Pursiainen, M. and Korhonen, J. (1978). *European Journal of Applied Microbiology and Biotechnology* **5**, 87.

Starkey, R. L. (1935). *Soil Science* **39**, 197.

Stephenson, M. and Stickland, L. H. (1931). *Biochemical Journal* **25**, 215.

Tehle, E. (1966). *Materials Protection* **5**(12), 21.

Temperley, T. G. (1965). *Corrosion Science* **5**, 581.

Thomas, A. R. and Hill, E. C. (1976a). *International Biodeterioration Bulletin* **12**, 87.

Thomas, A. R. and Hill, E. C. (1976b). *International Biodeterioration Bulletin* **12**, 116.

Thomas, A. R. and Hill, E. C. (1977). *International Biodeterioration Bulletin* **13**, 31.

Tiller, A. K. and Booth, G. H. (1968). *Corrosion Science* **8**, 549.

Wakerley, D. S. (1979). *Chemistry and Industry* 656.

Wan, K. C. (1979). M.Sc. Dissertation: University of Manchester.

Wanklyn, J. N. and Spruit, C. J. P. (1952). *Nature, London* **169**, 928.

Williams, M. E. (1972). M.Sc. Thesis: University of Manchester.

Willingham, C. A. and Quinby, H. L. (1971). *Developments in Industrial Microbiology* **12**, 278.

Wolzogen Kühr, C. A. H. von and Van der Vlugt, L. S. (1934). *Water, Den Haag* **18**(16), 147.

Wormwell, F. and Farrer, T. W. (1952).*Chemistry and Industry* 108.

7. Paintings and Sculptures

ALICJA B. STRZELCZYK

Laboratory of Paper and Leather Conservation, Institute of Conservation and Restoration of Antiquities, Nicolaus Copernicus University, Toruń, Poland

I. Introduction 203
II. Effect of Moisture on Microbial Deterioration of Paintings and Sculptures 205
III. Development of Microbial Communities on the Surface of Paintings . . . 207
IV. General Forms of Microbial Activity Affecting Paintings and Painted Sculptures 209
V. Microbial Deterioration of Paintings on Wood and Painted Sculptures . . 210
    A. Decay of the Wooden Support 210
    B. Decay of Paint Layers 212
VI. Microbial Deterioration of Paintings on Canvas 215
VII. Microbial Deterioration of Paintings on Paper 218
VIII. Microbial Deterioration of Mural Paintings 220
    A. Development of Heterotrophic Micro-Organisms 220
    B. Development of Autotrophic Micro-Organisms 222
X. Combating Micro-Organisms on Paintings and Sculptures 223
    A. Methods for Disinfecting Paintings and Sculptures 224
    B. Fungicides Used in Disinfection 226
References. 233

I. INTRODUCTION

Most paintings and painted sculptures needing conservation show traces of former or current microbial activity. They can be seen both on the face and on the reverse or support. The symptoms are various and depend on the kind of object, i.e. on the nature of the support (canvas, wood, paper or leather) as well as on the mode of painting and the kind of paints (oil paints, distemper or watercolours).

In works of art slightly attacked by micro-organisms, only some parts of the paintings may be damaged (on the front or back). In those badly overgrown, however, all of their components are subject to destruction. These components vary with time and the changing fashion, techniques and artistic taste.

The essential part on which the character of the painting depends is its support. This is either a wooden board, or canvas stretched in a frame, paper, parchment or a metal plate. The paintings generally consist of a number of layers. Over the canvas support sized with animal glue, a ground layer is usually laid, consisting of lime or gypsum with an addition of animal or plant glue. On the smoothed-out ground layer, the artist puts the drawing, and over this several or more layers of colour are laid down. All of these layers, by complementing one another, result in pictures admired throughout centuries. The colour layers consisting of finely ground natural pigments mixed with binders of oil and distemper (egg or glue distemper). The surface of the paintings was usually spread with a thin translucent layer of varnish. A similar multilayer structure is observed in painted sculptures.

Paintings on paper include, in the first place, watercolours, gouaches and pastels. All of them are distinguished by the lack of any underlayer or ground. The paint is laid directly on paper, which is the background for the paint, and its colour and state of preservation determine the appearance and durability of the whole painting. In watercolours, the paint forms a thin transparent layer. It contains a small amount of binder, usually gum arabic.

Pastels are made with crayons consisting of pigment with hardly any binder at all. Hence the fluffy texture of the pictures, which easily comes off. Therefore pastels are extremely difficult to store and preserve. Microbial attack is favoured when these paintings are kept under glass, where condensation water usually accumulates.

Mural paintings are made by a technique widely differing from those already described. In the old techniques of pure fresco, water paints were applied on wet plaster. Calcium carbonate, formed in the course of drying of the plaster, consolidated the pigment particles. Pure fresco is one of the most lasting techniques in mural painting. A mixed technique has also been used, consisting in complementary painting of the fresco with distemper paints (generally with casein binder). In conservation treatment of damaged frescoes, imperfections in the painting layer are repainted with distemper paints containing various binders (Doerner, 1971).

The organic components in paintings and sculptures (see Table 2) are good sources of nutrition and energy for a wide variety of micro-organisms which, under favourable conditions, begin to predominate in the works of art causing their irreversible damage (Strzelczyk, 1970).

II. EFFECT OF MOISTURE ON MICROBIAL DETERIORATION OF PAINTINGS AND SCULPTURES

High humidity is a prime factor causing microbial attack of paintings, sculptures and other works of art, such as mural paintings, easel paintings, sculptures, paintings on paper, parchment and leather etc. During their long existence, these objects have often come into contact with high humidity and thus been subject to destruction by agents favoured by it. Humidification of old works of art may arise from various sources. Some of them can be prevented, others are part of the permanent system of factors prevailing in the surroundings and are generally impossible to change. Construction faults of buildings due to neglect or war damage contribute to leakage and wall dampness, which favour development of micro-organisms. As a result, the excess water causes first swelling, then deformation and cracking of paintings, old wooden elements and sculptures. These conditions allow development of wood-rotting fungi, which attack timber in buildings. Since World War II, buildings of historical and artistic value have been badly damaged by the fungus *Coniophora cerebella* which thrives under conditions of excessive humidity.

Another agent of great importance in affecting microbial deterioration as well as physicochemical destruction of mural paintings is ground water permeating through walls inside the buildings. Salts contained in this water are transported from the soil on to the surface of paintings where they crystallize and assist development of micro-organisms. Stopping the circulation of ground water in the walls of old buildings is an extremely hard task and is often not feasible because of excessive cost.

The danger of microbial attack on paintings, sculptures and the like is greatly enhanced by condensation water. Objects displayed in air-conditioned museum rooms are not exposed to this danger. There are, however, still a huge number of objects of artistic value housed in old and equally valuable churches, in private collections or in stores with uncontrolled humidity and temperature. It is also known that condensation moisture favours development of micro-organisms on paintings stored for long periods of time in safes to protect them against theft (J. Smith, personal communication).

Condensation moisture settles on cold surfaces as they come into contact with warm air with a high water content. Under such conditions, a thin film of water forms over the surface of paintings, sculptures, books and furniture, and may persist for a long time. A similar phenomenon is effected by a lowering of the temperature inside buildings. The water capacity of air is then lowered, and excess water contained in it settles on the surface of objects as dew. Grzesikowa (1965) gives an analysis of this phenomenon in ancient buildings stating that the process occurs with diurnal, annual or any other change in temperature. It also takes place on cold and dark church walls, when the churches are aired on warm spring and summer days. This type of moisture appears on surfaces high above ground level in churches, and has nothing to do with water seeping up from the foundations. Condensation moisture disappears with a rise in ambient temperature, when the water capacity of air increases.

Ayerst (1968) found that not only particular micro-organisms differ in their requirements as regards ambient humidity and temperature, but also that, with varying temperature, their requirements regarding humidity become altered. It appears then that these agents are complementary to each other (Table 1).

Ayerst (1968) claims that bacteria and fungi at temperatures optimal for their growth show greater tolerance to extremely low or high humidity, and *vice versa*: at optimum humidity, micro-organisms are capable of tolerating extreme variations in temperature. The microorganism's response to low humidity is greatly affected by the content of nutrients in the environment. According to Ayerst (1968), in the presence of large concentrations of nutrients, micro-organisms show increased tolerance to low humidity. This fact is no doubt connected with increased production of metabolic water in respiration.

Tonolo and Giacobini (1958) demonstrated that components of paintings on canvas (supports, glues, painting layers) are attacked by micro-organisms at a relative humidity greater than 80%. It should

Table 1

The effect of temperature and relative humidity on the water content of air

Temperature (°C)	Water content (g/m³) of air for relative humidity values (%) of:			
	100	60	40	20
−5	3.24	1.94	1.30	0.65
0	4.84	2.90	1.94	0.97
+5	6.76	4.06	2.70	1.35
+10	9.33	5.60	3.73	1.87
+15	12.71	7.63	5.08	2.54
+18	15.22	9.13	6.09	3.04
+20	17.12	10.27	6.85	3.42
+25	22.80	13.68	9.12	4.56

be noted, however, that microbial attack may occur at a much lower relative air humidity in places in which water evaporation is obstructed.

III. DEVELOPMENT OF MICROBIAL COMMUNITIES ON THE SURFACE OF PAINTINGS

The surfaces of ancient objects bear a rich microflora from the air, which can survive for a long time in anabiosis. It settles down with dust. The studies of Kowalik and Sadurska (1957) and Gallo (1964), aimed at comparing the microflora of air with that developing on old books and prints, have shown that many members of the fungal microflora living in the air are able to develop on paper. It follows from the above studies, however, that not all kinds of fungi occurring in the air find adequate developmental conditions on books.

The author's extensive studies of the microflora growing on easel paintings and painted sculptures have demonstrated that, although the communities contain a large variety of micro-organisms, only some of them develop and damage these objects. This phenomenon has been demonstrated by comparing the microflora identified directly under the microscope with the microflora obtained from culture media inoculated with samples overgrown with micro-organisms. Gargani (1968) proposed sampling fungal growth directly from the paint layer by its impression on Scotch tape. The piece of Scotch tape removes the powdered paint together with fungal fruiting bodies. In this way, direct classification of micro-organisms seems to be easier. Bassi and Giacobini (1973) suggest using a scanning electron microscope for direct studies of micro-organisms destroying works of art.

In our studies, after inoculation of samples on different culture media, the rapidly growing and heavily spore-forming fungal genera *Penicillium*, *Aspergillus*, *Mucor* and *Stemphylium* were always predominant. Only occasionally did we find those fungi that attacked the object in question and could be distinguished and identified by direct microscopic examination. For this reason, we adopted as a principle to focus primarily on direct observations and only to confirm the results by indirect method, i.e. inoculating samples on various media.

The microflora attacking paintings and sculptures includes first of all species of microscopic fungi of the Fungi Imperfecti, some ascomycetes, fungi attacking wood (basidiomycetes and ascomycetes) and streptomycetes. The moisture of these objects does not as a rule favour development of bacteria, which are unable to utilize condensation water. This probably accounts for the fact that no bacterial colonies are found on these damaged paintings and sculptures. Bacteria may play a role in decaying processes of objects that suffered bad flooding, such as that in Florence in 1966.

Spores of most of the fungi mentioned are capable of utilizing condensation moisture. The relative humidity which favours their development, at least in the early stages, ranges from 75 to 95%. Under such conditions, the spores germinate into mycelium, which begins its enzymic activity. The organisms that can develop under these conditions, utilizing components of the paintings, are distinguished by their very high enzymic capacity and low nutritional requirements (Ważny and Rudniewski, 1972). If the humidity remains unchanged, developing hyphae produce spores within 48–72 hours. These disseminate over the surface of the object, and the fungi achieve domination over others that have settled incidentally with dust from the air. This accounts for the occurrence on paintings of numerous fungal colonies belonging to several genera or species. Their spores pose the greatest danger for other objects in their surroundings, being an inoculum of micro-organisms adapted to colonizing and attacking them. In microbial communities developing on these objects, changes are observed consisting of gradual overgrowth of the original microflora by other fungi and/or streptomycetes growing more slowly or decomposing more complex components.

Pietrykowska and Strzelczyk (1973), in their studies on the inter-action of streptomycetes and fungi destroying oil paintings, demon-strated that both streptomycetes and fungi were capable of producing substances that inhibited the other's development. This phenomenon depended on the kind of growth medium and on the humidity. In our opinion, factors that determine the microbial communities on paintings are availability of nutrient in the environment, adequate humidity and the capacity of some micro-organisms to produce substances that inhibit the development of others.

IV. GENERAL FORMS OF MICROBIAL ACTIVITY AFFECTING PAINTINGS AND PAINTED SCULPTURES

Invasions by micro-organisms of paintings, wood, canvas, plaster and other supports are somewhat similar in character and effect. They always result in irreversible deterioration of either the polychrome surface only or of the whole object. Microbial deterioration of these objects is made up of various mechanisms, which either complement or result from one another. They are as follows.

1. Microbial attacks on paintings usually result in deterioration of the whole of the object, generally starting from the support and then penetrating through all the layers. In painted sculptures or paintings on wood, deterioration often affects either the wood or the painting layers. Destruction of these parts, however, is not without effect on others, since following the decay of the wooden support by wood-rotting fungi and insects, the wood changes in volume and in sensitivity to water, which causes the painting layers to come off easily. In the case of deterioration affecting the polychrome only, the uncovered wood warps easily and deforms the object.

2. Development of micro-organisms on the surface of paintings results in formation of filaments spreading over the surface and masking their design and colour (compare Figs. 3 and 4 on pp. 216, 217).

3. Hyphae growing inside the painting cause the structure of the layers to become dislodged. A similar effect may be brought about by fungal fruiting bodies, which are frequently formed under the surface of the painting layers (Ważny and Rudniewski, 1972).

4. Extracellular enzymes enabling hyphae to penetrate inside the paintings cause deterioration of all components of the object. The

disintegration extends far beyond the range of the colonies. Enzymic activity causes the painting layers to loosen, crack and fall off the surface. Disintegration of the oil distemper binder in paints results in pigment particles being uncovered and liable to weathering. Advanced microbial activity brings about powdering of the surface of paintings (Ross, 1963).

5. Hyphae growing inside the paintings excrete metabolic products into the environment; among them may be organic acids which dissolve primarily the lime-ground. This accelerates the peeling of painting layers (Ross et al., 1968).

6. Fungal metabolism may also produce coloured compounds, which stain the paintings. The stains that appear as a result of fungal activity (especially the Dematiaceae) are almost impossible to remove from the surface of the paintings.

7. Development of fungi, particularly in their vegetative stages, promotes the settling of dust on paintings. Hyphae readily grow through the dust particles, consolidating them and utilizing them as nutrients (Ross, 1963).

V. MICROBIAL DETERIORATION OF PAINTINGS ON WOOD AND PAINTED SCULPTURES

A specific characteristic of these art objects is the considerable thickness of the support, which is a wooden board or sculpture. There are therefore two problems, namely the liability of wood to microbial attack and the liability of the paint layers to destruction. Development of micro-organisms may result in complete destruction of these art objects. Generally, however, their deterioration is due to the decay of the support or to the disintegration of the paint layers on an intact board (Table 2).

A. Decay of the Wooden Support

Wood decay is caused by members of the basidiomycetes. The source of infection of paintings and sculptures with these fungi is affected wooden elements of the rooms in which they are kept. Wood-rotting fungi are highly cellulolytic organisms. They damage the cell walls in the wood tissue, which results in a decrease in the weight of the wood, which becomes soft, dark or discoloured. In the active stage of fungal growth, the wood is also saturated with metabolic water. With time, the water dries out, and the wood badly shrinks and cracks into prismatic pieces. Destruction of the wooden support results in the paint layers peeling off the surface. Detachment of paint layers from paintings and painted sculptures is one of the most severe symptoms of fungal decay as is increased movement of wood effected by changing humidity. Among the basidiomycetes, soft-rot fungi, including *Gryophana lacrymans*, *Coniophora cerebella* and *Poria vaporaria*, have often been found on paintings and sculptures. These fungi are nearly always accompanied by insects of the family Anobiideae, which readily attack wood affected by wood-rotting fungi. Simultaneous development of fungi and insects often causes total disintegration of paintings on wood or sculptures. Mild-rot fungi, including species of *Chaetomium*, *Penicillium* and *Aspergillus*, which cause surface damage, are also often found on wooden supports (Ważny, 1970). The author has frequently found these fungi on wooden supports of ikons.

Non-painted wooden sculptures exhibited in the open air suffer from

Table 2

Some organic components of antique paintings and sculptures which are sources of nutrients for micro-organisms

Organic material	Occurrence and use
Cellulose	in canvas, timber and paper
Starch	in bookbindings as glue
	as relining paste
Gums	
arabic	in distemper paints
cherry	
tragacanth	
Sucrose	in distemper paints as plasticizer
Glucose	as plasticizer in watercolours (in honey)
Glycerine	as plasticizer in watercolours and emulsion paints
Gelatine	in paper as sizing glue,
	in canvas as sizing glue
	in grounds
	in mural paintings as paint binder
Linseed oil	in oil paints
Egg	in distemper paints as binder

so-called white-rot fungi. This kind of deterioration is manifested in the form of wood cracks running along the annual rings and in a change in the colour of the wood to silvery-grey.

B. Decay of Paint Layers

Wooden supports have been found to render paint layers more liable to microbial attack. Ross (1963) and Ross *et al.* (1968) demonstrated that paint coats on wooden supports become covered with fungal growth much more quickly than coats applied over masonry or metal substrates.

Development of micro-organisms may result in complete disintegration of the painting layers. According to Boustead (1963), the attacked paint layer becomes porous, less elastic and more easily dissolved in alkali. More importantly, it swells readily under humid conditions. This author, in studies on the liability of paint media to microbial attack, found that at a relative humidity of 68–98% the most liable to attack were casein, egg distemper, emulsion distemper and linseed oil in that order. He also pointed out the high sensitivity of dammar, mastic and calaphony.

Pigments contained in earths are not indifferent to microbial activity. It has been found that paints containing earth pigments (sienna, umber, boles) are particularly liable to be attacked. They are highly hygroscopic, rich in micro-elements, and hence are most frequently subject to mildewing. On the surface of paints made up of these pigments and water only, the author has found fungi of the genera *Penicillium*, *Trichothecium* and *Verticillium*.

Paints containing ions of heavy metals are generally resistant. Lead white is the most resistant, due to the presence of fungicidal lead ions and also to its great resistance to water absorption. Zinc oxide is also fairly resistant to microbial attack. Slansky (1960) claims that paints, such as lead white and chrome yellow, are darkened by atmospheric hydrogen sulphide. Our studies on destruction of oil paintings by streptomycetes confirm the above findings and have also shown that streptomycetes themselves may produce this gas. Micro-organisms found on the surface of paintings on wood comprise many fungal species and may also include streptomycetes, which have often been found by the author on paintings and sculptures.

In a small church at Barbarka near Toruń in Poland, there was a gothic sculpture of St. Barbara with well-preserved wood, but a badly damaged paint layer. The high humidity of the church in which it had rested for many years promoted mass development of species of *Trichoderma*, which grew out from beneath the painting layers to form fluffy white colonies along the cracks on the painted surface of the sculpture (Figs. 1 and 2). This fungus is known to grow readily on the surface layers of wood and to be antagonistic to other fungi. On the sculpture in question, it probably fed on components of glues and on substances contained in the surface layers of wood. The painting layer on this sculpture was cracked, peeling and falling off.

Fig. 1. A photograph of the wooden sculpture of St. Barbara showing the painting layer on the coat covered with fungal growth. The photograph was taken by W. Rasnowski.

Fig. 3. Photograph of the Crucifixion, front and reverse. Most intensive growth of strepto-mycetes took place where patches have been stuck. The photographs were taken by H. Sobolewska.

corresponding with the range of the glued on patches (Figs. 3 and 4).

In my experience, members of a number of fungal genera have been found on these objects. The most common are given below in the order of their frequency.

1. *Penicillium*. Many species of these fungi develop both on the surface and on the reverse side of paintings. They are very common fungi (oligotrophs and xerophiles) that quickly produce enormous numbers of spores and rapidly overgrow materials on which they develop. They may stain surfaces on which they grow, and mask them by producing numerous coloured spores.

2. *Aspergillus*. Strains of this genus develop on the surface of paints covering them with a deposit of olive-green spores. They disintegrate distemper and oil binders.

3. Species of *Stemphylium*, *Alternaria* and *Cladosporium* have been found on painting layers. They cause disintegration and staining. Particularly

Fig. 4. Photograph of the Crucifixion taken after disinfection and removal of microbial growth (in the course of conservation). The photograph was taken by H. Sobolewska.

bad stains were formed on skin colours on portraits. They usually avoid places painted with lead white. On the reverse sides they cause extensive black stains and severe weakening of the canvas.

4. *Mucor* and *Rhizopus* species develop particularly vigorously on the reverse sides of paintings, attacking glues. Their growth is often mistaken for cobwebs and removed without disinfection, which results in further dissemination of spores.

5. *Chaetomium* species attack the canvas of painting supports, forming olive-black perithecia. They stain the canvas brown–black, causing severe disintegration of cellulose.

6. *Aureobasidium* species form smooth black spots on the surface of paintings. They decompose oil binders and are extremely tolerant to desiccation.

7. *Geotrichum* species form white filaments on the surface of paintings, causing porosity of varnishes and oil paints. They develop particularly frequently in the presence of casein binders.

fungi formed fruiting bodies of the perithecia, pycnidia and stroma type under the painting layer, giving rise to formation of small blisters and craters. On frescoes in the castle of Karlstein in Czechoslovakia, Tonolo and Giacobini (1961) observed abundant subsurface growth of fungi which greatly contributed to loosening the binding between the paint layer and its support.

Development of streptomycetes, as described in Section VI (p. 215), causes destruction and masking of the painting surface by causing the spread of a light, powdery deposit (Bassi and Giacobini, 1973).

In all cases where development of bacteria, fungi or streptomycetes has been found, the surface of the paintings showed advanced degradation.

B. Development of Autotrophic Micro-Organisms

On damp areas exposed to sunlight abundant algal growth is observed. Unpublished studies from my laboratory have revealed that it may reach great intensity, covering large areas of plaster with a green coat, but sometimes avoiding places painted with paints toxic to algae. Sunlight, indispensable for development of algae, has a harmful effect on most heterotrophs because of its content of ultraviolet radiation. Algal growth on areas exposed to sunlight causes stains of various shades of green, from very dark, almost black, to yellow–green. The shade depends on the nature of the predominating algae.

Tonolo and Giacobini (1961) report that, among algae on mural paintings, green algae predominate. However, they do not list the genera represented. Sokoll (1977), who studied algal flora on plasters, found members of the blue–green algae, green algae and diatoms. A list of algal genera detected by this author is given in Table 4.

The species of algae found on paint surfaces largely depends on climatic conditions. Algae are extremely sensitive to temperature and it is this factor that mainly determines the composition of algal communities. Skinner (1971), in studies on the resistance of paints to algal growth, found that illuminated painted areas under temperate climatic conditions became covered with green algae (species of *Pleurococcus*, *Protococcus* and *Trentepholia*). Under tropical conditions, blue–green algae of the genera *Oscillatoria* and *Scytonema* developed on painted surfaces. The latter formed very dark, almost black stains.

Table 4

Algae isolated from mural paintings and plaster. From Sokoll (1977)

Tribe	Species
Chlorophyta	*Chlorella vulgaris, Ch. ellipsoidea, Stichococcus bacillaris, Ulothrix zonata, U. punctata, Protococcus viridis, Gleotila protogenita, Microthamnion strictissimum, Chlorosarcinopsis minor.*
Cyanophyta	*Oscillatoria pseudogeminata, O. irrigua, O. brevis, Phormidium autumnale, Ph. bohneri, Ph. valderianum, Microcystis parietina, Microcoleus delicatulus, Symploca muralis.*
Chrysophyta	*Botryochloris minima, Heterodendron, Pascheri sp., Fragilaria sp.*
Pyrrophyta	*Ceratium hirundinella, Pediastrum sp.*

The effect of algae on the surface of mural paintings is not limited
to masking their surface and digesting it by secretion of organic acids
(lactic, succinic, glycolic, acetic, oxalic and pyruvic acids; Levin,
1962). They are also capable of excreting large quantities of sugars
and amino acids, which promote development of bacteria and some
fungi. Among the fungi, members of the family Dematiaceae readily
accompany algae. Their colour makes them resistant to light. Thus
fungi in association with algae may participate in the destruction of
mural paintings.

IX. COMBATING MICRO-ORGANISMS ON PAINTINGS AND SCULPTURES

Paintings, sculptures and other objects attacked by micro-organisms
should be subjected to disinfection in order to kill the microflora living
on them and to prevent further microbial attack. The role of fungicides
is particularly important in the case of mural paintings, which are
constantly in danger of being subject to high humidity, either from
the soil or from atmospheric condensation on their surface. Modern
conservation demands very high standards for all materials used in
treating objects of art. Materials already used, or newly introduced in
conservation, are closely tested to preclude any possible noxious effect
on the objects treated.

Fungicides used for disinfection of sculptures, paintings on canvas
and board, mural paintings, as well as of water colours, pastels, prints
and books should comply with the following conditions. (1) A fungicide
should have a high fungitoxic value, so that it can be used in low

concentrations. (2) A fungicide must not adversely affect any of the components of the object of art, i.e. it must not affect the hue, the fastness of the paints, the binders, the glues in the ground, canvas, paper and wooden supports, or plaster. (3) A fungicide should be characterized by low volatility to ensure a prolonged protective effect on the object. This requirement does not refer to volatile fungicides used in fumigation. (4) A fungicide must not be liable to ageing associated with formation of noxious decay products. (5) A fungicide must not lose its biotoxic properties by combining with constituents of the object treated. (6) A fungicide should be easy to apply and have a low toxicity in man. As we know, not all fungicides meet all the above requirements. The restorer has therefore only a limited choice of materials for use.

World literature on conservation devotes little attention to the necessity of disinfecting objects of art infected by micro-organisms. There are also only a few reports of studies on the effect of fungicides on the components of paintings including paints, binders, grounds and supports, wood, canvas, paper and plaster. Hueck van der Plas (1966) gives a list of biocides and commercial preparations for combating micro-organisms on paints, paper, wood and textiles. In our laboratory studies are being carried out on the applicability of biocides for disinfection and protection of paintings on canvas, wood, paper and murals against microbial attack.

A. Methods for Disinfecting Paintings and Sculptures

Objects showing only trace growth of microflora are subjected to disinfection. This precedes any further conservation treatment. The choice of fungicide and the method of disinfection depends on the kind of object, on its condition and on the intensity of the microbial attack.

1. Fumigation

This is one of the best methods for disinfecting painted and non-painted sculptures, paintings on wood, canvas and paper, textiles, as well as books and old prints badly attacked by fungi or insects (Borecki et al., 1965; Gallo and Gallo, 1971). Application of the fungicide in gaseous form ensures good penetration inside the object and cuts down the necessity of using fungicides in solutions. Fumigation is performed in vacuum chambers filled with toxic gas after the air has been pumped out.

Fumigation destroys fungi and insects in all developmental stages. The effect of the gas, however, does not protect the object against another attack after taking the object out of the chamber. Unfortunately, as a vacuum chamber is an expensive piece of equipment, not all conservation laboratories can afford it. Cunha and Cunha (1971) describe a model of a portable fumigation chamber.

2. Treatment with Fungicidal Vapours

This method is commonly used for disinfecting books and prints. It is also recommended for disinfecting watercolours and pastels (Strzelczyk and Rosa, 1975). It has the advantage of being easy to use in laboratories, since it does not require sophisticated equipment. The object is placed in a polyethylene bag between sheets of filter paper saturated with a fungicide solution and then dried. Air-tight cabinets with shelves permitting free circulation of the fungicide vapours are also used. Plenderleith and Werner (1974) recommend the use of thymol in this method. The fungicidal vapour treatment may also be applied for disinfection of paintings on wood and painted sculptures attacked superficially by moulds, but not by wood-rotting fungi or insects.

3. Spraying with Fungicidal Solutions

Fungicide solutions in organic solvents are used. It is of prime importance that neither the fungicide nor the solvent should have any noxious effect on the varnishes, paints, binders or other components of the antique object. This method of disinfection is used for objects showing bad deterioration of the surface layer, e.g. for easel pictures with a badly peeling painting layer or with mural paintings whose paint layer is powdering off.

Disinfecting large areas, such as murals, may constitute a danger to the restorer's health, as it involves prolonged inhalation of the spray. For this reason, the operation must be carried out in perfectly ventilated rooms, or with the use of protective masks.

The spraying of the fungicide must not be too vigorous, since living spores may be introduced into the air and infect other objects in the vicinity.

For complete destruction of all of the microflora overgrowing the object, the spraying should be performed three times at 3–7 day intervals (Strzelczyk, 1970).

4. Impregnation by Brushing

The same fungicide solutions used for spraying may also be applied to paintings or sculptures by means of a brush. The fungicide solution has then a better contact with the surface of the object and penetrates deeper through its layers. Fungicidal treatment by brushing is used to disinfect very mouldy canvas supports and painting surfaces badly attacked by micro-organisms, but not yet destroyed to a degree that would bring a risk of damage to the painting from strokes of a brush spreading the fungicide. This method of applying disinfectants is less harmful to the worker.

As in the spraying method, for complete destruction of the micro-organisms, three applications of the disinfection at 3–7 day intervals are recommended. The amount of disinfectant applied each time should be such that it would not permeate the object throughout, and the solvent should evaporate from its surface as quickly as possible.

5. Injecting

This method is used to destroy wood-rotting insects and fungi in paintings, sculptures and beam-framed floors. The fungicides and insecticides used for injections are commercial preparations. The disinfectant is introduced into the worm holes by means of a medical syringe. After a few days the holes are sealed with wax paste (Plenderleith and Werner, 1974).

B. Fungicides used for Disinfection

In practice, restorers of ancient objects have at their disposal only a few fungicidal compounds or complex commercial preparations that comply with their demands. Some of these disinfectants have properties precluding them from use with certain objects, while, with others, they may be applied with good effect. In order to use the disinfectants effectively without causing any damage to the objects, it is essential to have a good knowledge of their properties.

1. Disinfection of Canvas Paintings

Treatment aimed at destroying micro-organisms on these objects is applied on their face or reverse sides and, in the case of particularly strong microbial attack, even on both sides at the same time. The painting is then sprayed over the face, and brushed over the reverse side. Objects that are mouldy only on one side, and not very badly, are disinfected only on the attacked side.

It has been found that the most effective disinfectant is a solution of p-chlor-m-cresol and phenylmercuric acetate in turpentine and acetone in the following proportions: p-chlor-m-cresol, 0.3 g; phenyl-mercuric acetate, 0.01 g; turpentine, 67 cm³; acetone, 33 cm³. The author has found that a small addition of mercuric acetate enhances the fungitoxic properties. The turpentine solvent does not dissolve varnishes or soften paints, and addition of acetone accelerates evaporation of the solvents.

Voronina (1968), testing various fungicides for disinfection of easel paintings, found that p-chlor-m-cresol is active in disinfecting oil and distemper paintings, except for egg distemper. Accelerated ageing for 50 hours under a quartz–mercury lamp did not cause any change in the solubility and hue of the painting layers. Studies carried out in my laboratory have confirmed that it is innocuous to all of the layers making up canvas paintings (Stachowiak, 1975). Slansky (1960) recommends wax with pentachlorophenol dissolved in white spirit for protecting the canvas in old paintings against microbial attack. For new canvas used for doubling paintings, he recommends sodium fluoride, sodium fluoro-silicate, salicylanilid or cadmium soap. Voronina and Araktcheyeva (1977) worked out a method of protecting canvas supports of paintings against microbial attack by treatment with nipagin (methyl ether of p-aminobenzoic acid).

Among the above fungicides, pentachorophenol cannot be considered for disinfecting canvas since, under conditions of high humidity, it easily dissociates to provide chloride ions. Canvas, being a cellulose-containing material, is extremely sensitive to acid hydrolysis. For reasons that will be discussed later, sodium fluoride must also not be used for paintings.

Plenderleith and Werner (1974) and Waterer (1973) recommended a
0.5% o-phenylphenol solution for disinfection of textiles and leather.
Hueck (1965) mentions a number of commercial preparations for combating insects in old textiles.

2. Disinfection of Paintings on Wood and Sculptures

Painted wood is often subject to wood-rotting fungus and insect attack.
Since these pests generally develop deep inside the wood, the disinfectants for destroying them must penetrate inside the object.

The best sterilizing effects are obtained by fumigation in vacuum
chambers with ethylene oxide as the toxic fumigant. It destroys fungi
and insects in the wood, the latter in all developmental stages. Commercial products, such as Cartox (Poland), Carboxide or Ethoxide (Great
Britain) consist of 10% ethylene oxide and 90% carbon dioxide, which
acts as a diluent and eliminates the explosive properties of ethylene
oxide. It also enhances the penetrating capacity of ethylene oxide.
Similar properties are shown by the gas under the trademark 'Cryoxide'
consisting of 11% ethylene oxide and 89% Freon (Plenderleith and
Werner, 1974). Gallo and Gallo (1971) give a list of firms producing
mixtures of ethylene oxide with neutral gases. The mixture has been
demonstrated (Tonolo and Giacobini, 1958) to have no noxious effect
on paints in easel paintings and murals. The application of all of the
above-mentioned gases requires expensive equipment, such as vacuum
chambers. These are made by a number of European firms, such as
Mallet S.A., Paris, France (Cunha and Cunha, 1971).

Fumigation with gases containing ethylene oxide is used for paintings,
sculptures, furniture, wooden fragments of interior decorations and
ethnographical objects. After treatment with ethylene oxide, protection
of the objects against further microbial attack is provided by one of
the following methods. Museum store rooms are disinfected by introducing the gas after sealing them for a period of two to several
days depending on the nature of the gas. Plenderleith and Werner
(1974) recommended fumigation of ethnographical objects with hydro-
cyanide, as is done in the British Museum, or with methyl bromide.
Fumigation of large objects is done professionally by Rentokil Laboratories Ltd of 7 Morocco Street, London, S.E.1. (Plenderleith and
Werner, 1974).

Fumigation with carbon disulphide vapour is used with good effect in disinfecting and destroying insects in paintings on wood. The fumigant penetrates deep inside the object, does not damage the polychromy and can be used in chambers made from polyethylene bags. It is applied at 100 cm³ of carbon disulphide per m³ of air in the chamber. The object is left in the chamber for 2–3 weeks. It must be borne in mind that carbon disulphide vapours are heavy and accumulate near the bottom of the chamber. The object must therefore be placed rather low: carbon disulphide mixed with air becomes explosive and inflammable, it should be handled only by experienced staff.

Paintings and painted sculptures are also disinfected with fungicides in organic solvents or with commercial preparations made specially for disinfecting antique objects. The latter are both fungicidal and insecticidal. Either group of disinfectant prevents microbial and insect attack for some time. They are more labour-consuming but often less troublesome in use, as they do not require special equipment. They are applied by spraying, brushing or injecting.

The choice of method depends on the kind of infection. In superficial fungal infection, spraying or brushing is used. Injections are used with objects attacked by wood-rotting fungi and insects (see Sections IX.A.2.3 and 4).

I recommend for disinfection *p*-chlor-*m*-cresol and phenylmercuric acetate solutions in turpentine and acetone (see Section IX.B.1). The solution is applied as for disinfection of canvas paintings. Several repeated treatments have been found to give good results, i.e. complete disinfection without any undesirable side effects.

A Polish commercial preparation designed for disinfection of paintings, sculptures and non-painted wood is Antox (INCO, Warszawa, ul. Salezego 6). It is commonly used for combating fungi and insects in paintings, wood and sculptures. The preparation contains chlorinated phenol derivatives (pentachlorophenol) in aliphatic and chlorinated aromatic solvents. It is applied by injection into the worm holes or by brushing. In addition, INCO makes several other preparations designed specially for ancient wood. These are: Soltox 5F (waterborne preservatives), Xylamit Żeglarski (Nautical) and Xylamit Destylowany Stolarski (distilled for woodwork/oil-type preservatives). The preparations are in common use in Poland in conservation of old wooden elements (altars, beam-framed floors, interior decorations) and ethnographical objects (Czajnik *et al.*, 1970).

Sodium fluoride in aqueous solution, which has been proposed by some authors (Slansky, 1960; Plenderleith and Werner, 1974), is suitable only for saturation of timber. Considering that it is applied in aqueous solution, it is not recommended for use with paintings and sculptures, since it may cause irreversible changes in the condition of the object. Moreover, sodium fluoride reacts readily with calcium salts, which results in formation of insoluble and non-fungicidal calcium fluoride.

A list of commercial preparations made for wood disinfection in Britain is found in the Wood Preservation Register published by Timber and Plywood (194–200 Bishopsgate, London E.C.2) (Plenderleith and Werner, 1974).

3. Disinfection of Paintings on Paper

The most commonly used, traditional fungicide for disinfection of paper objects is thymol, recommended by Plenderleith and Werner (1974) and Plumbe (1964). According to these authors, thymol may be safely used for watercolours, pastels, prints, drawings, books, parchment and vellum. Disinfection in thymol vapours is carried out in heated cabinets for 14 days. These authors warn that thymol softens oil paints and varnishes. Clapp (1974) warns against treating parchment with thymol. Likewise, Smirnova (1962), after a close investigation of the effectiveness of thymol solutions for the disinfection of parchment, concludes that this fungicide is not suitable for this purpose.

My own observations (Strzelczyk, 1969) showed, however, that thymol vapours in sublethal doses promoted germination of fungal spores destroying old paper. With higher doses, the response of spores of various fungi varied widely. It has also been found that, due to its high volatility, thymol did not prevent further fungal attacks.

In studies on the possibility of applying ethylene oxide (with carbon dioxide) for sterilization of watercolours and pastels, it has been found that it is completely neutral to watercolour paints and crayons (Strzelczyk and Rosa, 1975). It may therefore be recommended for sterilization of these paintings in vacuum chambers (the time of fumigation with Cartox is 24 hours).

In the absence of a fumigation chamber, a reliable method is sterilization of these paintings in p-chlor-m-cresol or phenylmercuric acetate vapours. Before suggesting this method of disinfection, the fungicides were tested for their effect on paper, 56 watercolour paints and 96 crayons from various Polish and foreign firms including Winsor, Newton and Rowney in the United Kingdom, Talens in Holland, Eberhard in the U.S.A., Lefranc in France and Toison D'or in Czechoslovakia. The paints and crayons did not respond to the fungicides by change of hue or penetration inside the paper. Vapours of either fungicide showed highly fungitoxic effects and protected the objects against microbial attack for a long time.

Infected pastels and watercolours are disinfected in the following way. (a) Prepare sheets of filter paper to fit the size of the object, and a polyethylene bag of similar size. (b) Saturate the filter paper with 10% p-chlor-m-cresol solution (or 1% phenylmercuric acetate) in ethanol and leave until quite dry. (c) Put the object to be disinfected between the disinfectant-saturated sheets, put it in the polyethylene bag and seal. (d) Pastels with a particularly fluffy painting layer should be separated from the filter paper with a thick cardboard frame. (e) The duration of sterilization in p-chlor-m-cresol vapours is seven days at room temperature, and in phenylmercuric acetate three days at 40°C.

The method has been applied many times with good results for disinfection of watercolours and pastels in my laboratory.

4. *Disinfection of Mural Paintings*

Restorers have a limited choice of disinfectants suitable for mural paint-ings. The preparations must comply with very high demands, that is, they should be highly biotoxic and have a lasting effect. Most of those tried so far have proved to have short-duration effects. An instance of this was the attempt at inhibiting invasion of heterotrophic micro-organisms on frescoes in Florence by means of the antibiotic nystatin (Gargani, 1968). The effect of the antibiotic proved to be of short duration. Gargani (1968) also presumes that, since it acts selectively, it may not be capable of destroying all the microflora attacking the frescoes. This was shown by rapid development of new fungi on the disinfected surfaces.

Highly recommendable for disinfection of mural paintings are p-chlor-m-cresol and phenylmercuric acetate. Both of these fungicides have been thoroughly tested in our laboratory (Strzelczyk, 1979). They were found to be highly fungitoxic but had no noxious effect on binders and pigments used in murals. They also did not cause the plasters to darken.

Roznerska (1973) found out that phenylmercuric acetate could be used as a fungicide to treat paints used in painting murals. Hueck van der Plas (1966) recommends p-chlor-m-cresol for disinfecting paints, glues, binders, leather and paper. Both of these disinfectants are of limited duration, as pointed out also by Payne (1963), but still they are superior to nystatin in this respect. The author (Strzelczyk, 1978) recommends a solution consisting of 0.3% p-chlor-m-cresol and 0.1% phenylmercuric acetate in ethanol. The object should be sprayed three times at 3–7 day intervals. Precautions should be taken, as the solution is highly toxic to humans. Tonolo and Giacobini (1958, 1961) stress the desirability of keeping mural paintings under air conditioning at constant humidity and temperature. They suggest that murals should be first disinfected with ethylene oxide and then kept in a nitrogen atmosphere in gas-tight chambers. In view of what has already been stated on water migration from the ground and water condensation on murals, it seems that ways of rendering mural paintings resistant to the development of heterotrophic micro-organisms are still to be sought.

Combating autotrophic micro-organisms—algae—on mural paintings is a very difficult task. Algae find favourable condition for mass development on open and illuminated areas of these objects. Factors promoting their development also promote the decay and evaporation of disinfectants from the surface of the murals.

The literature cites a number of algicides that destroy algae in water environments, for example, in swimming pools (Hueck van der Plas, 1966; Fitzgerald, 1971). They are generally water-soluble compounds and therefore easily washed out of the porous surfaces, e.g. for mural paintings. Few of them can be selected for testing their applicability in combating algae on murals. It has been found, however, that in cases where emergency measures must be taken against algal invasion on elevations or other open spaces, treatment with an algicide gives positive results. They destroy algal growth and delay their reappearance.

Tests carried out in my laboratory have demonstrated that the preparation called Lastanox TA (made by Lachema, Brno, Czechoslovakia) was the most effective in combating algae on plasters and on surfaces of stone monuments (Czerwonka, 1976; Staszewska, 1977). The preparation did not have a noxious effect on any of the components of mural paintings, i.e. on the ground, the binders and the pigments. It also showed considerable resistance to ageing in the air-conditioned chamber, where it did not lose its algicidal properties. The preparation was effective when used as a 1% solution in ethanol. It should be sprayed three times at 3–7 day intervals. Studies aimed at the best choice of algicides of long duration to be used on mural paintings are still in progress.

REFERENCES

Ayerst, G. (1968). *In* Proceedings of the First International Biodeterioration Symposium, Southampton, Elsevier, Amsterdam, pp. 223–241.

Bassi, M. and Giacobini, C. (1973). *International Biodeterioration Bulletin* **9** (3), 57–68.

Borecki, Z., Czerwińska, E., Eckstein, Z. and Kowalik, R. (1965). 'Chemiczne środki grzybobójcze (fungicydy), Państwowe Wydawnictwo Rolnicze i Leśne'. Warszawa.

Boustead, W. (1963). *In* 'Recent Advances in Conservation' (A. H. Walters and J. S. Elphick, eds.), pp. 73–78. Butterworths, London.

Clapp, A. F. (1974). 'Curatorial Care of Works of Art on Paper', Intermuseum Conservation Association, Oberlin, Ohio, pp. 46–48.

Cunha, G. M. and Cunha, D. G. (1971). 'Conservation of Library Materials, I'. The Scarecrow Press Inc., Metuchen, N.J.

Czajnik, M., Lehnert, Z., Lerczyński, S. and Ważny, J. (1970). 'Impregnacja i odgrzybianie w budownictwie'. Arkady, Warszawa.

Czerwonka, M. (1976). M.Sc. Thesis: Copernicus University, Toruń, Poland.

Doerner, M. (1971). 'Malmaterial und seine Verwendung im Bilde'. Ferdinand Enke Verlag, Stuttgart, pp. 169–188.

Fitzgerald, G. P. (1971). *Algicides. The University of Wisconsin, Literature Review*, No. 2, pp. 1–50.

Gallo, F. (1964). *Bolletino dell'Istituto di Patologia del Libro 'Alfonso Gallo'*, Roma I–IV, pp. 1–18.

Gallo, F. and Gallo, P. (1971). *Bolletino dell'Istituto di Patologia del Libro 'Alfonso Gallo'* **XXX**, I–II, pp. 35–69.

Galloway, L. D. (1954). *Journal of Applied Bacteriology* **17**, 207–212.

Gargani, G. (1968). *Proceedings of the First International Biodeterioration Symposium, Southampton*, pp. 252–257.

Grzesikowa, H. (1965). *Biblioteka Muzealnictwa i Ochrony Zabytków* **XI**, 156–164.

Hueck, H. J. (1965). *TNO-Nieuws* **20**, 301–307.

Hueck van der Plas, E. (1966). *International Biodeterioration Bulletin* **2** (2), 69–120.

Kathpalia, Y. P. (1960). *Indian Pulp and Paper, July*, pp. 117–125.

Kowalik, R. and Sadurska, I. (1957). *Acta Microbiologica Polonica* **5**, 277–284.

Krumperman, P. H. (1958). *American Paint Journal* **42** (38), 72–78.

Kuritzina, D. S. (1968). *Vestnik Moskovskovo Universiteta Botanika* **4**, 31–41.

Kuritzina, D. S. and Sizowa, T. P. (1967). *Mycologya y Phytopatologya* **1** (4), 342–344.

Levin, R. A. (1962). 'Physiology and Biochemistry of Algae'. Academic Press, New York and London.

Nuksha, J. P. (1960). *Microbiologya XXIX* **1**, 132–136.

Payne, H. F. (1963). 'Organic Coating Technology', vol. II. John Wiley, New York and London.

Pietrykowska, S. and Strzelczyk, A. (1973). *Acta Universitatis N. Copernici, V* **52**, 203–216. English summary.

Pietrykowska-Leźnicka, S. (1979). Ph.D. Thesis: N. Copernicus University, Toruń, Poland.
Plenderleith, H. J. and Werner, A. E. A. (1974). 'The Conservation of Antiquities and Works of Art'. Oxford University Press, London.
Plumbe, W. J. (1964). 'The Preservation of Books in Tropical and Subtropical Countries'. Oxford University Press, Hong Kong, London.
Rosa, H. and Strzelczyk, A. (1979). *Acta Universitatis N. Copernici*, *VII*, 119–128. English summary.
Ross, R. T. (1963). *Advances in Applied Microbiology* **5**, 217–234.
Ross, R. T., Sladen, J. B. and Wienart, L. A. (1968). *Proceedings of the First International Biodeterioration Symposium, Southampton*, pp. 317–325.
Roznerska, M. (1973). *Acta Universitatis N. Copernici*, *V* **52**, 183–201.
Skinner, C. E. (1971). *Proceedings of the Second Biodeterioration Symposium, Luntheren*, pp. 1–18.
Slansky, B. (1960). Technika Malarstwa, Arkady, Warszawa.
Smirnova, B. J. (1962). Voprosy Conservacyi y restavracyi bomagyi y pergamena, Academya Nauk SSSR, Moscov-Leningrad, pp. 49–59.
Sokoll, M. (1977). M.Sc. Thesis: N. Copernicus University, Toruń, Poland.
Stachowiak, E. (1975). M.Sc. Thesis: N. Copernicus University, Toruń, Poland.
Staszewska, G. (1977). M.Sc. Thesis: N. Copernicus University, Toruń, Poland.
Strzelczyk, A. (1969). *Acta Mycologica* **V**, 213–218. English summary.
Strzelczyk, A. (1970). *Biblioteka Muzealnictwa i Ochrony Zabytków* **XVII**, 109–118. English summary.
Strzelczyk, A. (1978). *Ochrona Zabytków* **2** (121), XXXI, 128–131. English summary.
Strzelczyk, A. (1979). *Acta Universitatis N. Copernici* (in press).
Strzelczyk, A. and Pietrykowska, S. (1969). *Ochrona Zabytków* **4**, 263–272. English summary.
Strzelczyk, A. and Rosa, H. (1975). *Ochrona Zabytków* **1** (108), XXVIII, 61–66. English summary.
Tonolo, A. and Giacobini, C. (1958). *Bolletino dell'Istituto Centrale del Restauro, Roma 36*, pp. 191–196.
Tonolo, A. and Giacobini, C. (1961). *Recent Advances in Conservation*. Contributions to the JJC. Rome Conference, pp. 62–64. Butterworths, London.
Voronina, L. J. (1968). *So-obshtchenya* **20**, 57–65.
Voronina, L. J. (1969). *So-obshtchenya* **24–25**, 103–104.
Voronina, L. J. and Araktcheyeva, D. Z. (1977). Hudojestvennoye Nasledye **33** (3), Moskva 32–40.
Waterer, J. W. (1973). 'A Guide to the Conservation and Restoration of Objects Made Wholly or in Part of Leather'. G. Bell, London.
Ważny, J. (1965). *Biblioteka Muzealnictwa i Ochrony Zabytków* **XI**, 151–155.
Ważny, J. (1970). 'Zeszyty Naukowe SGGW, Leśnictwo XIV', pp. 51–64.
Ważny, J. and Rudniewski, P. (1972). *Material und Organismen* **7** (2), 81–92.

8. Tobacco

T. G. MITCHELL and P. C. STAUBER

*Group Research and Development Centre, British-American Tobacco
Co. Ltd., Southampton SO9 1PE, England*

I. Introduction	235
II. Types of Biodeterioration	237
III. Barn Rots	240
IV. Marketing, Storage and Shipping of Cured Leaf	246
V. Manufacturing	251
VI. Distribution	253
VII. Economic Aspects	254
VIII. Acknowledgements	256
References	256

I. INTRODUCTION

Tobacco has a long history as a commodity of significance in international commerce. As a high-value crop of seasonal production, but year-round utilization, the requirements for preventing microbial deterioration were developed empirically long before any information on the micro-organisms and principles involved became available. Essentially, these consisted of methods of drying the leaves in a standardized manner, known as curing, and a recognition of the need for subsequent storing and shipping the leaf sufficiently dry to avoid damage, but without its shattering.

Additional sorting, cleaning and drying of the leaf normally occurs in a process known as redrying, which ensures further against deterioration in long-term storage or shipping. Threshing, in which the leaf lamina is separated from the midrib, is frequently carried out in the redrying plant.

The types and patterns of microbial deterioration of tobacco are determined by moisture content (and therefore relative humidity) more than any other factor. In spite of the wide variation in moisture content of leaves from harvesting (about 90%) to ultimate utilization (about 12%), there are only a few distinctive microbiological patterns of deterioration. This reflects the fact that, after curing, variations in moisture content are relatively small and high values generally occur for short periods only.

The moisture–humidity relationship of unprocessed, cured tobacco is shown in Figure 1. Considerable variations occur in the chemical composition of tobacco, between leaves of the same variety from different positions on the plant (Bacot, 1960), between different types

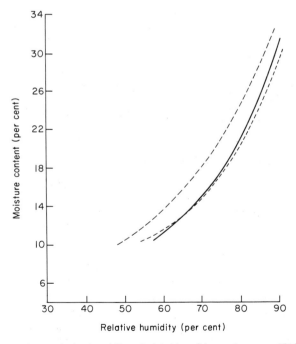

Fig. 1. Moisture content—relative humidity relationships of three tobaccos at 22°C. — — — — — indicates relationship for flue-cured tobacco (sample A), – – – – – – – for flue-cured tobacco (sample B) and —— for air-cured Burley-type tobacco.

Table 1

Composition of tobacco samples shown in Figure 1

	Composition (per cent dry weight)					
Sample	Nicotine	Total sugars	Reducing sugars	Total nitrogen	Protein nitrogen	Ash
Fluc-cured 'A'	2.3	23.7	18.7	2.0	1.0	11.2
Fluc-cured 'B'	1.3	5.2	5.2	2.9	1.6	14.6
Air-cured, Burley	3.3	0.6	0.6	3.7	1.8	23.7

of tobacco (Akehurst, 1968) and, to some extent, as a result of differences in curing practice. For example, the compositions of the three tobaccos described in Figure 1 are summarized in Table 1. It would be expected that such large differences, for example in sugar content, would be reflected by differences in moisture–relative humidity relationships, and this is borne out by Figure 1. The figure also demonstrates that the differences between tobaccos nominally of the same type can be as great as those between different tobacco types. Some of these variations are of considerable practical importance in affecting margins of safety when shipping or storing tobacco.

II. TYPES OF BIODETERIORATION

Micro-organisms cause losses to the tobacco industry by a variety of mechanisms. The most important of these are as follows. (a) Rotting, in which there is a complete disintegration of leaf structure. This occurs particularly in curing, but may also result from prolonged microbial growth at any stage. (b) Production of off-odours due to growth of micro-organisms on tobacco. This is usually due to fungi (Lucas and Pounds, 1973; Welty and Vickroy, 1975), but occasionally may involve actinomycetes (Johnson, 1924). Musty taint due to formation of chloro-anisoles, which are known to arise from fungal modification of chloro-phenols (Cserjesi and Johnson, 1972), has been identified as a problem both on unmanufactured tobacco (Bemelmans and Ten Noever de Brauw, 1974) and on tobacco products (Curtis et al., 1974). (c) Com-positional changes causing a decrease in value by weight loss, and also a change in quality of the leaf, e.g. low sugar content, and increases

Table 2

Micro-organisms that cause biodeterioration in tobacco curing

Micro-organism	Effect	Reference
FUNGI		
Alternaria alternata	Minor damage in flue-curing.	Gayed (1978)
	Freckle rot of air-cured leaf (Connecticut).	Anderson (1948)
	Shed-burn in air-curing.	Johnson (1924)
Aspergillus flavus	Pole rot of flue-cured (minor cause) in Rhodesia.	Hartill (1967), Cole (1975)
Aspergillus spp.	Minor role in damage in both flue- and air-curing.	Anderson (1948), Gayed (1978)
Botrytis cinerea	Stem-rot and shed-burn of air-cured tobacco.	Johnson (1924), Anderson (1948)
Cercospora nicotianae	Barn spot in flue-curing.	Lucas (1975)
Fusarium spp.	Stem-rot and shed-burn of air-cured tobacco. Minor cause in Connecticut, U.S.A.	Johnson (1924), Anderson (1948)
Penicillium spp.	Minor cause of damage to flue-cured tobacco.	Gayed (1978)
Peronospora tabacina	Damage during air-curing in Europe.	Todd (1961)
Phytophthora parasitica var. *nicotianae*	Large nectrotic lesions in air-curing of cigar wrapper leaf or dark brown zones contrasting with normal lemon colour of flue-cured leaf.	Lucas (1975)
Pythium spp.	Flue-curing in South Carolina, U.S.A.	Holdeman and Burkholder (1956)
Rhizopus arrhizus	Barn rot in flue-curing in Canada, Australia, Rhodesia, New Zealand and South Africa.	Gayed (1972a, b; 1978), Lucas (1975), Stephen (1955), Cole (1975), Hartill (1969), Reinecke (1962)
Rhizopus nigricans	Barn rot in flue-curing in Spain.	Llanos (1978)
Sclerotinia sclerotiorum	Stem and stalk rot of air-cured leaf.	Anderson (1948)
Trichothecium sp.	Stem rot of air-cured leaf.	Johnson (1924)

Table 2—*contd.*

Micro-organism	Effect	Reference
ACTINOMYCETES, BACTERIA		
Actinomycetes	Minor damage in flue-curing.	Gayed (1978)
Bacterial decay	Secondary invaders in stem rot in air-curing. Rotting of bulk-cured and box-cured leaf.	Johnson (1924), Watkins (1976a, b)
Erwinia carotovora	Rots of flue-cured in Canada.	Gayed (1978)
Erwinia aroideae	Rot of flue-cured, associated with hollow stalk in field.	Holdeman and Burkholder (1956)
Erwinia spp. *Pseudomonas* spp. *Xanthomonas* spp. *Phytobacterium* spp.	Soft rots in curing following wounding of leaf.	Durán-Grande (1974)

in contents of sterols and protein nitrogen (Keller *et al.*, 1969; Tancogne, 1977). (d) Changes in appearance of the leaf which will be associated with low quality and value. These include colour, usually darkening, and formation of tears and holes. Together with loss of elasticity, these are particularly important in relation to cigar-wrapper leaf.

Mycotoxins, whose formation is frequently associated with bio-deterioration, have not been a major factor in tobacco losses to date. Studies have failed to identify aflatoxin in either sound or mould-damaged tobaccos (Tso and Sorokin, 1967), although it has been shown that it can be formed in pure-culture studies in the laboratory (Pattee, 1969). However, even if it occurs occasionally in commercial tobaccos, two careful studies have shown that there is little possibility of it being transferred to cigarette smoke (Tso and Sorokin, 1967; Kaminski *et al.*, 1970). Other, currently less important, mycotoxins such as penicillic acid may occur in mouldy tobacco and transfer to smoke in small amounts (Snow *et al.*, 1972). The most appropriate means of avoiding them and possibly other unknown mycotoxins is to avoid development of mould-damaged leaf.

III. BARN ROTS

Significant losses can occur during curing. Collectively, these un-desirable changes are known as barn or pole rots, and may vary from superficial mould growth on part of the leaf to extensive rotting affecting considerable parts of a fully loaded barn. Micro-organisms that have been recognized as causing loss in this way are listed in Table 2.

Curing of tobacco is a farm process with two objectives, namely development of specific desirable characteristics concerned with leaf quality and usability, and change of the leaf from a biologically unstable one containing approximately 90% water into a product of about 10% water content stable on long-term storage. Two main types of curing are used, namely flue- and air-curing, and the differences between them are important in determining the types of microbial activity and bio-deterioration that occur. The effects of different curing processes on leaf and smoke attributes are described by Wolf (1967).

In traditional flue-curing, which is used for 'bright' or Virginia-type tobacco (Akehurst, 1968), the leaf is picked by hand or mechanically when judged to be ripe. It is then tied or stitched to wooden laths or placed in metal racks for hanging in the barns. The heating and ventilation of the barns can be controlled, the first phase of the cure (yellowing conditions) being carried out at high humidity (90%) and raised temperatures (35°C). It is in this period that losses can occur in flue-curing. In subsequent stages, the air temperature is raised to 'fix' the colour of the leaf and it is then dried out. The midrib is the most difficult part of the leaf to dry and, if not done thoroughly, leads to mould development on this part after curing.

In air-curing, the process is slower and heat is not generally used. The leaf is harvested either by cutting off the whole plant just above the ground and hanging the complete plant in the barn, or by picking individual leaves which are then attached to laths as in flue-curing. Instead of well-defined phases, there is a progressive wilting and drying of the leaves over a period of several weeks. In the absence of ancillary heating in the barn, the rate of moisture loss from the leaves is very dependent on weather conditions.

Despite the variety of micro-organisms implicated in biodeterioration in curing (Table 2), most authorities are agreed that only a few are of major importance. In addition, some spoilage in the barn is due to continued growth of plant pathogens following infection initiated in the field. Examples of this include some or most of the damage by *Alternaria alternata*, *Cercospora nicotianae*, *Phytophthora parasitica* var *nicotianae*, *Sclerotinia sclerotiorum* and strains of *Erwinia* spp. Small amounts of damage by *C. nicotianae* may be considered acceptable, or even desirable, as many buyers associate barn spot with leaves fully mature at harvest, which are most in demand.

Fig. 2. Development of rot spreading from midrib in Virginia tobacco leaf. The leaf was artificially inoculated with *Rhizopus arrhizus*, and held at 37°C and above 90% humidity for 48 hours.

In flue-curing, the main type of barn rot identified in many parts of the World has been found to be caused by *Rhizopus arrhizus*. It grows rapidly in the conditions optimal for yellowing the leaves, and appears to act as a weak pathogen of the senescent tissue (Hopkins, 1956). Losses are particularly associated with injury to the butts of leaves during handling when tying or stitching them to wooden laths for hanging in the barn (Gayed, 1972a, b). Yellowing conditions are close to the optimum for the organism regarding temperature (Weimer and Harter, 1923) and cell sap stimulates spore germination (Gayed, 1972b). The fungus causes a rot which spreads out from the midrib causing the tissue to soften and discolour (Fig. 2). The mechanism involved has not been studied, but it is presumed that pectolytic enzymes are involved in the rotting action, whereas the brown colour is probably caused by polyphenol oxidation following cell death and release of polyphenol oxidase. Affected leaves tend to disintegrate around the point at which they are tied to the lath, and they may fall to the barn floor. They are usually a complete loss as the rotted areas do not dry out and the leaves are of

Fig. 3. Fragment of flue-cured leaf. The dark-coloured upper fragment, with scattered growth of actinomycete colonies, contrasts with the lighter coloured and blemish-free normal leaf below.

unacceptable appearance. Control is largely dependent on minimizing mechanical injury to the leaf, although it may be partially controlled by avoiding very high humidities in the yellowing phase (Reinecke, 1962). Some success has been reported by using dichloran(2,6-dichloro-4-nitroaniline) either sprayed on the leaves or for treating the twine used to tie them (Cole, 1975; Lucas, 1975).

Mechanical injury is also considered an important factor in development of a bacterial decay associated with modern developments in harvesting and flue-curing. Mechanical harvesters cause more leaf breakage than hand picking and, combined with curing leaves in boxes instead of tied to laths, this can lead to extensive rotting if there is surplus moisture present on the leaves or if they are packed too tightly (Watkins, 1976b). A similar problem can occur in bulk-curing barns, in which the leaves are held in racks (Watkins, 1976a). Apart from injury, this type of spoilage is probably dependent on the presence of free water on the leaf surface at the time the barns are loaded. When combined

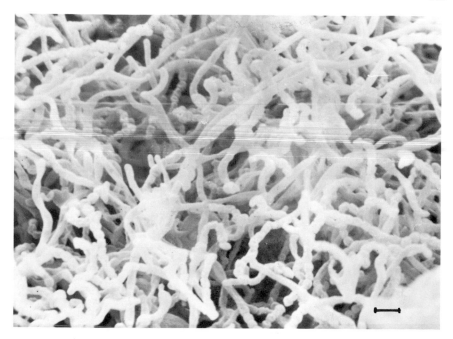

Fig. 4. Scanning electron micrograph of actinomycete growth on the tobacco leaf shown in Figure 3. The bar indicates a distance of 2.0 μm.

with close packing in the barn, it may be difficult or impossible to remove the excess water and induce wilting in the early phases of the curing cycle, and the conditions are then very favourable for rapid bacterial growth.

Brown-spot damage by *Alternaria alternata* has been shown to increase in the yellowing phase of curing (Main and Chaplin, 1969), but the importance of field infection is shown by an optimum temperature for initiation of infection of about 21°C (Spurr, 1973) against the 35°C in the yellowing phase of curing.

Damage by actinomycetes in flue-curing is generally of only minor significance (Gayed, 1978) because their occurrence and growth are associated, not with extensive rotting, but with superficial disfigurement which may be more important as an indicator of poor curing practice (Figs. 3 and 4). Growth of actinomycetes on leaves in flue-curing is accompanied by an increase in numbers of bacteria, whereas fungi show neither quantitative nor qualitative changes when compared with

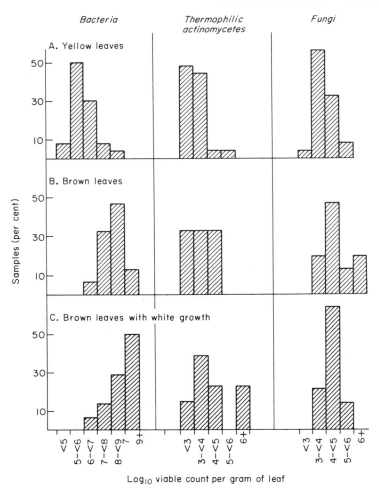

Fig. 5. Microbial population of sound leaves and development of superficial actinomycete growth in flue-curing of tobacco.

unaffected leaves (Fig. 5). The actinomycetes involved in this type of deterioration are of mixed types from their appearance when examined *in situ*. At least some are thermophilic, such as *Thermoactinomyces vulgaris*, which is a useful indicator organism as it can be recovered and counted selectively in the presence of a large excess of bacteria and fungi (Cross, 1968; Mitchell and Stauber, 1975). It is clear that conditions leading to growth of bacteria and actinomycetes in flue-curing are different from those giving rise to pole-rot. Observation has indicated that the problem occurs mainly with sand leaves (i.e. from the base of the plant), and is associated with long yellowing periods and overpacking of barns. The need for long yellowing times is particularly associated with picking leaf before it is fully ripe, and the low dry-matter content of sand leaves encourages overpacking of barns in an effort to lower curing costs. Evidence that thermophilic actinomycetes grow under these conditions suggests that the temperature, at least within the tightly packed bunches of leaf, rises considerably above the nominal 35°C–40°C for yellowing.

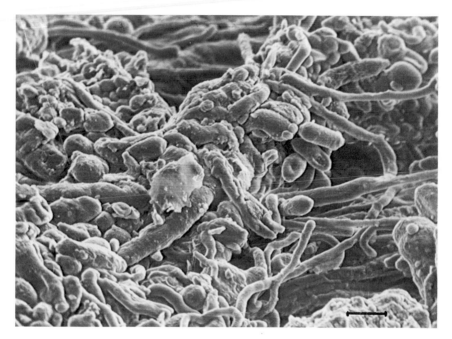

Fig. 6. Scanning electron micrograph of growth of *Cladosporium* sp. on the surface of dark air-cured tobacco leaf. The bar indicates a distance of 10 μm.

In air-curing, there should be a long slow decline in the moisture content of the leaf, and any interference with this through development of periods of high humidity may allow some mould growth to occur. As a result, it is not uncommon to find small amounts of superficial mould growth on otherwise sound air-cured tobaccos, particularly in the butts and along the midrib. The moulds which arise in this way range from dematiaceous types such as species of *Alternaria* and *Clado-sporium* (Fig. 6), continuing to grow from their presence as phylloplane fungi in the field, to the saprophytic fungi of stored commodities typified by *Aspergillus* and *Penicillium* species, along with a variety of other hyphomycetes. Such superficial fungal growths are important in causing a loss in market value of the affected leaf, off-odours and changes in chemical composition (Table 3), but seldom cause complete rotting of the leaf structure. Changes associated with mould growth, shown in Table 3, would all tend to reflect a loss in quality of the leaf. Recently, the occurrence of sterols not associated with tobacco has been found to be associated with mould growth (Tancogne, 1977).

Table 3

Effect of mould growth, the dominant type being *Cladosporium* sp., on composition of dark air-cured tobacco

	Composition (per cent dry weight)			
Sample	Nicotine	Total sugars	Reducing sugars	Protein nitrogen
Affected by mould growth (see Fig. 6)	1.7	2.1	1.5	1.3
Normal	2.4	6.6	6.0	1.0

Stem and stalk rot of air-cured tobaccos were found by Anderson (1948) to occur mainly as a continuation of infection initiated before harvest. The fungus *Sclerotinia sclerotiorum* causes rotting of the main stem (stalk) of the plant, and this may cause the leaves to fall when curing stalk-cut tobacco. Subsequently, the infection may spread into the midribs of individual leaves, causing a rot to develop even when primed from the plants.

Botrytis cinerea can be considered a weak pathogen associated with high humidities after harvesting. Infection can be initiated by dead blossoms falling on leaves (Anderson, 1948), or it may develop if freshly harvested leaves are left in piles for too long before hanging in the curing shed. It causes a rapidly spreading rot of the leaf tissue.

IV. MARKETING, STORAGE AND SHIPPING

After curing, the spoilage pattern of tobacco is dominated by the storage moulds characteristic of many commodities held at restricted moisture levels for long periods. Despite steps such as redrying to obtain an even and safe moisture level, damage can occur as a result of deliberate or accidental addition of water at the farm, market, store or in shipping, while sweating, condensation and storage in high humidity climates can all cause the moisture level to increase above the critical equilibrium relative humidity value for mould growth.

Existing evidence suggests that the critical equilibrium relative humidity value for mould growth to occur on tobacco is similar to that found for other substrates, i.e. about 70%, but it is not easy to relate experimental results from laboratory studies (Welty and Nelson, 1971; Welty and Weeks, 1975) to practical problems. This is because, in the packages of leaf used in commerce, normally of 250 kg or more in weight, moisture contents of 17–20% (corresponding to an equilibrium relative humidity value of 67–77%) may cause sweating to ensue which, in its turn, will force a redistribution of water leading to patches of much higher moisture content. Mould growth will then occur in areas of high moisture content much more rapidly than predicted from the original equilibrium relative humidity value of the tobacco.

There have been several studies of the microflora of mould-damaged cured tobacco, including some that have compared it with the same leaf type unaffected by mould. Species of *Aspergillus* and *Penicillium* are

Table 4

Changes in the mycoflora of flue-cured leaf as a result of threshing and redrying

Stage of process	Total viable count of fungi (per g of tobacco)	Percentage of recovered mycoflora identified as:			
		Aspergillus glaucus group	*Aspergillus niger*	Other storage fungi[a]	Field fungi[b]
Sound cured leaf, ex market	3,890	9	7	0.5	83.5
After desanding	4,074	33	25	0.5	41.5
After moisture addition ('ordering')	5,129	20	33	0.1	47
After threshing	10,470	48	37	0.6	14.4
After redrying	6,918	47.9	45	7	0.1

[a] This group comprises *Aspergillus flavus*, *A. tamarii*, *A. ochraceus*, *A. nidulans* and *Penicillium* species.
[b] This group includes *Alternaria* spp., *Phoma insidiosa*, *Chaetomium* spp., *Trichoderma viride*, *Myrothecium verrucaria*, and species of *Mucor*, *Scopulariopsis*, *Syncephalastrum* and *Absidia*.

usually predominant by incidence or by plate count in damaged flue-cured leaf (Welty and Lucas, 1969; Florczak, 1971; Maeda *et al.*, 1971), and this contrasts with the incidence of *Aspergillus* spp. before curing when it is difficult to recover significant numbers (Welty *et al.*, 1968). In comparison, undamaged cured leaf from commercial sources is frequently dominated by *Aspergillus* species, reflecting the cross con-tamination that can occur in sorting and redrying (Table 4). The diminution in field fungi, such as *Alternaria* spp., as a result of processing is due to a combination of physical removal along with surface sand and dirt, and probably destruction by heat. Counts of individual species of *Aspergillus* made on commercial tobaccos in this way can give a false impression of their significance as deteriogens. *Aspergillus niger* has a higher relative humidity requirement for germination and growth than species of the *A. glaucus* complex (Ayerst, 1969; Pitt and Christian, 1968), and this is reflected by the latter being associated with more incidents of tobacco spoilage than the former. In one examination of the causes of spoilage of individual samples of flue-cured leaf at a redrying plant in the U.S.A. (T. G. Mitchell, unpublished observations), direct study indicated 78% to be due to the *A. glaucus* group, 19% to *A. niger* and 3% to other fungi.

Damage due to mixtures of *Aspergillus* and *Penicillium* species is more common in piles of tobacco at marketing, in which small parts of the bulk may be of much higher moisture content than the remainder but, in boxes of redried leaf, growth of fungi is usually limited to species of the *A. glaucus* complex, particularly *A. repens*. The regular occurrence of the latter is to be expected, but it is more difficult to explain the rarity of other fungi able to grow under limiting-moisture conditions. For example, Pitt and Christian (1968) found *Xeromyces bisporus, Chryso-sporium fastidium* and *C. xerophilum* as significant parts of the spoilage flora of prunes, and have shown that these are capable of growth at lower water-activity levels than members of the *A. glaucus* group. There appear to be no records of their isolation from spoiled tobacco. At the present time, it remains uncertain whether this reflects different methods of examination and isolation or aspects of the chemical com-position or pH value of the substrate (Pitt and Hocking, 1977). Similarly, *Aspergillus restrictus* does not appear to be an important deteriogen of tobacco despite its known ability to grow under moisture-limiting conditions (Raper and Fennell, 1965). In our own studies, use of malt-salt agar medium, which is satisfactory for growth of this mould, has resulted in only occasional isolation of the species from either damaged or sound, cured tobaccos. Possibly its occurrence as a deteriogen in other situations is a reflection of some selective property of the substrate or its alleged resistance to antifungal agents (Smith, 1969) which are not routinely used on tobacco. In contrast with *A. restrictus*, we have found the related *A. penicilloides* to occur regularly on some cigar (air-cured) tobaccos, and on products made from them, in the absence of significant populations of the *A. glaucus* type. Laboratory studies indicate an optimum temperature for *A. penicilloides* of about 30°C compared with 20°C–25°C for *A. restrictus*, and an ability to grow in the presence of higher concentrations of salt. These properties would, on the one hand, favour its growth on commercial tobaccos originating from the tropics and subtropics, while the possibility of it growing at lower relative-humidity values than the common *A. glaucus* types could be important in allowing this slow-growing fungus to develop.

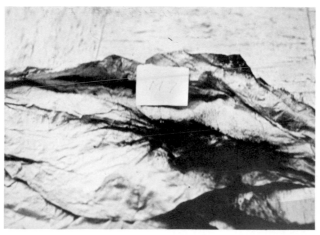

Fig. 7. Growth of *Aspergillus niger* and rot development along the midrib of a flue-cured tobacco leaf from the auction floor.

At relative humidity values of 80% or more, it is not clear why *A. niger* (Fig. 7) should occur as the major spoilage fungus. *Aspergillus flavus*, *A. ochraceus* and *Penicillium* spp. are found frequently and in association with *A. niger*, but it is the last species which is usually predominant. Whether small differences in moisture tolerance are important, or if selection is dependent on some physiological property allowing *A. niger* a competitive advantage, are points that require clarification.

Fig. 8. Corner of a case of flue-cured tobacco in which heating and mould growth occurred during shipping. The tobacco has caked together in layers. The white areas are caused by superficial growth of *Scopulariopsis* sp.

Types of *Scopulariopsis* spp. can be recovered from cured tobaccos, particularly air-cured varieties, and there are indications that they grow on the leaf during the air-curing process. In addition, they can form significant growths on tobacco at high moisture contents (> 30%) under some circumstances (Fig. 8). In the example shown, growth occurred on the faces of cases of flue-cured leaf during shipping. Growth was associated with considerable heating of the tobacco bales, causing moisture migration and condensation on the outer wrappings in contact with the bale surface. When tobacco 'heats' or 'ferments' in this way, considerable changes in nitrogenous constituents can occur (Frankenburg, 1950) with evolution of ammonia and other volatile compounds. Under such conditions, the tolerance of *Scopulariopsis* sp. to ammonia (Bothast *et al.*, 1975) may be an important factor in its selective colonization of the leaf. In the example shown, major changes occurred in the chemical composition of the damaged tobacco (Table 5); it was darkened and had an obnoxious odour, but these changes were only partly due to the *Scopulariopsis* sp.

V. MANUFACTURE

Under modern conditions, few opportunities exist for microbial activity to occur on tobacco during processing, as it is normally held at elevated moisture levels for only a few hours. Exceptions to this occur in the processing of stems (midribs) which are very hard and require higher moisture levels for cutting, traditional fermentation and reconstitution of waste tobacco into sheet form.

Table 5

Changes in composition of flue-cured leaf associated with growth of *Scopulariopsis* sp.

| | Percentage content on dry-weight basis in | |
Component	Undamaged leaf	Leaf with growth of *Scopulariopsis* sp.
Nicotine	1.35	1.47
Total sugars	8.0	0.6
Reducing sugars	7.2	0.5
Total free amino acids	1.5	0.12
Aspartic acid	0.16	0.05
Asparagine	0.41	0.001
Glutamine	0.12	0.001
Proline	0.37	0.008

Use of moisture levels of 30% or more in stem-processing leads to an abundance of water on the surfaces of the equipment. This encourages growth of yeasts and bacteria which may be transferred to the stems in large numbers unless the surfaces are regularly cleaned. The presence of relatively large inocula of yeasts on the stems will cause a decrease in sugar content, and development of off-odours (winey-stem) in bulks held at elevated moisture levels for 18 hours or more.

Traditional tobacco fermentations are applied particularly to cigar tobaccos. The agents involved have been the subject of considerable debate, particularly regarding the role of micro-organisms. The work of Frankenburg (1950) identified the changes undergone during fermentation, while his listing of the various moisture levels and other conditions involved indicates that the significance of leaf enzyme or microbial action will vary with the type of fermentation employed.

Microbial spoilage of fermentations can arise in two distinct ways. Control of large bulks of fermenting tobacco is effected by remaking them at intervals to prevent an excessive rise in temperature. Failure to judge events correctly can lead to rotting in the centre of the bulks due to the action of thermophilic cellulolytic clostridia (Reid *et al.*, 1938b). Alternatively, it is possible for mould growth to develop on and in the superficial parts of the bulk if the 'fermentation' is slow. The cause of this may be a lack of enzymic activity in the leaf or absence of the appropriate micro-organisms to initiate the fermentation. Some superficial mould growth is not uncommon on bulks of fermenting tobacco, but unrestrained growth of moulds is undesirable since even-

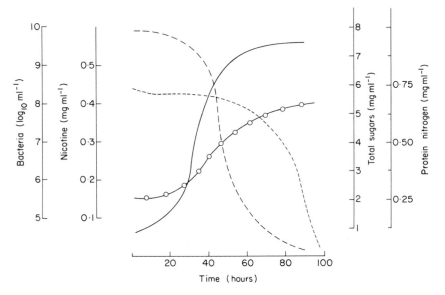

Fig. 9. Changes in the composition of an aqueous tobacco slurry (5% solids from a flue-cured waste) held at 22°C, stirred and aerated. ——— indicates changes in the content of bacteria, ———————— of nicotine, ———————— of total sugars and O—O—O—O of protein nitrogen.

tually it leads to development of off-odours (Reid *et al.*, 1938a) and a deterioration in the physical properties of the leaf. 'Natural' control of this is probably normally due to a combination of volatiles from the tobacco and the temperature rise. For cigar tobaccos, the presence of significant populations of bacteria may be positively correlated with good quality (Reid *et al.*, 1937; English *et al.*, 1967), whereas fungi are associated with poor-quality tobaccos.

Forced or chamber fermentations have been introduced to attain the same effects as traditional methods in a shorter time. These can also be subject to undesirable microbial activity as Corbaz (1974) has reported growth of thermophilic actinomycetes on poorly controlled chamber fermentations of air-cured leaf. The actinomycetes identified were *Micropolyspora faeni*, *Thermomonospora viridis* and *Thermoactinomyces vulgaris*.

Reconstitution of tobacco into sheet form involves preparing a slurry of the tobacco in water and casting it in the form of a sheet on a steel band or forming a paper structure on a wire. During preparation and casting of the sheet, the aqueous tobacco slurry is at a relative humidity close to 100%. Any delay in processing will permit microbial growth resulting in progressive modification of tobacco constituents (Fig. 9). This should not be a problem in a properly regulated plant, but spoilage can also occur after development of wet spots due to uneven drying. In addition, some users and uses require the sheet to be at a higher moisture content than is normally safe, and the *A. glaucus* group of fungi may then grow. Antifungal agents, such as thiabendazole (Moshy *et al.*, 1968) or sorbic acid (Harvey *et al.*, 1975), are incorporated in the sheet when required to prevent mould growth.

VI. DISTRIBUTION

Problems of microbial biodeterioration in tobacco products are rare, because cigarettes and cigars are usually packed at a safe moisture content. The packing normally employed does not provide a complete moisture barrier, but more problems of loss of quality occur through loss of moisture than of water uptake leading to mould growth. In most instances, uptake of atmospheric water would lead to an unacceptable product through, for example, staining of the paper, before mould growth had developed.

Occasionally, growth of *A. glaucus* types or *A. penicilloides* may occur on cigars through uneven distribution of moisture in the various components during manufacture. Equilibration may occur only slowly, and small patches of mould growth can develop on the cigar surface after packing. A related problem from contaminated glue leading to mould growth on cigars was described by True (1914). Uneven distribution of moisture in the product can also lead to mould growth on tobacco

Fig. 10. Growth of a member of the *Aspergillus glaucus* group on pigtail tobacco after packing, showing that growth is limited to overlapping regions where a high content of moisture has been trapped in the 'wrapper'.

products of minor importance today, such as plug and twist (Fig. 10).

Pipe tobaccos are usually packed at relatively high moisture contents (e.g. 18–25%), although there will be considerable variation in the equilibrium relative humidity value between products at the same moisture level depending on the extent to which sugars and humectants are added during manufacture. However, many are at risk from mould growth so that preservatives, such as benzoate and sorbate, are employed routinely to prevent this. In come cases, vacuum-sealed tins have been employed for pipe tobaccos to prevent mould growth, but failures have occurred which in some cases were probably due to yeast action.

VII. ECONOMIC ASPECTS

No information is available on average losses due to microbial deterioration of tobacco. They are probably less than 1% of production based on weight, but its significance is greater than this value suggests.

The total monetary value of losses, even at a fraction of 1%, is considerable as it has been estimated that about a quarter of the world tobacco-leaf production enters international trade, and that export value including manufactured goods amounts to 2.5% of World trade (United Nations Conference on Trade and Development, 1978). However, its importance lies less in any gross figure for value of goods lost than in its occurrence as major incidents, involving significant charges on the profitability of an undertaking, whether farm or manufacturer.

Losses at the farm level are frequently associated with climate. In air-curing, the unpredictability of weather may make it problematical whether to use any heating except in special cases such as cigar-wrapper leaf. In 1947, it was estimated that the loss from pole rot in Connecticut, U.S.A. was U.S. $1.5 million, representing 10% of the value of the cigar-filler crop (Anderson, 1948). Bacterial barn rot in flue-curing in the U.S.A. has been reported to cause losses of up to 5,000 kg for some growers, but little or none to others (Watkins, 1976b). At 1977 auction prices in the U.S.A., this would represent a loss of about U.S. $12,000 to a farmer. Yearly losses from pole rot during flue-curing in Canada have been estimated to average $1–3 million over the period 1973–8 (Gayed, 1978).

Losses in the period between curing and redrying vary with the country, depending on climate. In countries with high humidities and temperatures at the time when leaf is sold by farmers, it is necessary to arrange for rapid sorting and redrying of the leaf to prevent mould growth. This usually means that the equipment can be used only for part of the year, and costs are correspondingly higher than if the processing could take place through the entire year. Cool storage facilities and vacuum-cooling units are sometimes employed as adjuncts at this stage to spread the season and minimize damage. In Rhodesia, losses at the time of marketing have been estimated at U.S. $2–3 million in 1977 (Davis, 1977). Although many of the losses and extra costs at marketing are borne by processors, those associated with changes in grading and valuation (Everette et al., 1972; Watkins, 1976a) represent a considerable loss to the farmer.

When mould growth develops in storage or shipping, a very large volume of leaf may be involved as the same defect is likely to affect the whole batch. Direct losses can then amount to U.S. $200,000 or much more, but to this has to be added the considerable costs of assessment, separating and recovering unaffected material and interruptions to production. Serious losses in the distributed product are rare, although inadequate packaging can result in mould growth on the product particularly in humid tropical climates. The repercussions are much greater than the cost of materials lost, as there is a considerable short-term risk of loss of market share, production problems and potential loss of confidence in a branded product.

VIII. ACKNOWLEDGEMENTS

The authors are most grateful to Mr. F. Hunt for examining samples by scanning electron microscopy and the provision of the photographs used in this article.

REFERENCES

Akehurst, B. C. (1968). 'Tobacco'. Longmans, London.
Anderson, P. J. (1948). *Connecticut Agricultural Experimental Station Bulletin No. 517.*
Ayerst, G. (1969). *Journal of Stored Products Research* **5**, 127.
Bacot, A. M. (1960). *The Chemical Composition of Representative Grades of the 1952 and 1954 Crops of Flue-Cured Tobacco.* Technical Bulletin No. 1225. U.S.D.A., Washington.
Bemelmans, J. M. H. and Ten Noever de Brauw, M. C. (1974). *Science of the Total Environment* **3**, 126.
Bothast, R. J., Lancaster, E. B. and Hesseltine, C. W. (1975). *European Journal of Applied Microbiology* **1**, 55.
Cole, J. S. (1975). *Rhodesian Journal of Agricultural Research* **13**, 15.
Corbaz, R. (1974). *Annales du Tabac, Section 2* **11**, 217.
Cserjesi, A. J. and Johnson, E. L. (1972). *Canadian Journal of Microbiology* **18**, 45.
Cross, T. (1968). *Journal of Applied Bacteriology* **31**, 36.
Curtis, R. F., Dennis, C., Gee, J. M., Gee, M. G., Griffiths, N. M., Land, D. G., Peel, J. L. and Robinson, D. (1974). *Journal of the Science of Food and Agriculture* **25**, 811.
Davis, P. A. (1977). *Tobacco Today* **5** (13), 18.
Durán-Grande, M. (1974). *Microbiología Española* **27**, 103.
English, C. F., Bell, E. J. and Berger, A. J. (1967). *Applied Microbiology* **15**, 117.
Everette, G. A., Davis, D. L. and Teague, G. S. (1972). *University of Kentucky College of Agriculture Extension Service Leaflet* AGR-9.
Florczak, K. (1971). *Biuletyn Centralnego Laboratorium Przemyslu Tytoniowego* (1–2), 73.
Frankenburg, W. G. (1950). *Advances in Enzymology* **10**, 325.
Gayed, S. K. (1972a). *Canadian Journal of Plant Science* **52**, 103.
Gayed, S. K. (1972b). *The Lighter* **42** (3), 29.
Gayed, S. K. (1978). *The Lighter* **48** (4), 16.
Hartill, W. F. T. (1967). *Rhodesia, Zambia and Malawi Journal of Agricultural Research* **5**, 61.

Hartill, W. F. T. (1969). *New Zealand Tobacco Growers Journal (November)*, 21.

Harvey, W. R., Baker, P. G. and Fountaine, W. B. (1975). *Tobacco Science* **19**, 133.

Holdeman, Q. L. and Burkholder, W. H. (1956). *Phytopathology* **46**, 69.

Hopkins, J. C. F. (1956). *Tobacco Diseases, with Special Reference to Africa*. Commonwealth Mycological Institute, Kew.

Johnson, J. (1924). *Tobacco Diseases and their Control*. U.S.D.A. Bulletin No. 1256.

Kaminski, E. J., Lazanas, J. C., Wolfson, L. L., Fancher, O. E. and Calandra, J. C. (1970). *Beiträge zur Tabakforschung* **5**, 189.

Keller, C. J., Bush, L. P. and Grunwald, C. (1969). *Journal of Agricultural and Food Chemistry* **17**, 331.

Llanos, M. C. (1978). *CORESTA Information Bulletin* (1), 24.

Lucas, G. B. (1975). *Diseases of Tobacco*, 3rd edition. Biological Consulting Associates, Raleigh, U.S.A.

Lucas, G. B. and Pounds, J. R. (1973). *Tobacco Science* **17**, 167.

Maeda, S., Shiroki, C. and Udagawa, S. (1971). *Bulletin of the Hatano Tobacco Experimental Station* (70), 133.

Main, C. E. and Chaplin, J. F. (1969). *Tobacco Science* **13**, 17.

Mitchell, T. G. and Stauber, P. C. (1975) *In* 'Microbial Aspects of the Deterioration of Materials' (D. W. Lovelock and R. J. Gilbert, eds.), pp. 203–211. Academic Press, London, New York and San Francisco.

Moshy, R. J., Fiore, J. V. and Halter, H. M. (1968). *Abstracts 22nd. Tobacco Chemists Research Conference*, Richmond, U.S.A.

Pattee, H. E. (1969). *Applied Microbiology* **18**, 952.

Pitt, J. I. and Christian, J. H. B. (1968). *Applied Microbiology* **16**, 1853.

Pitt, J. I. and Hocking, A. D. (1977). *Journal of General Microbiology* **101**, 35.

Raper, K. B. and Fennell, D. I. (1965). 'The Genus Aspergillus'. Williams and Wilkins, Baltimore.

Reid, J. J., Haley, D. E., McKinstry, D. W. and Surmatis, J. D. (1937). *Journal of Bacteriology* **34**, 460.

Reid, J. J., McKinstry, D. W. and Haley, D. E. (1938a). *Pennsylvania Agricultural Experimental Station Bulletin* No. 356.

Reid, J. J., McKinstry, D. W. and Haley, D. E. (1938b). *Pennsylvania Agricultural Experimental Station Bulletin* No. 363.

Reinecke, J. (1962). *Farming in South Africa* **38**, 41.

Smith, G. (1969). 'An Introduction to Industrial Mycology', 6th edition. Edward Arnold, London.

Snow, J. P., Lucas, G. B., Harvan, D., Pero, R. W. and Owens, R. G. (1972). *Applied Microbiology* **24**, 34.

Spurr, H. W. (1973). *Tobacco Science* **17**, 145.

Stephen, R. C. (1955). *Rhodesian Tobacco* (11), 5.

Tancogne, J. (1977). *Annales du Tabac, Section 2* **14**, 197.

Todd, F. A. (1961). *Plant Disease Reporter* **45**, 319.

True, R. H. (1914). *The Molds of Cigars and Their Prevention*. U.S.D.A. Bulletin No. 109.

Tso, T. C. and Sorokin, T. (1967). *Beiträge zur Tabakforschung* **4**, 18.

United Nations Conference on Trade and Development (1978). Marketing and Distribution of Tobacco. United Nations, Geneva.

Watkins, R. (1976a). *Flue-Cured Tobacco Farmer* **12** (6), 14.

Watkins, R. (1976b). *In* '1977 Tobacco Information', pp. 60–63 (by North Carolina Agricultural Extension Service). N.C. State University, U.S.D.A. Raleigh.

Weimer, J. L. and Harter, L. L. (1923). *Journal of Agricultural Research* **24**, 1.
Welty, R. E. and Lucas, G. B. (1969). *Applied Microbiology* **17**, 360.
Welty, R. E. and Nelson, L. A. (1971). *Applied Microbiology* **21**, 854.
Welty, R. E. and Vickroy, D. G. (1975). *Beiträge zur Tabakforschung* **8**, 102.
Welty, R. E. and Weeks, W. W. (1975). *Tobacco Science* **19**, 77.
Welty, R. E., Lucas, G. B., Fletcher, J. T. and Young, H. (1968). *Applied Microbiology*
    16, 1309.
Wolf, F. A. (1967). *In* 'Tobacco and Tobacco Smoke' (E. L. Wynder and D. Hoffman,
    eds.), pp. 5–45. Academic Press, New York and London.

Note Added in Proof

Since preparing this contribution, two publications have appeared which have extended our knowledge of tobacco biodeterioration.

Gayed (1978) identified *Botrytis cinerea, Rhizopus reflexus, Alternaria alternata* and *Fusarium tricinctum* as being the most common fungi associated with affected leaf. The most virulent in attacking uninjured leaf in laboratory experiments was *Botrytis cinerea*.

A variety of themophilic fungi have been isolated from cured tobacco leaves and cigarettes in Nigeria by Ogundero (1979). They were considered to be responsible for spoilage of tobacco products in Nigeria, but it is not clear under which conditions losses arise from this group of fungi.

In our own laboratory, it has now become clear that *Xeromyces bisporus* does occur on tobacco and is capable of growing on it under appropriate conditions. Use of the glucose–fructose medium of Pitt and Hocking (1977) led to its recovery from mould-damaged tobacco from which no fungi had been isolated using standard media such as malt salt agar, variants of Czapek-Dox agar and potato dextrose agar. Thus the infrequent recognition of this fungus is probably due to a combination of its requiring a complex medium containing high concentrations of sugar and the ease with which it is overgrown in the presence of other 'storage' moulds.

References

Gayed, S. K. (1978). *Canadian Plant Disease Survey* **58**, 104.
Ogundero, V. W. (1979). *Mycopathologia* **69**, 131.
Pitt, J. I. and Hocking, A. D. (1977). *Journal of General Microbiology* **101**, 35.

9. Fuels and Oils

C. GENNER and E. C. HILL

Department of Microbiology, University College of Wales, Cardiff, Wales, U.K.

I. Introduction	260
A. High Volume Ratio of Hydrocarbon to Water	260
B. High Ratio of Water to Hydrocarbon	260
II. Fuels	261
A. Conditions for Microbial Growth	261
B. Types of Fuels Involved	261
C. Consequences of Microbial Growth	263
D. Growth Associated with Subsonic Aircraft	264
E. Fuel Baffles	269
F. Supersonic Aircraft	269
G. Detection of Microbial Infections in Fuel Systems	271
H. Antimicrobial Procedures	272
I. Infections in Light Fuel Oils	278
J. Natural History of *Cladosporium resinae*	280
III. Lubricating and Hydraulic Oils	280
A. Introduction	280
B. Characteristics of Oils Involved	281
C. Consequences of Microbial Growth	282
IV. Soluble Oil Emulsions	288
A. Ecological Aspects of 'Soluble' Oil Emulsions in Machine Tools	288
B. The Spoilage Process	291
C. Biocides in Cutting Oils	292
D. On-Site Test Methods	293
E. Physical Methods of Controlling Micro-Organisms	293
F. Rolling Mill Emulsions	294
G. Other Oil-in-Water Emulsions	294
H. Health Aspects of Emulsion Spoilage	295
V. Water-in-Oil Emulsions	300
VI. Conclusions	300
References	301

and these spores can remain viable for several months (Thomas, 1973). Spores of *C. resinae* and *Aspergillus fumigatus* can remain viable after exposure to 80°C or −32°C (Thomas and Hill, 1977a). Therefore the conditions encountered during flight, i.e. cooling to 0°C or lower, do not seriously affect the viability of fungal spores suspended in the fuel.

As mentioned above, problems caused by microbial growth in fuel are many. Those caused by the physical presence of organisms are fairly self evident. However, the contribution of the organisms to corrosion has been greatly researched in the last 15 years. Gorog *et al.* (1970) reported the main mechanisms of microbial corrosion to be as follows.

(1) Corrosive chemicals produced, e.g. carbon dioxide, hydrogen sulphide, sulphuric acid, ammonium and organic acids.

(2) Electrochemical corrosion, due to changes in oxidation–reduction equilibrium, i.e. depolarization.

(3) Indirect attack, due to deposit formation whereby surface aeration is different at various points, i.e. differential aeration-cell formation.

(4) Indirect attack through the action of microbes on protective coatings.

Microbial corrosion of aluminium fuel tanks generally results in formation of pits in the metal surface. This pitting can eventually result in perforation of the tank wall. The pits are usually found beneath mats of microbial material (Hendey, 1964). During microbial growth and corrosion, increased concentrations of aluminium may be found in the water phase in contact with the metal, and at the water–fuel interface (Hedrick *et al.*, 1964). Parberry (1968) found that percentage weight loss due to corrosion in the aluminium could be correlated with growth of *C. resinae*, and the drop in pH value of the aqueous phase. Miller *et al.* (1964) presented evidence which supported the theory that variations in potential are important in corrosion of aluminium. They measured the voltage between areas covered by microbial film and those where the film was absent, and found potential differences of up to 60 mV. Various mineral elements affect corrosion of aluminium, e.g. chlorine, iron and calcium increase corrosion (Blanchard and Goucher, 1965), whereas borate, phosphate, nitrite and nitrates inhibit corrosion (Rowe, 1957). Microbes can increase corrosion by removing corrosion-inhibiting components, such as phosphate and nitrate, but leaving the corrosive elements.

The materials used in many aircraft tanks may be alloys of aluminium and metals such as magnesium and copper. Evidence suggests that micro-organisms can use the magnesium from these alloys thereby increasing corrosion and leaving the alloy depleted in magnesium (Hedrick et al., 1967). The metal can be treated and made more corrosion resistant, but this treatment results in decreased structural strength.

E. Fuel Baffles

To help minimize movement of fuel in the fuel tanks (particularly when they are only part full), baffling systems are sometimes employed. These must be made of a light-weight material which has structural strength, and must resist microbial attack. Substances used for this application include polyurethane foam and polyesterurethane foam. However, Hedrick and Crum (1968) reported that polyesterurethane foam gave increased microbial cell counts and oxygen uptake when incubated with a bacterial isolate. When exposed to growth of a fungal isolate, extensive matting occurred and the tensile strength of the foam was greatly lowered. Similar results were found with polyurethane foam. They concluded that inclusion of baffles constructed of these materials would contribute to increasing activity of fuel-utilizing microbes.

F. Supersonic Aircraft

Although no serious incident of microbial growth and corrosion has been reported in supersonic aircraft-fuel systems, extensive research has been conducted on the susceptibility of these systems to microbial attack. As stated above, the fuel aboard subsonic aircraft is cooled during flight and may reach subzero temperatures. Microbial growth occurs in the aircraft during grounding. The microbial environment is different in supersonic aircraft, since the fuel in the integral wing tanks is heated by friction caused by passage of air over the wing surface during supersonic flight. The fuel is also used as a heat sink for some of the aircraft systems (see Fig. 2). The fuel may reach temperatures as high as 90°C to 100°C in some outboard tanks. In order to keep the aircraft trimmed to the correct altitude in flight, it is moved from tank to tank. In this way the fuel may at some times be at around 40°C. At this temperature the main problem organism in subsonic

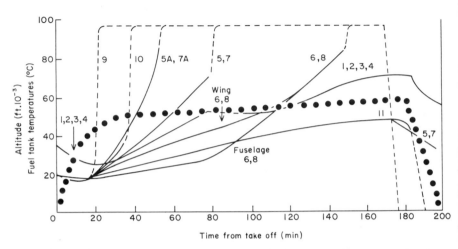

Fig. 2. Predicted temperatures in the various fuel tanks of a supersonic transport aircraft as a function of time from take-off (——, temperature in fuel-containing tanks; ----, temperature in empty tanks) and altitude (●). Tank 11 is the tail fuselage trim tank, and tanks 1, 2, 3 and 4 are collector tanks from which fuel is delivered to the engines. From Hill and Thomas (1975).

aircraft, *Cladosporium resinae*, cannot grow well. If growth does occur, *C. resinae* is a poor competitor to other organisms better suited to growth under these conditions. One such organism that has been found to grow well under simulated conditions similar to those found in these fuel tanks is an albino strain of *Aspergillus fumigatus* (Hill, 1971; Thomas and Hill, 1976a). This organism was able to withstand the high and low temperatures that might be encountered during operation of a supersonic aircraft and to accelerate corrosion (Thomas and Hill, 1976b). The organisms sporulate freely in a fuel-and-water medium and the spores can remain viable in dry aviation kerosine for prolonged periods (Thomas and Hill, 1977a). They can survive heating to 80°C and cooling to −32°C for periods long enough to cover a supersonic flight. When suspended in the water phase, the spores are less heat resistant and 60°C is probably the maximum temperature through which they can survive for prolonged periods (Upsher, 1976; Thomas and Hill, 1977a).

Scott and Forsyth (1976) carried out an extensive survey of the fuel pick-up points throughout the World, isolating and identifying the microbial contaminants present. They found that *Aspergillus fumigatus* was as common a contaminating organism as *Cladosporium resinae*. Therefore, it appears that supersonic aircraft taking on fuel at these pick-up points might be inoculated with some organisms capable of growth in their fuel tanks. Thus a potential exists for extensive microbial growth in supersonic aircraft. The threat has been met by meticulous attention to design, so that water cannot accumulate in the wing structure, and 'housekeeping' of a very high order. Should these measures be relaxed, we should have the unusual situation where microbes would grow in some parts of an aircraft fuel system in flight rather than when on the ground.

G. Detection of Microbial Infections in Fuel Systems

Microbial infections in aircraft fuels can be detected by close inspection of the inside of the fuel tanks for 'mat' production (Scott and Hill, 1971). Pipe or filter blockage also suggests that a microbial presence might be the cause. However, before these problems occur various tests can be used which will demonstrate microbial activity. One such test was reported by Russell et al. (1968) which involved the use of radio-active phosphorus (^{32}P). The test involved incubation of the fuel sample with $Na_3{}^{32}PO_4$ for about two hours, followed by membrane (Millipore) filtration of the fuel and subsequent estimation of the radioactivity retained on the filter. Micro-organisms rapidly take up the phosphate and so the radioactivity can be directly related to microbial presence on the filter. This test, however, could hardly be used for 'onsite' testing. Various other tests have been developed which are designed for field testing. One such test, reported by Hill (1970b), depends on detection of phosphatase secreted by fungi during metabolism. The rapid test may be successfully used by unskilled personnel to detect microbial activity before serious problems occur. It depends on obtaining a suitable water sample from below the fuel. Similarly, water bottoms can be tested using the various 'Dipslide' test kits at present available (Hill, 1975b; Genner, 1976). These use selective growth media which culture any organisms present and give a semi-quantitative estimate of the infection. There is a time-lag of 2–3 days between sampling and obtaining the result. These slides have been widely used in industry for detection of infections in various industrial fluids, and will be referred to in more detail below. An alternative approach, which has met with some success, is to analyse wing-drain samples for aluminium ions as an indirect indication of corrosion.

H. Antimicrobial Procedures

In order to prevent the adverse effects of microbial growth in aircraft fuel there are several lines of approach which may be used, namely good housekeeping practices, biocides to prevent growth and tank linings to prevent corrosion. The best approach appears to be a combination of all three approaches, good housekeeping and biocide use to minimize growth in the first place, and tank coatings to prevent corrosion if contamination and growth should occur.

1. Good Housekeeping

The first method for minimizing the incidence and effects of microbial activity is to apply constant vigilant administration of good housekeeping procedures. These primarily involve regular removal of free water from ground-storage tanks and aircraft-wing tanks thereby minimizing the likelihood of microbial growth becoming established (Lansdown, 1965). Mill scale and rust contamination in fuel, if transferred from storage tank to aircraft, will increase growth and corrosion. Most problems appear to occur at the smaller refuelling terminals where standards of fuel quality can fall unnoticed (Scott, 1971). Filtering the fuel during refuelling is the best method of preventing this contamination. However, even with the most vigilant care, water gets into aircraft fuel tanks by other methods already mentioned. This water must be regularly removed via the wing-drainage cocks. This may be difficult in many aircraft since water droplets, once formed, tend to stick to the wing surfaces even when these are downward facing, and they may not run to the drainage points. In new aircraft, particularly supersonic aircraft, however, the design of the integral fuel tanks is such that there are few corners or horizontal surfaces for water to lie; consequently, water removal is much more efficient than in older aircraft. Most fuelling installations incorporate coalescer units at fuel-transfer points to strip out free water.

Older ground-fuel installations were often underground and hence free water could never be entirely removed. Modern installations are above ground and incorporate inclined tanks to allow maximum water drainage. Attempts at water removal are, however, sometimes frustrated when infected fuel is handled, as the surface activity induced by microbial growth interferes with water separation. Although ample time for water settlement in ground tanks is advisable, it should never be prolonged to the extent that aerobic, followed by anaerobic, microbes can grow. Kerosine contains over 300 p.p.m. dissolved oxygen and if tank turnover is adequate, aerobic conditions are maintained (Scott and Hill, 1971). Prolonged stagnation allows the aerobic flora to deplete this oxygen, and then sulphate-reducing bacteria can flourish. Severe corrosion problems are associated with fuel containing sulphide.

Another area where vigilance can prevent extensive problems is in visual inspection of the aircraft tank interiors and fuel system. As soon as it is evident that an infection has occurred, corrective measures can immediately be implemented. Signs, such as the appearance of slime on surfaces, of damage to protective coatings, pitting and blistering, all indicate that corrective measures should begin. Fuel-gauge malfunction and excessive pump screen or filter plugging also indicate a developing microbial presence.

2. Biocides

Microbial growth in any medium may be prevented by the use of biocides. The choice of potential biocides for application to jet fuel is very limited due to the stringent specifications enforced by engine manufacturers. Nevertheless, many reports have been published which assess various biocides for this application. Rogers and Kaplan (1968) screened 97 water- and fuel-soluble materials, but concluded that the three classes of biocide already in use, namely various chromates, organoborates and ethylene glycol monomethyl ether (EGME) were the most acceptable and effective for this application. Park (1973) has categorized those properties of biocides important to the airframe manufacturer.

The use of chromates, which are water soluble and fuel insoluble, does have limitations. Strontium chromate is an effective biocide against various fungi as a saturated solution in water (Hedrick and Carroll, 1966). In addition, it is a recognized inhibitor of metal corrosion, being extensively incorporated into paints and rubber sealants used in aircraft tanks. However, strontium chromate is insoluble in fuel and, as within the fuel tanks water may lie in discrete droplets, it is difficult to envisage how all of the droplets within the tank can be treated with biocide. Rubidge (1975) gives evidence suggesting that the 'binding' chemicals used in conventional strontium chromate tablets may actually stimulate fungal growth. For a biocide to be really effective in aircraft tanks, it must be at least sparingly fuel soluble so that, once the fuel has been treated, all water droplets are exposed to the biocide. If the biocide has a high partition coefficient from fuel to water, all water droplets will contain an effective concentration of biocide (Hill, 1970a). Chromate biocides are still effectively used in storage tank bottoms where the water phase tends to form a complete layer over the tank bottom (London et al., 1965).

The other two classes of biocide mentioned earlier, namely organo-borates and ethylene glycol alkyl ethers, are sparingly soluble in fuel and freely soluble in water and have high partition coefficients from water to fuel. Both have been in use since about 1960. Originally, EGME was added to jet fuel as an anti-icing additive to prevent water droplets freezing in the fuel at the very low temperature attained during flight. Ice particles can cause fuel-flow problems just like microbial sludge.

After the American Air Force introduced EGME into their fuel in 1962, it was noted that storage-tank bottoms below this treated fuel were sterile. This finding prompted investigations which showed that 0.05% (by volume) in fuel was an effective antimicrobial concentration. The additive also contained 0.4% glycerol which increased the biocidal activity and also acted as a protectant for the tank lining materials (Hitzman et al., 1963). Since the additive had already been approved by engine and airframe manufacturers and the Federal Aeronautics Administration and found to be compatible with materials of aircraft construction, its immediate introduction as a simple means of microbe control was possible. Since 1962 it has been widely used by military air forces since its cost can be justified by its dual role as an anti-icing and biocidal additive.

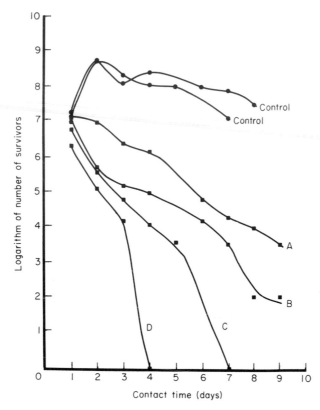

Fig. 3. Survival of bacteria with time in the aqueous phase of an aviation kerosine–aqueous mineral salts mixtures in the ratio 100:1. The kerosine contained ethylene glycol monomethyl ether at concentrations of 0.1 (A), 0.15 (B), 0.20 (C) and 0.25% (v/v) (D). From Hill (1970a).

Many reports have been published on the activity of EGME and 0.1%–0.15% is the most commonly recommended concentration in fuel (Dayal *et al.*, 1971) (see also Fig. 3). Hill (1970a) reported that if the concentration is allowed to fall much below 0.1% EGME can be actively growth promoting, and high volume ratios of water to fuel also lower its effectiveness. Thomas and Hill (1977b) showed that temperature plays an important part in the activity of biocides. At or below 4°C fungicidal activity of EGME is very low, but as the temperature rises, its activity is increased so that at the temperatures encountered in supersonic aircraft tanks its activity is greatly enhanced. However, Hitzman (1965) showed that the partition coefficient is also greatly affected by temperature. The coefficient of EGME at 18°C is 800:1 whereas at 27°C it is 200:1. Thus, at lower temperatures, EGME becomes more concentrated in the water phase, and hence its anti-icing properties are enhanced. However, the increase in biocidal activity with increased temperature compensates for this change in partition and the net result is an increase in antifungal activity as temperature rises (Thomas and Hill, 1977b).

Ethylene glycol alkyl ethers have been known to be bactericidal for many years (Berry and Michaels, 1950), and ethylene glycol methyl ether (EGME) is the form that has become widely accepted for application to jet fuels and it is now commonly used without the original glycerol. However, many reports have been published of work designed to find a more suitable anti-icing additive of the glycol ether type, with greater antimicrobial activity and less effect on the flash point of fuel. Neihof and Bailey (1978) reported that, of seven compounds tested, di-ethylene glycol monomethyl ether was the most suitable replacement since it was effective at 1–2% (in aqueous phase) compared with the 10–17% required for EGME. However, most of the other compounds tested had less favourable partition coefficients than EGME and it would be more difficult to ensure that the required anti-microbial concentration accumulated in the water phase. Despite their short-comings, in particular their ability to stimulate growth at low concentration, the ethylene glycols are widely accepted as they combust completely, are freely fuel miscible, are non-toxic and are readily available. There is some current concern that their antibacterial activity falls short of their antifungal activity.

The other biocidal compounds in wide use since the early 1960s are the organoborates. Borates have long been known to inhibit microbial and enzymic activity (Zittle, 1951) and this activity is greater for organic borate compounds. The activity of organoborates has been investigated by DeGray and Fitzgibbons (1966). Some organoborate compounds are known to precipitate boric acid when treated fuel comes into contact with water and this may cause filter plugging. One product which is widely used, namely Biobor-JF, was found by Saunders *et al.* (1966) not to cause this precipitation. The recommended concentration for use of Biobor-JF is between 135 and 270 p.p.m. (equivalent to 10–20 p.p.m. boron in fuel). Hostetler and Powers (1963) found this to be an effective concentration. In 1965, Boggs reported that, after 280 flight hours of testing in a Jet Star, the fuel tanks remained sterile and no adverse effects were observed on the engines, finishes, sealants or other equipment. He found that a 20 p.p.m. boron concentration used was distributed well throughout the fuel. Low concentrations do not stimulate microbial growth but a residue of ash persists after combustion. Biobor-JF is now widely used by civil airline operators to maintain acceptable microbial populations in their fuel tanks. Engine builders tend to allow 135 p.p.m. for continuous use and 270 p.p.m. for shock treatment. The antimicrobial activity of both EGME and Biobor-JF is, however, quite slow, a long exposure time being required for effective use. Even when used at 270 p.p.m. as a shock treatment, Biobor-JF must have a contact time of several days, and EGME must be used continuously as a constant component of the aircraft fuel. A more desirable method of controlling the problem would be an intermittent treatment at the regular servicing times, or as required, to combat any specific problems.

Recently some new biocides have been developed which are very effective in the fuel environment. One of the most biologically active is RH 886 (Rohm and Hass (U.K.) Ltd.) (Thomas and Hill, 1977b). When used in combination with EGME it was active even at very low concentrations (about 1 p.p.m.). The RH 886 biocide is based on thiazolin. Other compounds that have been tested for this kind of application include esters of hydroxybenzoic acid, derivatives of dioxane and various quaternary ammonium compounds. Quaternary ammonium compounds are particularly suitable, some resulting in sterilization after two hours exposure, and they also have good cleaning properties. However, they may lower the interfacial tension between fuel and water and compromise water separating equipment. In addition to the problem of finding a biocide with all the desirable biological properties, there are problems of acceptability by engine and airframe manufacturers fuel suppliers and aviation safety authorities.

3. Tank Coatings

One means of preventing the corrosion problems associated with microbial growth in aircraft fuel tanks is to treat the inside of the tank with a protective coating. Some form of coating is obligatory to prevent leaks. This approach was quite successful in prejet aircraft using aviation gasoline. The treatment generally consisted of an application of a thin coating of a protective sealant such as an epoxy- or polysulphide-based primer covered with a layer of elastomeric coating such as Buna-N (based on butadiene nitrile rubber). However, the situation is more difficult with jet aircraft using a kerosine-type fuel, since microbial growth is more prevalent in this type of fuel. Hazzard (1965) reported that hyphae of *C. resinae* can penetrate some of these protective coatings. In some cases, penetration occurred within four or five days of application. London *et al.* (1965) found that some tank coatings stimulate microbial growth as they may supply nutrients, in particular carbon and nitrogenous compounds. In fact, Kereluk and Baxter (1963) reported that a water extract of Buna-N provided a better growth medium than the Bushnell–Hass mineral salts medium.

Hazzard and Kuster (1963) studied several types of tank coating and concluded that epoxy and polyurethane coatings provided better protection against fungal attack than nitrile- or polysulphide-based coatings. Miller *et al.* (1964) also found that microbes could utilize Buna-N-type sealants as a nutrient source but found that polyurethane was a much poorer source of nutrients. However, they did demonstrate that microbes could break down polyurethane coatings even though they were much more resistant to attack than Buna-N, and operating experience with polyurethane confirms this. A survey of the coatings available was conducted by Boggs (1965).

To date, it appears that no wholly satisfactory tank lining material has been developed. Most of the materials available have inadequacies. The surface must be wholly covered, even a pin prick may be sufficient to allow corrosion to begin. Also, the coating may mask corrosion and make it more difficult to identify since corrosion can spread beneath the lining material. The incorporation of biocides into coatings appears to have been only partially successful (Scott, 1971).

I. Infections in Light Fuel Oils

It might be assumed that the large numbers of publications on microbial growth in aviation kerosine and the few publications on infections in other light fuel oils reflect the incidence of such occurrences. This is by no means so, and a comprehensive survey by Liggett (1976) of marine, rail and road diesel fuels, tractor vapourizing oil, paraffin, gas oils and central-heating fuels gave a clear indication of the widespread nature of the phenomenon. Unfortunately, the information is in a 'limited access' thesis. The consequences of these infections are of less obvious significance that those in aircraft, so there is no associated dramatic corrosion of aluminium alloys. However, there are associated system malfunctions, particularly filter plugging and lowering of interfacial tension between fuel and water, which Liggett (1976) has quantified. Hostetler and Powers (1963) have demonstrated that injector fouling in diesel locomotives can be attributed to infection of the fuel and they quote figures indicating that biocides minimize formation of 'gums', sediments and filter plugging in heating oils and rail and road vehicle fuels.

The technical innovation of using gas turbines as power units for naval vessels (and a few merchant vessels) has been accompanied by predictable microbial problems. It is impossible to keep sea water out of the fuel and, as the fuel is normally carried in the bottom of the ship, it is also impossible to drain this water completely. Indeed, in some vessels, the fuel is displaced by sea water as it is used, in order to maintain a consistent 'trim'. Even though the water is sea water, *Cladosporium resinae* is still the dominant infecting fungus; Klemme and Leonard (1971) have emphasized the ˙ignificance of sulphate-reducing bacteria. The fuel temperature reflects the external sea temperature and hence can be expected to vary from 4 to 25°C. The fuel is pumped from the main storage tanks to 'service' tanks and from there, via coalescers and filters, to the gas turbines. Functional problems occur when infection is heavy. Coalescers malfunction, filters plug, and 'slugs' of water activate engine shut-down devices. The high ratio of water to fuel is a factor in planning biocide regimes. As EGME has been largely unsuccessful (and even stimulatory occasionally), organoborates are most commonly used. The latter have been criticized by Klemme and Neihof (1976b) as 20,000 p.p.m. is needed in the water phase to control sulphate-reducing bacteria. If sea-water displacement of fuel is practised, any biocide leaching into or added to the water phase must be environmentally acceptable as this water is discharged when the ship is refuelled. Several candidate biocides have been proposed by Klemme and Neihof (1976a), and Neihof et al. (1976) have investigated photodegradation as a property endowing environmental acceptability. Houghton and Gage (1979) have described microbial growth in diesel fuel in Royal Navy ships but have reservations about the effectiveness of Biobor-JF. It seems likely that acceptable antimicrobial methods will eventually be evolved based on better housekeeping, better biocides and also on physical methods such as centrifuging.

J. Natural History of *Cladosporium resinae*

It is somewhat surprising that *Cladosporium resinae* has achieved such dominance in aviation-fuel infections in the light of the large variety of moulds and bacteria that can metabolize this class of hydrocarbons. The organism is sometimes referred to as the creosote fungus because of its ability to colonize creosote-treated timber. Its natural history has been extensively described by Sheridan *et al.* (1972), following work by Parberry (1968) who demonstrated a wide geographical distribution of this fungus in soil. The dominance of *C. resinae* in fuel is probably due to a combination of characters which combine to exert a substantial selective pressure. These are: (i) the ability to metabolize aliphatic hydrocarbons in the C_8–C_{20} range; (ii) the ability to sporulate freely when growing in kerosine (Hendey, 1964); (iii) the long-term viability of the spores when suspended in kerosine; (iv) the ability of the organism to tolerate the successive freezing and thawing cycles which occur in the aircraft wing (Hill and Thomas, 1975). With only an occasional hiccup to re-assess the correct nomenclature of this fungus, it seems likely that it will continue to dominate our thoughts on fuel infections.

III. LUBRICATING AND HYDRAULIC OILS

A. Introduction

Another class of hydrocarbon product use where the hydrocarbon : water ratio is very high is lubrication and hydraulic oil systems. Some of these systems have recently been found to be susceptible to microbial attack under certain conditions. In factories, 'straight' oils (i.e. not oil emulsions) are used in metal working, bearing lubrication and in hydraulic systems. In motor vehicles, straight oils are used for bearing lubrication and cooling, and for some hydraulic functions. Similar uses are found in ships engines, control systems and cargo-handling equipment. In all of these applications, these oil systems are designed to be free from water. By many mechanisms, some similar to those referred to above, for fuels, water contaminates virtually all of these systems. These mechanisms include condensation due to differences in temperature between air and surfaces, fluctuations in temperature of the oil causing precipitation of dissolved water, and general leakage of water into the systems from cooling systems or from other sources. Once water gets into these systems, if suitable organisms are present and other factors such as temperature, nutrients and pH value are suitable, microbial growth can occur.

B. Characteristics of Oil Involved

Mineral lubricating oils consist mainly of hydrocarbons of a wide variety of structural types of which the main components are alkanes and naphthenes, although small amounts of aromatics may be present. Some synthetic lubricating oils are used consisting of esters and methyl silicones. Lubricating oils are divided into different types as a function of their kinematic viscosity (a measure of the resistance to gravity flow), namely spindle oils (35 centistokes), light machine oils (35–80), heavy machine oils (80–110) and cylinder oils (110). The low viscosity oils may contain a higher aromatic content of up to 25% (Fowle, 1973).

For many applications, hydraulic and lubricating oils may be modified by addition of additives. These may include anti-oxidant, antirust, antiwear, extreme pressure and antifoam additives, together with corrosion inhibitors, pour-point depressants, viscosity index extenders, dispersants, detergents and acid neutralizers. The chemical composition of these additives, of course, covers a wide range of structures and elements and any of these may or may not be present in a particular oil. Their effect on microbial growth may be one of enhancement or inhibition (Hill and Al-Haidary, 1976; see Fig. 4). The total additive package may comprise 20% of the oil product, and the vital nitrogenous and phosphorus-containing compounds are commonly incorporated.

In some applications, the temperature of the oil during use is too high to allow microbial growth and the systems are virtually self-sterilizing. The normal motor car engine operates at 80°–90°C, with the lubricating oil probably averaging about 90°C during use. The oil may reach much higher temperatures on its cycle through the engine. Similarly, in high-speed marine diesel engines, the average temperature of the lubricating oil is too high to allow microbial growth. However, the crankcase oil of slow-speed marine diesel engines is maintained at 35–45°C, an ideal temperature range for microbial growth.

C. Consequences of Microbial Growth

Microbial growth in lubricating oils and hydraulic oils can cause many problems some of which are similar to those experienced in fuel systems. They can be similarly divided into two categories, namely problems caused by the metabolism of the organisms and problems caused by the physical presence of the organisms.

1. Hydraulic Oils

Generally, hydraulic oils are used for transmitting measured information to control movements. This function relies on predictable flow of the oil through pipes, valves and orifices. If microbial growth occurs, blockage of the small orifices, pipelines and filters may cause system malfunction. Hydraulic systems are comparatively static, and growth tends to occur in the water bottom of the header tank and in discrete water droplets throughout the system. Biological material becomes dispersed through the oil after agitation. Growth in hydraulic systems

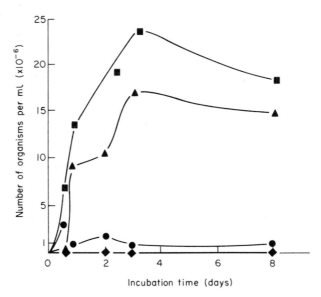

Fig. 4. Time-course of growth of a mixed bacterial population in shake flasks at 37°C in media containing mineral salts and colloidal silica coated with additives. ■, indicates growth in the presence of additive 1; ▲, of additive 2; ◆, of additive 3; ●, growth in the absence of an additive. From Hill and Genner (1980).

may be of aerobic or anaerobic organisms. Aerobic growth is possible, particularly in high-pressure systems, since the high pressure increases the dissolved oxygen content available to the micro-organisms. The

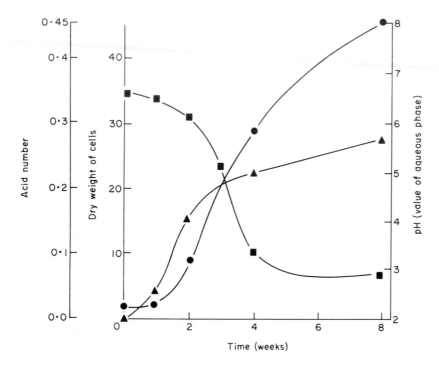

Fig. 5. Time-course of growth of *Candida guilliermondii* in hydraulic oil-aqueous mineral salts mixture in a 1:1 ratio at 30°C. ■, indicates the pH value of the aqueous phase; ▲, the dry weight of organisms (mg per 100 ml) in the aqueous phase; ●, acid number (mg of potassium hydroxide per g of oil) as determined by potentiometric titration (Institute of Petroleum Method 177/74). From Hill and Al-Haidary (1976).

high pressure in these systems does not inhibit microbial activity (Hill, 1968). Problems can also be caused by metabolism of the organisms. Growth can result in depletion of additives from the oil; the kinematic viscosity can be increased by selective removal of lighter hydrocarbons; the surface activity may be changed promoting stable water in oil emulsions; the total acid number may be increased due to acid production (see Fig. 5); and an increase in the corrosiveness of the oil to metals may develop (Hill and Al-Haidary, 1976). The increase in corrosiveness results mainly from the changes listed.

2. Lubricating Oils

As previously mentioned, the operating temperatures of lubricating oils are usually too high to allow microbial growth. One exception to this is the slow-speed marine diesel engine where the operating temperature varies from 35°C–45°C. Since 1969, several serious incidents have occurred where heavy microbial infections have been implicated as the cause of extensive corrosion and, in at least one case, engine seizure (King and McKenzie, 1977; Hill, 1978b).

As with fuels, one prerequisite for microbial growth in a hydrocarbon-containing system is the presence of free water. Growth initially occurs in any water which might be present at the bottom of the engine sump. The water most frequently enters the oil from the cooling systems of the engine and as such may contain coolant additives, such as corrosion inhibitors or 'soluble' oil. These coolant additives may supply nutrients and so increase growth. Most also contain nitrite, and as bacteria isolated are frequently nitrite reducers, we may deduce that this is stimulatory. The coolants themselves may be already infected with organisms and so provide a ready inoculum. The initial growth usually consists of Gram-negative bacteria of the *Pseudomonas* sp. type. On some occasions, species of *Aspergillus* may be the initial infecting organism. This growth can result in water-in-oil emulsification due to surface-active metabolic products such as fatty acids and proteins. Corrosion can occur throughout the system due to the various mechanisms already outlined in Section II.A (p. 26). These include chemical corrosion caused by metabolic byproducts such as acids and sulphides, electrochemical corrosion and the effect of differential aeration cells. Indirect mechanisms include removal of protective coatings such as paint, destruction of protective additives and simple stabilization of a water-in-oil emulsion.

Once high concentrations of aerobic organisms are present, there may be a considerable lowering of the dissolved oxygen content throughout the system. This lowered oxygen content encourages growth of anaerobic and facultative bacteria. This anaerobic growth results in production of more emulsifying metabolic products and the aggressive byproduct hydrogen sulphide. The presence of hydrogen sulphide in such a system aggravates the corrosion and occurs particularly in laid-up ships. The microbial activity usually results in depletion of the various additives which may be present in the oil.

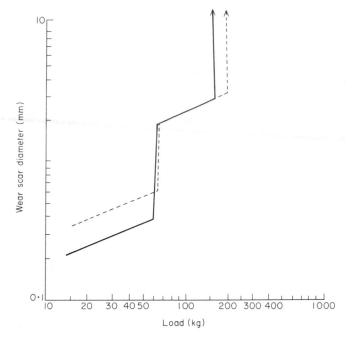

Fig. 6. Shell Four-Ball Wear Test (Institute of Petroleum Method 239/73) on a fresh crankcase oil (——) and an infected ship's sample (----). The wear scars on steel balls immersed in the lubricant were measured using increased loads. After Hill (1978b).

Removal of any of these components can again lower the effectiveness of the oil. Hill (1978b) showed that a lubricating oil after microbial attack had lowered load-carrying capacity (Fig. 6) and Hill and Al-Haidary (1976) reported that certain of these additives act as nutrients and so can stimulate growth. King and McKenzie (1977) reported that the water-in-oil emulsions formed from surface-active components produced by microbial growth markedly affect the lubricating properties of the oil.

All of these factors combine to lower the efficiency of the oil. This results in increased wear of bearing surfaces and eventually suspension of fine metal particles throughout the oil. These particles increase the aggressiveness of the oil, and oxides formed from these particles result in further stabilization of an emulsion. The eventual conclusion of this train of events has, in many cases, been seizure of the crankshaft and other working parts of the engine. A comprehensive report including case histories was published by the General Council of British Shipping (Hill, 1978c).

a. *Detection of microbial infection.* The early detection of an extensive microbial presence in an oil system may allow corrective measures to be used and thus prevent possible disastrous consequences. Initial signs that suggest microbial activity in a ship's lubricating oil were listed by Hill (1978b) and include a stable water content in the oil, increased acidity, smells, slime and sludge production, together with the early corrosion signs on journals and bearings and general corrosion throughout the engine. In many cases the serious consequences have been avoided by on-board use of 'Dipslides' in the oil, either routinely or after several macroscopic signs have been noted. Once a serious microbial presence has been detected, corrective measures should be employed without delay.

b. *Preventive measures.* Microbial problems could be avoided if water did not enter the sump or if any water that did enter the oil system could be efficiently and quickly removed. However, in many cases, due to the design of the oil systems, it is impracticable to operate water-free

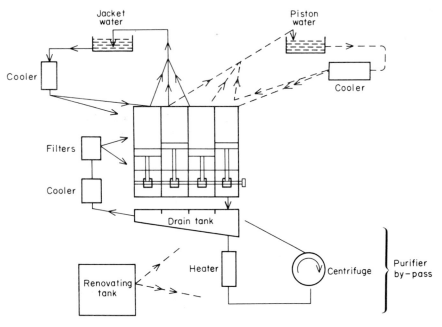

Fig. 7. A simplified block diagram of the crankcase oil and cooling-water systems in a marine crosshead diesel engine. From Hill and Genner (1980).

systems. Where engines have water-cooled pistons, eventual leakage at the glands is almost inevitable.

Virtually all ships have lubricating-oil purification systems consisting of a heat exchanger (to heat the oil in order to lower its viscosity) and a continuous centrifuge (purifier) to remove water and any particulate matter which may be present. This system operates continuously on a slip-stream of the main oil charge (Fig. 7). However, in most cases, the system design allows some water to remain in the very bottom of the sump, and a fraction of a per cent of water remains after centrifugation. Good housekeeping can be very rewarding, however, if the water content and general condition of the oil are constantly monitored, the purifier is used effectively and extra vigilance is used following a major water leak. If this housekeeping should fail and an extensive infection occurs, there are various corrective measures which may be employed. These can be divided into (a) chemical and (b) physical methods.

(i) *Chemical measures.* There are many biocides on the market which can be used in lubricating-oil systems to combat an infection. If the oil is not too badly damaged by infection, a biocide may be added. Possibly an 'additive package' can also be added to the oil to bring it back to specification. The choice of biocide should, however, be made carefully since some biocides have been found to be incompatible with certain oils. Biocides which are available and have been recommended for application to oils embrace many chemical types. These include mixtures of compounds based on triazines, amines, acetals, chlorinated phenols, cresols, quaternary ammonium compounds, aldehydes (particularly formaldehyde) and amides. The antimicrobial activities of these various compounds have been investigated and some results published (Lipson *et al.*, 1969; Hill, 1972). Hill (1978b) reported that the biocide benzylcresol markedly lowered the load-carrying capacity of a lubricating oil. In 1979, he also reported that some biocides are corrosive and should be avoided in lubricating oils. If the oil is badly damaged by the infection, it should be discarded, the system should be sterilized and a fresh charge of oil used. It may be prudent to use a biocide in the fresh oil charge in order to prevent a recurrence.

(ii) *Physical measures.* If the oil is not too seriously affected by the infection, heat may be used to sterilize it. The oil may be pumped out of the sump into the renovating tank, a storage tank large enough to take the entire charge of oil and in which the oil can be heated to about

90°C for 24 or 48 hours. This should result in sterilization. In the meantime, the sump and oilways should be thoroughly cleaned and chemically sterilized before the oil is returned to the system. Again it may be prudent to add a biocide to prevent recurrence. A very substantial research programme on thermal sterilization has been completed by Hill and Genner (1980). Its main conclusions are that most slow-speed diesel engines have an incipient microbial infection in the sump bottom. In the majority of cases the purifier system continuously sterilizes or partly sterilizes a proportion of the oil (Fig. 8) and this is sufficient to suppress a growth explosion. In a few instances, the purifier system, due to inadequacies in temperature, holding time capacity relative to the crankcase oil charge volume, or merely location, cannot cope with the potential growth rate of infecting microbes and the population progressively escalates (Hill and Genner, 1980).

IV. 'SOLUBLE' OIL EMULSIONS

A. Ecological Aspects of 'Soluble' Oil Emulsions in Machine Tools

A major category of petroleum-product spoilage is that of infections in 'soluble' oil emulsions (Fig. 9). The aqueous phase is the continuous and major phase, comprising about 85–99% of the petroleum product in use, the other components being mineral oil and emulsifying chemicals and a host of various minor ingredients which could include coupling agents, anticorrosive agents, extreme pressure additives, natural oils and fats, antifoams, dyes and biocides. The mineral oil containing all of the additives is marketed as a cutting-oil concentrate and added to water by the end-user to make emulsions of various strengths. The 'oil' phase is dispersed in droplets measuring from less than one to a few micrometres in diameter, and hence an enormous oil–water interface exists. The components distribute themselves between the two phases according to their relative solubilities as modified by the influence of other ingredients such as coupling agents. Although it has been suggested that microbes arraign themselves around oil droplets, this does not appear to be the distribution in 'field' samples in which organisms seem to move freely through the water phase.

These petroleum products are used in metal working to cool machine tools and work pieces as well as to provide lubrication to the machining

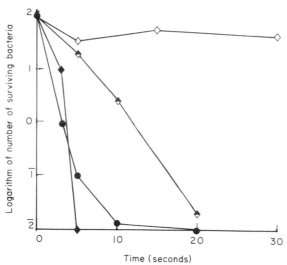

Fig. 8. Time-course of survival of bacteria in infected lubricating oils heated in capillary tubes for up to 30 seconds. ◆, indicates use of oil T at 80°C, ◈, at 70°C, ◇, at 60°C. ●, indicates use of oil E at 70°C. From Hill and Genner (1980).

Fig. 9. A photograph of bottles containing oil-in-water hydraulic oil emulsions. A fresh emulsion is shown on the left, and a grossly degraded field sample on the right.

operation, and hence the bulk temperature is a few degrees above ambient. Natural cooling occurs in the sump; a few large installations may utilize heat exchangers, particularly cooling towers, to cool the fluid. Aeration occurs when the oil emulsions flows over the tool and work piece and the fluid in the sump rapidly becomes anaerobic.

derivatives less so. Some attempts to quantify decreases in emulsion stability have been made by Hill (1975a) using a Coulter Counter (Fig. 10), and by Byrom and Hill (1971) using capacitance measurements. Increases in corrosiveness due to microbial spoilage have been partially quantified by Hill and Gibbon (1969). In practice, it is easier to determine the number of organisms present and assume that numbers in excess of 10^6, 10^7 or 10^8 per ml, depending on the product in use and the conditions of use, are indicative of spoilage. In economic terms, spoilage increases machining costs by causing: (i) short life for 'cutting' oil; (ii) high disposal costs of spoiled emulsion; (iii) increased tool wear and fouling of grinding wheels; (iv) deterioration of surface finish of machined parts; (v) increased corrosion; (vi) objectionable smells adversely affecting work force; (vii) possible health hazards.

Although these affects can rarely be quantified, they have been sufficient justification for stimulating much research on the use of antimicrobial chemicals in cutting fluids.

C. Biocides in Cutting Oils

Bennett (1974) reviewed the use of biocides and published many individual papers. Several test methods have been used to simulate machine-tool conditions (Pivnick and Fabian, 1953; Bennett, 1974; Hill et al., 1976; Rossmore, 1977). There are no clear conclusions from the host of publications, as different biocides vary in effectiveness in different formulations. There is a general concensus of opinion that adding a biocide to the 'soluble' oil concentrate as sold is of limited value, as the biocide potency may decrease on storage and will certainly be progressively lowered after the emulsion has been made up. The formulator has the difficult task of selecting an appropriate level of addition to the cutting-oil concentrate, as the user may make up the emulsion over a wide range of dilutions. There is then the possibility that the biocide may be non-effective if diluted too much or irritant if diluted too little. Nevertheless, formulators do attempt to protect most concentrates with a preservative.

An alternative is to preserve the emulsion by making repeated biocide additions. This is also a difficult regime in practice through inadequacies of gauging the timing and quantity of successive biocide additions.

There is a very strong consensus that cleaning and sterilizing machine tool systems between successive charges of cutting fluid is of prime importance to controlling spoilage, and various commercial detergent–sterilizers are available for this. Some of them can be applied to the old 'cutting oil' charge a few hours before it is discarded.

D. On-Site Test Methods

The key to a good antimicrobial regime is to monitor either the concentration of biocide in the cutting oil or the numbers of micro-organisms present. Some attempts have been made to introduce on-site tests for spoilage organisms but the assay of biocides has proved too complex for routine plant use. Rossmore (1971), proposed a methylene blue reduction test for in-plant use and Hill *et al.* (1967b) gave a simple procedure based on reduction of a tetrazolium salt, namely the 'Red-Spot' test. Both tests had their shortcomings and the most commonly used method in recent years has been the 'Dip-slide' (Hill, 1975b; Genner, 1976). Dip-slides can distinguish between aerobic bacteria, sulphate-reducing bacteria, yeasts and fungi; they can be used by semiskilled staff, and will give a semiquantitative answer in from one to five days.

E. Physical Methods of Controlling Micro-Organisms

A variety of physical methods have from time to time been proposed for controlling cutting-fluid spoilage. Magnetic fields and ultraviolet radiation have been tried without success, and there has been some more detailed work conducted on gamma irradiation (Rossmore and Brazin, 1969; Mixer *et al.*, 1969). There are, however, unavoidable side effects which have not been satisfactorily resolved and a cost analysis by Mixer *et al.* (1969) clearly established that a biocide regime should be cheaper than irradiation.

Various combination processes of radiation and heat or biocides, have also been investigated (Heinrichs and Rossmore, 1971). Symes and Cowap (1975) have proposed a method for continuously removing spoilage microbes by a specific filter media but the field effectiveness of the technique has not always substantiated laboratory predictions. Ultrasound can lower microbial populations (Rossmore, 1974) but is probably more effective when used in combination with a low concen-

demonstrated. It is common to hear of unusual outbreaks of dermatitis and there is often talk of eye infections or unspecified respiratory malfunction. Such incidents are rarely investigated in depth and, if they are, it is usually politically expedient not to publish the results. Unfortunately, even where there are good works' medical services, any investigations tend to be superficial and rarely attempt to correlate medical symptoms with specific spoilage situations.

The potential areas of risk given by Hill and Al-Zubaidy (1979) are: (i) the emulsion passively carrying pathogenic bacteria; (ii) the emulsion changing chemically due to microbial action; (iii) inhalation of living or dead bacteria invoking a response leading to lung malfunction; (iv) bacteria and their enzymes and toxins being irritant to skin or lung; (v) secondary infections due to opportunistic pathogens. These will be considered separately but, in practice, the boundaries are blurred.

2. Pathogenic Bacteria in Emulsions

It is conceivable that pathogenic bacteria from the work force handling oil emulsions could survive in oil emulsions and be passed on to new hosts. Bennett and Wheeler (1954) added 18 pathogenic Gram-negative bacteria and 12 Gram-positive bacteria to two sterile oil emulsions and noted that Gram-positive bacteria died out rapidly (most in less than one day) whereas some Gram-negative bacteria survived for much longer (e.g. *Shigella* sp. 76 days, *Salmonella* sp. 33 days). Rossmore and Williams (1967) confirmed that *Staphylococcus aureus* survives poorly in coolants, particularly those with a normal spoilage contamination, and it is general experience that, despite the undoubted repeated contamination of oil emulsions in use with *Staph. aureus*, it is rarely isolated. Isolation of Gram-negative pathogens from field samples is not uncommon, and Pivnick and Englehard (1954), Lloyd *et al.* (1975) and others have reported good survival and even growth of Gram-negative bacteria such as species of *Salmonella* and *Klebsiella*.

There is doubt about the significance of *Escherichia coli* as an indicator of faecal pollution. Coliform bacteria have frequently been isolated from emulsions and have even been grown in emulsions, and there have been various stances taken on their role as significant spoilage organisms. Our personal experience is that many coliform bacteria isolated are atypical when detailed biochemical tests are conducted and may be a type of spoilage organism with the incidental ability to ferment lactose in the presence of bile salt. Even though some pathogens may survive and even grow in oil emulsions there is no published work which positively links incidence of human infection to a known source in an oil emulsion.

3. Chemical Changes in Emulsions

It is conceivable that some of the byproducts of microbial cutting-oil spoilage, such as fatty acids, could be irritant to body tissues. There is as yet little published evidence for fatty-acid accumulation in cutting fluids, although fatty acids have been extensively studied in other oil–water systems (McKenzie et al., 1976). Fatty acids are much more likely to accumulate in fungal infections than in bacterial infections. Of more concern is the generation of sulphide in cutting oils. The suggested threshold limit value for hydrogen sulphide in the United Kingdom is 10 mg per m^3 of air; 20–150 mg per m^3 causes eye irritation, 500 mg per m^3 causes headaches and dizziness and 800–1,000 mg per m^3 kills in 30 minutes. Hydrogen sulphide is a known skin irritant, and spot checks in the authors' laboratories reveal solution levels in cutting oils of 0–5 mg per litre.

4. Inhalation of Bacteria

Al-Zubaidy (1978) has reported respiratory responses in animals inhaling an aerosol from a cutting-oil emulsion infected with Ps. aeruginosa and indicated that, as expected, the presence of oil appears to stimulate antibody production. Unfortunately, as far as is known, no work has yet been published on levels of antibodies to cutting-oil spoilage organisms in metal workers, and hence a vital piece of information is still missing. We simply do not know if any machinists have developed such antibodies. It is important to note that microbes need not be viable to invoke a serological response, and it has been a feature of humidifier fever outbreaks that appropriate air sampling has detected relatively small numbers of living organisms. Hill and Al-Zubaidy (1979) quote a maximum figure of 70,308 viable bacteria per m^3 in air 0.6 m from a grinding machine, but this number diminished dramatically further from the machine, presumably due to dilution as well as loss of viability.

There is good evidence from other industries that repeated inhalation

of the biocide, which would have a high pH value, and not from solutions in oil emulsions. Nevertheless, it is prudent to be well aware of the environmental impact of cutting oil biocides and to establish guide lines based on skin and inhalation toxicity.

V. WATER-IN-OIL EMULSIONS

The last group of petroleum products which can be delineated as physically and ecologically distinct are the water-in-oil emulsions, which have a water:oil ratio of about 40:60. They are mostly sold as finished products and are used instead of mineral oils when the product is required to be flame-proof. Typically they are used as hydraulic oils in rolling mills. Most field experience is that they suffer few spoilage problems, probably because potential spoilage organisms have difficulty passing through the continuous oil phase to colonize the finely distributed water droplets. Chemically they are not unlike 'soluble' oils, as they must have a substantial emulsifier pack to maintain stability of the invert emulsion. Rossmore and Szlatky (1977) isolated several species of fungi, and species of *Pseudomonas*, *Proteus* and *Bacillus*, from hydraulic systems experiencing filter plugging and system malfunction. There were technical difficulties in the isolation procedures and the sample was preferably first inverted to an oil-in-water emulsion using Arlacel 80 and Tween 60. Growth studies in fresh fluid were conducted, but difficulty was experienced in establishing a spoilage flora. These authors forecast an increase in this type of spoilage as legislation is compelling wider use of these petroleum products.

VI. CONCLUSIONS

Microbial spoilage of petroleum products and associated corrosion and malfunction has been known and studied for 40 years. The technology for this study is now largely developed and widely used, but the product user is still only partially educated to use the products in ways that minimize microbial growth. The literature is copious, but spread through a wide range of microbiological, engineering and petroleum orientated journals. A reasonably complete list of those publications is available from the Biodeterioration Information Centre, Birmingham, United Kingdom, in their Specialised Bibliography, 5B9, Fuels, Hydrocarbons and Lubricants (over 500 references cited).

REFERENCES

Abbott, B. S. and Gledhill, W. E. (1971). *Advances in Applied Microbiology* **14**, 24.

Allen, F. H. (1945). *Journal of the Institute of Petroleum* **31**, 9.

Al-Zubaidy, T. S. (1978). Ph.D. Thesis: University of Wales, Cardiff.

Anon (1952). *Chemistry Research*, D.S.I.R. Chemical Research Laboratory, H.M.S.O., 1953.

Anon (1956). *Chemistry Research*, D.S.I.R. Chemical Research Laboratory, H.M.S.O., 1957.

Anon (1961). *North American Aviation Inc.*, Report presented in co-ordinating Research Council Symposium of Fuel Corrosion, February 1961.

Bakanauskas, S. (1958). *United States Air Force, Wright Air Development Center, Technical Report 58.*

Banaszak, E. F., Thiede, W. H. and Fink, J. N. (1970). *New England Journal of Medicine* **283**, 271.

Bennett, E. O. (1956). *Lubrication Engineering* **13**, 310.

Bennett, E. O. (1972). *Society of Manufacturing Engineers U.S.A.* Technical Paper MP 72–226.

Bennett, E. O. (1974). *Progress in Industrial Microbiology* **13**, 121.

Bennett, E. O. and Wheeler, H. O. (1954). *Applied Microbiology* **2**, 368.

Berry, H. and Michaels, I. (1950). *Journal of Pharmacy and Pharmacology* **2**, 27.

Blanchard, G. C. and Goucher, C. R. (1965). *Developments in Industrial Microbiology* 6, 95.

Bohdanowicz-Murek, K., Zyska, B. J., Zankowicz, L. P., Wawrzyniak, E. M. and Terech, I. (1980). *In* Proceedings 4th International Biodeterioration Symposium, Berlin (T. A. Oxley, H. C. Gunther-Bekker, D. Allsopp, eds.), pp. 357–362. Pitman Publishing Ltd., London.

Boggs, W. A. (1965). *Society of Automotive Engineers, Business Aircraft Conference Proceedings*, May 1965.

Born, M. J., Vasil, M. L., Sadiff, J. C. and Iglewski, B. H. (1977). *Infection and Immunity* **16**, 326.

Bushnell, L. D. and Haas, H. F. (1941). *Journal of Bacteriology* **41**, 653.

Byrom, D. and Hill, E. C. (1971). *In* 'Microbiology 1971' (P. Hepple, ed.), pp. 42–59, Institute of Petroleum, London. Applied Science Publishers Ltd., London.

Cooney, J. J. and Kula, T. J. (1970). *International Biodeterioration Bulletin* **6** (3), 109.

Darby, R. T., Simmons, E. G. and Wiley, B. J. (1968). *International Biodeterioration Bulletin* **4** (1), 39.

Dayal, H. M., Uniyal, B. P., Agarwal, P. N. and Nigam S. S. (1971). *Journal of Science and Technology* **9-B** (1), 24.

DeGray, R. J. and Fitzgibbons, W. O. (1966). *Developments in Industrial Microbiology* **7**, 384.

DeGray, R. J. and Killian, L. N. (1960). *Industrial and Engineering Chemistry* **52** (12), 58.

Duffet, N. D., Gold, S. H. and Weirich, C. L. (1943). *Journal of Bacteriology* **45**, 37.

Ellis, L. F., Samuel-Maharajah, R., Mendelow, M., Ruth, L. and Pivnick, H. (1957). *Applied Microbiology* **5**, 345.

Engel, W. B. and Swateck, F. E. (1966). *Developments in Industrial Microbiology* **7**, 354.

Feisal, E. V. and Bennett, E. O. (1961). *Journal of Applied Bacteriology* **24**, 125.

Fowle, T. I. (1973). *In* 'Handbook of Tribology' (M. J. Neale, ed.), Section B2. Butterworths, London.

Ganser, P. (1940). *Pipe Line Gas Journal* **231**, 271.

Garrard, S. D., Richmond, J. B. and Hirsch, M. M. (1951). *Pediatrics, Springfield* **8**, 482.

Genner, C. (1976). *Process Biochemistry* **11**, 39.

Gorog, J., Ronay, D. and Vamos, E. (1970). *Proceedings of Corrosion Week, Manifestation European Federation of Corrosion, 41st, Budapest*, pp. 7–12. Akademiai, Kiado.

Green, R. H., Olson, R. L., Gustan, E. A. and Pilgrim, A. J. (1967). *Developments in Industrial Microbiology* **8**, 227.

Guynes, G. J. and Bennett, E. O. (1959). *Applied Microbiology* **7**, 117.

Harke, H. P. (1977). *Contact Dermatitis* **3**, 51.

Hazzard, G. F. (1961). *Australian Defense Standards Laboratory*, Report 252, Part 1.

Hazzard, G. F. (1965). *Australian Defense Standards Laboratory*, Report 252, Part 4.

Hazzard, G. F. and Kuster, E. C. (1963). *Australian Defense Standards Laboratory*, Report 252, Part 2.

Hedrick, H. G. and Carroll, M. T. (1966). *Developments in Industrial Microbiology* **7**, 372.

Hedrick, H. G. and Crum, M. G. (1968). *Applied Microbiology* **16** (12), 1826.

Hedrick, H. G., Miller, C. E., Halkias, J. E. and Hildebrand, J. E. (1964). *Applied Microbiology* **12**, 197.

Hedrick, H. G., Reynolds, R. J. and Crum, M. G. (1967). *Developments in Industrial Microbiology* **8**, 267.

Hedrick, H. G., Reynolds, R. J. and Crum, M. G. (1968). *Developments in Industrial Microbiology* **9**, 415.

Heinrichs, T. F. and Rossmore, H. W. (1971). *Developments in Industrial Microbiology* **12**, 341.

Hendey, N. I. (1964). *Transactions of the British Mycological Society* **47** (4), 467.

Hill, E. C. (1967). *Metals and Materials* **1**, 294.

Hill, E. C. (1968). *Fluid Power International* **33**, 56.

Hill, E. C. (1970a). *Journal of the Institution of Petroleum* **56** (549), 138.

Hill, E. C. (1970b). *Aircraft Engineering*, July 1970, p. 24.

Hill, E. C. (1971). *Proceedings of Global Impacts of Applied Microbiology III*, pp. 203–204. Examiner Press, Bombay.

Hill, E. C. (1972). *Journal of the Institute of Petroleum* **58**, 248.

Hill, E. C. (1975a). *In* Proceedings of the Third International Biodeterioration Symposium, Rhode Island (J. M. Sharpley and A. M. Kaplan, eds.), pp. 243–258. Applied Science Publishers Ltd., London.

Hill, E. C. (1975b). *In* 'Microbial Aspects of the Deterioration of Materials' (R. J. Gilbert and D. W. Lovelock, eds.), pp. 127–136. Applied Science Publishers Ltd., London.

Hill, E. C. (1978a). *In* 'Developments in Biodegradation of Hydrocarbons—1' (R. J. Watkinson, ed.), pp. 201–225. Applied Science Publishers Ltd., London.

Hill, E. C. (1978b). *Transactions of the Institute of Marine Engineers* **90**, 165.

Hill, E. C. (1978c). *Report No. TR/069*, General Council of British Shipping, London.

Hill, E. C. (1979). *Society of Manufacturing Engineers U. S. A.*, Technical Paper MF 79–391.

Hill, E. C. and Al-Haidary, N. K. (1976). *Proceedings of a Symposium on Microbial Corrosion Affecting the Petroleum Industry, Sunbury, U.K.*, p. 51. Institute of Petroleum IP 77–001.

Hill, E. C. and Al-Zubaidy, T. S. (1979). *Tribology International* **12**, 161.
Hill, E. C. and Genner, C. (1980). *In* Proceedings 4th International Biodeterioration Symposium, Berlin (T. A. Oxley, H. C. Gunther-Bekker and D. Allsopp, eds.), pp. 37–43. Pitman Publishing Ltd., London.
Hill, E. C. and Gibbon, O. M. (1969). *Metal Finishing* **15**, 395.
Hill, E. C. and Penberthy, I. O. (1968). *Metals and Materials* **2**, 359.
Hill, E. C. and Thomas, A. R. (1975). *In* Proceedings of the Third International Biodeterioration Symposium, Rhode Island (J. M. Sharpley and A. M. Kaplan, eds.), pp. 157–174. Applied Science Publishers Ltd., London.
Hill, E. C., Evans, D. A. and Davies, I. (1967a). *Journal of the Institute of Petroleum* **53**, 280.
Hill, E. C., Davies, I., Pritchard, J. A. V. and Byrom, D. (1967b). *Journal of the Institute of Petroleum* **53**, 275.
Hill, E. C., Graham Jones, J. and Sinclair, A. (1967c). *Metals and Materials* **1**, 407.
Hill, E. C., Gibbon, O. M. and Davies, P. E. (1976). *Tribology International* **9**, 121.
Hitzman, D. O. (1965). *Developments in Industrial Microbiology* **6**, 105.
Hitzman, D. O., Shotton, J. A. and Alquist, H. E. (1963). *Society of Automotive Engineers and American Society of Naval Engineers*, Joint National Aero-Nautical Meeting, Paper SAE 683D, Washington D.C. April 8–11.
Hodgson, G. (1979). *Industrial Medicine and Surgery* **39**, 16.
Holdom, R. S. (1976). *Tribology International* **9**, 271.
Hostetler, H. F. and Powers, E. J. (1963). 28th Mid-year Meeting of the American Petroleum Institute's Division of Refining, Philadelphia, May 13.
Houghton, D. R. and Gage, S. (1979). *Transactions of the Institute of Marine Engineers* **91**, 189.
Iglewski, B. H., Burns, R. P. and Gipson, I. K. (1977). *Investigative Ophthalmology and Visual Science* **16**, 73.
Isenberg, D. L. and Bennett, E. O. (1959). *Applied Microbiology* **7**, 121.
Jacobson, J. A., Hoadley, A. W. and Farmer, J. J. (1976). *American Journal of Public Health* **66**, 1092.
Jerusalimsky, N. D. and Skryabin, G. K. (1966). *Zeitschrift für Allgemeine Mikrobiologie* **6**, 23.
Keczkes, K. and Brown, P. M. (1976). *Contact Dermatitis* **2**, 92.
Kereluk, K. and Baxter, R. M. (1963). *Developments in Industrial Microbiology* **4**, 235.
King, R. and McKenzie, P. (1977). *Transactions of the Institute of Marine Engineers* **89**, 37.
Klemme, D. E. and Leonard, J. M. (1960). *Naval Research Laboratory Memorandum Report*, 5501, Washington, D.C.
Klemme, D. E. and Leonard, J. M. (1971). *Naval Research Laboratory Memorandum Report*, 2324, Washington, D.C.
Klemme, D. E. and Neihof, R. A. (1976a). *Naval Research Laboratory Memorandum Report*, 3212, Washington, D.C.
Klemme, D. E. and Neihof, R. A. (1976b). *Naval Research Laboratory Memorandum Report*, 3259, Washington, D. C.
Komagata, T., Nakase, T. and Katsuya, N. (1964). *Journal of General and Applied Microbiology* **10**, 313.
Lansdown, A. R. (1965). *Journal of the Royal Aeronautical Society* **69**, 763.
Leathen, W. W. and Kinsel, N. A. (1963). *Developments in Industrial Microbiology* **4**, 9.
Lee, M. and Chandler, A. C. (1941). *Journal of Bacteriology* **41**, 373.
Liberthson, L. (1945). *Lubrication Engineering* **1**, 103.

Liggett, S. (1976). M.Sc. Thesis: University of Wales, Cardiff.
Lipson, D., Sadurska, I. and Kowalik, R. (1969). *Postepy Mikrobiologii* **8**, 169.
Lloyd, G., Lloyd, G. I. and Schofield, J. (1975). *Tribology International* **8**, 27.
London, S. A., Finefrock, V. H. and Killian, L. N. (1965). *Developments in Industrial Microbiology* **6**, 61.
Markovetz, A. J. and Kallio, R. E. (1964). *Journal of Bacteriology* **87**, 968.
McKenzie, P., Akbar, A. S. and Miller, J. D. (1976). *Proceedings of a Symposium on Microbial Corrosion Affecting the Petroleum Industry, Sunbury, U.K.*, p. 37. Institute of Petroleum, IP 77–001.
Miller, R. N., Herron, W. C., Krigrens, A. G., Cameron, J. L. and Terry, B. M. (1964). *Material Protection* **3**, 60.
Mixer, R. Y., Cavallo, J. J. and Hart, W. C. (1969). *Society of Tool and Manufacturing Engineers, U.S.A.*, Technical Paper MR 69–261.
Miyoshi, M. (1895). *Jahrbücher für Wissenschaftliche Botanik* **28**, 269.
Neihof, R. A. and Bailey, C. A. (1978). *Applied and Environmental Microbiology* **35** (4), 698.
Neihof, R. A., Patouillet, P. J., Hannan, P. J. and Klemme, D. E. (1976). *Naval Research Laboratory Memorandum Report*, 3320, Washington, D.C.
Nyns, E. J., Auquiere, J. P. and Wiaux, A. L. (1968). *Antonie van Leeuwenhoek* **34**, 441.
Parberry, D. G. (1968). *International Biodeterioration Bulletin* **4**, 79.
Park, P. (1973). *International Biodeterioration Bulletin* **9**, (3), 79.
Pivnick, H. and Englehard, W. E. (1954). *Applied Microbiology* **2**, 140.
Pivnick, H. and Fabian, F. W. (1953). *Applied Microbiology* **1**, 204.
Pollock, M., Taylor, N. S. and Callahan, L. T. (1977). *Infection and Immunity* **15**, 776.
Powelson, D. M. (1962). *Stanford Research Institute Report*, Project No. B-3658.
Prince, A. E. (1961). *Developments in Industrial Microbiology* **2**, 197.
Rogers, M. R. and Kaplan, A. M. (1965). *Developments in Industrial Microbiology* **6**, 80.
Rogers, M. R. and Kaplan, A. M. (1968). *Developments in Industrial Microbiology* **9**, 448.
Rossmore, H. W. (1971). *International Biodeterioration Bulletin* **7**, 147.
Rossmore, H. W. (1974). *Society of Manufacturing Engineers, U.S.A.*, Technical Paper MR74–169.
Rossmore, H. W. (1977). *In* 'Biodeterioration Investigation Techniques' (A. H. Walters, ed.), pp. 227–241. Applied Science Publishers Ltd., London.
Rossmore, H. W. and Brazin, J. G. (1969). *In* Proceedings 1st International Biodeterioration Symposium, Southampton (A. H. Walters and J. Elphick, eds.), pp. 386–401. Elsevier Publishing Co. Ltd., London.
Rossmore, H. W. and Holtzman, G. H. (1974). *Developments in Industrial Microbiology* **15**, 273.
Rossmore, H. W. and Szlatky, K. (1977). *International Biodeterioration Bulletin* **13**, 96.
Rossmore, H. W. and Williams, B. W. (1967). *Health Laboratory Science* **4**, 160.
Rossmore, H. W., De Mare, J. and Smith, T. H. F. (1972). *In* Proceedings 2nd International Biodeterioration Symposium, Lunteren (A. H. Walters and E. H. Hueck-Van der Plas, eds.), pp. 286–293. Applied Science Publishers Ltd., London.
Rowe, L. C. (1957). *Corrosion* **13**, 750.
Rubidge, T. (1975). *International Biodeterioration Bulletin* **11** (4), 133.
Russell, D. R., Crum, M. G. and Hedrick, H. G. (1968). *Developments in Industrial Microbiology* **9**, 426.
Sabina, L. R. and Pivnick, H. (1956). *Applied Microbiology* **4**, 171.
Saunders, J. C., Wotring, W. T. and Taylor, J. F. (1966). The Standard Oil Co. (Ohio). Technical Report 20 October, 1966.
Scott, J. A. (1971). *Society of Automotive Engineers*, Paper No. 710438.

Scott, J. A. and Forsyth, T. J. (1976). *International Biodeterioration Bulletin* **12** (1), 1.
Scott, J. A. and Hill, E. C. (1971). *Microbiology 1971. Proceedings of the Institute of Petroleum*, pp. 25–41. Applied Science Publishers Ltd., Amsterdam.
Sheridan, J. E., Nelson, J. and Tan, Y. L. (1972). *Tuatara (New Zealand)* **19** (2), 70.
Singh, V. (1976). *Current Medical Practice* **20**, 214.
Smith, T. H. F. (1969). *Lubrication Engineering* **8**, 313.
Symes, W. R. and Cowap, D. (1975). *In* Proceedings 3rd International Biodegradation Symposium, Rhode Island (J. M. Sharpley and A. M. Kaplan, eds.), pp. 233–242. Applied Science Publishers, Ltd., London.
Tausson, W. O. and Aleshina, W. A. (1932). *Mikrobiologia* **1**, 229.
Thaysen, A. C. (1939). *Journal of the Institute of Petroleum* **25**, 411.
Thomas, A. R. (1973). Ph.D. Thesis: University of Wales, Cardiff, 150 pp.
Thomas, A. R. and Hill, E. C. (1976a). *International Biodeterioration Bulletin* **12** (3), 87.
Thomas, A. R. and Hill, E. C. (1976b). *International Biodeterioration Bulletin* **12** (4), 116.
Thomas, A. R. and Hill, E. C. (1977a). *International Biodeterioration Bulletin* **13** (1), 1.
Thomas, A. R. and Hill, E. C. (1977b). *International Biodeterioration Bulletin* **13** (2), 31.
Toussen, T. A. (1939). *Mikrobiologia* **8**, 828.
Upsher, F. J. (1976). *Paper Number I.P. 76–001*. Institute of Petroleum, London.
Vermooten, C. A. L. and Binnington, C. D. (1975). *The South African Mechanical Engineer* **25**, 68.
Wort, M. and Lloyd, G. I. (1979). *Proceedings of a Symposium on the 'Physical Methods of Microbial Control in Oils and Coolants'*. Institute of Petroleum, London.
Zittle, C. A. (1951). *Advances in Enzymology* **12**, 493.
ZoBell, C. E. (1945). *Science* **102**, 364.
ZoBell, C. E. (1946). *Bacteriological Reviews* **10**, 1.
ZoBell, C. E. and Prokop, J. V. (1966). *Zeitschrift für Allgemeine Microbiologie* **6**, 143.

Note Added in Proof
New sources of kerosine fuels may become available. May and Neihof (1980) have reported that aviation fuel JP-5 derived from coal and two types of oil shale supported growth of anaerobic sulphate-reducing bacteria, but the coal-derived fuel and one of the shale-derived fuels were inhibitory to *Cladosporium* sp.

Meybaum and Schiapparelli (1980) have used an electrochemical technique to predict the corrosiveness of water associated with fungal growth in aircraft fuel and have stated that it correlates with visual assessments. The same workers (Schiapparelli and Meybaum, 1980) have indicated that dodecanoic acid produced by *Cladosporium resinae* plays an important role in aluminium alloy corrosion.

There are indications that current fuel biocides may act selectively. Williams and Lugg (1980) have shown that oxidase-positive Gram-negative bacteria can flourish in water bottoms of fuel containing EGME in conditions which inhibited germination of spores of *Cladosporium resinae*. Potentially corrosive carboxylic acids were produced and the fuel/water system emulsified.

There are always problems in matching the appropriate preservative to a specific cutting-oil formulation. Onyekwelu *et al.* (1981) have published data on the compatability of 10 commercial biocides with 12 cutting fluid formulations. Much better long-term performance is expected nowadays from such biocides, and this is an actively researched area.

Physical methods of controlling microbial infections are receiving more attention. Hill and Genner (1981) have given guide-lines for designing the lube-oil system of

II. BACTERIAL ATTACK

In the liquid state, the principal causative microbial agents of deterioration are bacteria. Both Gram-positive and Gram-negative organisms have been isolated from spoiled latex paints. Buono *et al.* (1973) and Ross (1964, 1972) have isolated both Gram-positive spore-forming bacteria as well as Gram-negative bacteria. Winters (1972) and Goll (1972) have found that pseudomonads are the most commonly isolated organisms. The principal results of bacterial attack are a decrease in the viscosity of the paint, gas evolution and production of odours.

Bacteria have nutritional requirements that include the presence of compounds supplying the elements carbon, nitrogen, sulphur, oxygen and traces of potassium, magnesium, calcium and iron. Miller (1973) points out that cellulosic ethers used as thickening agents in latex formulations are the most likely carbon-energy source for growth of bacteria. Bacterial utilization of the cellulosic ether is thought to cause a decrease in the viscosity of the paint to a level where it cannot be applied to a surface. Loss in the viscosity of latex paints seems to be the greatest economic effect confronting paint manufacturers. Latex paint in a can, in which a significant decrease in viscosity has occurred, represents an unsaleable product. These paints must either be discarded or rethickened, which is both timely and costly. It is not uncommon for a paint manufacturer to experience this problem. Some production batches which have exceeded 45,000 litres have been discarded due to this problem of viscosity. Holmes (1969), in a survey across the United States, found that about 50% of the paint manufacturers had experienced viscosity problems in their latex paints; many of them believed that the main cause of their viscosity problems was due to the presence of cellulases.

The multiplicity of cellulases from many micro-organisms has been studied on the basis of different fractionation procedures. Most microbial cellulases occur as a typical extracellular enzyme. Extracellular enzymes are synthesized inside the cell and secreted. They do not generally appear in the external environment as a result of autolysis. These extracellular enzymes enable the micro-organism to metabolize large molecules found outside the cell.

Cellulases are produced by insects, molluscs, protozoa, bacteria and fungi. Cellulase is an adaptive enzyme and it is only produced when the micro-organism is grown on cellulose. Therefore, cellulolytic micro-organisms would secrete cellulases either in a solution of cellulose ether or in a latex paint. Cellulases would not be secreted by micro-organisms in any raw material.

Both fungi and bacteria have been implicated as producing cellulases in latex paints (Lalk, 1964; Winters, 1972). Floyd and others (1966) have shown that oxidizing agents, quaternary ammonium compounds, phenylmercuric acetate, N-halo compounds and lactones are effective cellulase inhibitors when they are used in latex formulations. Wienert and Vanderstraeten (1973) claimed that modified barium metaborate is an excellent cellulase inhibitor.

Winters and Goll (1974) have discussed the role of cellulases in deterioration of latex paints. The hydrolytic effect of these cellulases on the cellulose ether at the β-1–4 linkage, which results either in release of reducing sugar or in production of glucose, is described. Floyd et al. (1966) have also shown that the degree and type of substitution on the cellulose chain are the most important parameters regulating cellulase activity on the substrate. If the substitution occurs in a regular pattern, so that every glucose unit has been chemically etherized, then the cellulose ether is protected from cellulase attack. This type of product has been synthesized and is called the 'cellulase resistant' cellulosic ether.

Although earlier studies by Winters (1972) and Winters and Goll (1974) indicated that cellulases were the main and predominant cause for viscosity decrease in latex paints, Winters (1980) and Miller (1973) have shown that the majority of micro-organisms isolated from latex paints and raw materials are non-cellulolytic. These studies suggest that the presence of cellulases is an over-emphasized phenomenon. The reasons why many paint manufacturers believed that cellulases were the dominant cause for viscosity decrease in latex paints was that cellulases were known to be produced by micro-organisms, and that latex paints were susceptible to growth by these micro-organisms.

The majority of latex paints that have experienced a decrease in viscosity have been devoid of micro-organisms. In examining other causes for viscosity decrease in latex paints, Winters (1980) found that both oxidizing and reducing agents would lower the apparent viscosity of cellulosic ethers. Subcommittee DO1.28 (Biodeterioration) of the American Society for Testing Materials (Rosenberg, 1978) also found that the majority of viscosity problems in latex paints were chemical in nature rather than due to cellulases.

Oxidizing agents, such as potassium persulphate, t-butylhydroperoxide and hydrogen peroxide, will cause a decrease in the apparent viscosity of cellulosic ethers. These oxidants seem to affect the apparent viscosity of cellulosic ethers through formation of cellulosic radicals by hydrogen abstraction with either hydroxyl or sulphate free radicals. The decrease in apparent viscosity of cellulosic ethers, therefore, may be the result of disruption of supramolecular aggregates. The presence of oxidant in emulsions is usually due to the use of an excess of catalyst in the polymerization process.

Sulphur-containing reducing agents, such as sodium metabisulphite, sodium sulphite, sodium formaldehyde sulphoxylate and sodium bisulphite will also cause a decrease in the apparent viscosity of cellulosic ethers. These reducing agents may be present in latex emulsions when they are used as a redox couple for initiation of the polymerization process. Other sulphur-reducing agents, such as hydrosulphite, may be present in extender slurries as bleaching agents. The presence of these reducing agents will cause a lowering of the redox potential of the raw material. Polymerization of acrylamide by these reducing agents is possible if a redox couple is formed with molecular oxygen. Therefore, polymerization of acrylamide by addition of raw materials would indicate the presence of reductant, as well as oxidant.

Direct presence of a reductant in a sample of raw material can be determined by a redox procedure. Raw material (10 g) is added to a 1.0% hydroxyethyl cellulose solution containing 1,000 p.p.m. of potassium persulphate. The apparent viscosity is recorded immediately and after 24 hours incubation at 25°C. A decrease in the apparent viscosity of the cellulosic ether solution which is greater than the control (1,000 p.p.m. persulphate and no raw material) suggests the presence of reductant. The rate of viscosity decrease is accelerated with all of the reductants. The increase in the rate of viscosity decrease is assumed to be due to formation of redox couples which accelerates the rate of free radical formation.

The presence of reductants in raw materials can also be determined by adding 1,000 p.p.m. potassium persulphate to a 25% (w/v) acrylamide solution containing 1.0 g of raw material. The mixture is incubated at 50°C and gel-time formation is recorded. The persulphate control (no raw material) will cause polymerization of acrylamide in

40 minutes. The presence of either sodium metabisulphite or sodium formaldehyde sulphoxylate increases the rate of polymerization, which lowers the gel time to 18 minutes for metabisulphite and 12 minutes for the sulphoxylate. Therefore the presence of a reductant in a raw material sample is indicated by an increase in the rate of free-radical formation which causes an increase in the rate of polymerization.

All cellulosic ethers are susceptible to attack by persulphate. Differences in rates of viscosity decrease are observed among the various types of cellulosic ethers. The highly substituted hydroxyethyl cellulose, methylhydroxypropyl cellulose and methylhydroxyethyl cellulose show the slowest rate in viscosity decrease. These variations may be due to the differences in cross-linking that occurs in formation of supramolecular aggregates. Although some of these cellulosic ethers are termed 'cellulase resistant', all of them are susceptible to attack by both persulphate and sulphur-containing reductants.

It has been the practice of paint formulators to use latex emulsions which contain oxidants and no decrease in the apparent viscosity of paints was observed. This is due to the fact that sulphate free radicals, which possess a very short half-life, will react with other paint components before they can react with the cellulosic ether molecules. Unless the oxidant concentration is high enough (greater than 250 p.p.m.), no adverse viscosity decrease will result. However, the redox reaction between persulphate and sulphur-containing reductants does result in formation of other sulphur-containing free radicals, which have a longer half-life than sulphate free radicals, and seem to have a higher affinity for the cellulosic ether molecules in latex formulations.

It is often a common practice of paint manufacturers to use two or more sources of a particular latex emulsion. However, although the monomer may be the same, two latex emulsions can differ in their mode of synthesis. One latex emulsion may be synthesized by a straight reflux by persulphate, while the other may be produced through a redox-couple mechanism. One may still contain an excess of persulphate while the other emulsion may contain an excess of reducing agent. If two different latex emulsions are pumped into the same latex-storage tank, formation of sulphur free radicals can result. Use of the material from this storage tank will result in a decrease in the apparent viscosity of the formulated latex paint. It is suggested that the practice of putting two latex emulsions, one containing oxidant and the other reductant, into the same tank be discontinued.

There is still a feeling by many that the presence of cellulases is the main cause of viscosity decrease in latex paints. Although this concern may be warranted, there is sufficient data to show that the presence of oxidants (reductants) in raw materials may be the dominant cause of a viscosity decrease in latex paints, not the presence of cellulases. Oxidants and reductants have been isolated from raw materials and low-viscosity latex paints and, when raw materials containing oxidants and reductants are formulated into latex paints, a decrease in apparent viscosity results.

Since the viscosity problem may not be microbial, proper housekeeping would not be the solution to the problem. Solution of the redox problem can be accomplished either by excluding the use of raw materials containing oxidants and reductants or by the use of oxidant–reductant inhibitors.

The presence of cellulases is detected biochemically by assaying for the presence of either reducing sugar or free glucose. These products are not formed when cellulosic ethers are attacked by oxidants or reductants. The presence of oxidants and reductants is detected by the methods described previously. Therefore, it is a simple task to determine the cause of viscosity decrease in any latex paint. Paints attacked by cellulases cannot be rethickened because of the continuous catalytic action of cellulase, whereas paints that have been affected by the presence of oxidants and reductants can be rethickened to a stable viscosity.

Amines at a 1% (w/w) concentration inhibited persulphate activity by 99.1–65.0%. Persulphate activity was inhibited by 91.8–47% at the 0.1% (w/w) concentration. Whenever the amine was substituted by an additional methyl, propyl or butyl group, the apparent inhibitory effect was lowered. Isopropylamine, methylethanolamine and the fatty amine oxethylate were equally inhibitory. Ammonia was only slightly lower in its inhibitory effect. The effectiveness of ammonia and these amines in inhibiting the activity of persulphate on cellulosic ethers is probably the reason why latex paints containing these agents rarely show viscosity instability. My practical experience is that over 90% of latex paints which show viscosity instability do not contain these agents.

Practical experience has demonstrated that those latex paints formulated with amines invariably never experience viscosity instability. It should be borne in mind, however, that a high concentration of persulphate coupled with a low concentration of amines can lead to viscosity loss in latex paints. However, once it is recognized that viscosity instability in latex paints is caused by the presence of oxidizing and reducing agents, proper formulation can prove an effective control. Since certain polyvinylacetate latex systems must avoid the use of ammonia and amines, elimination of excess oxidants and reductants in raw materials is strongly recommended.

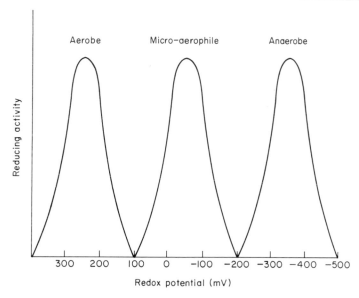

Fig. 1. Reducing activities of aerobic, micro-aerophilic and anaerobic bacteria at different redox potentials.

Changes in redox potential during bacterial growth are complex parameters not at present clearly understood. However, its importance in regulating growth of bacteria has been stressed by many workers (Wimpenny and Nicklen, 1971). Winters and Goll (1974) have shown the importance of redox potential in its effect on the viscosity of cellulosic ethers, as well as its potential effect on bacterial growth in latex paints and emulsions. Redox potential in latex systems depends on the presence of oxygen, oxidant and reductant. Oxygen is well recognized as one of the important regulators of bacteria growing both aerobically and anaerobically. The amount of dissolved oxygen in latex systems has never been measured but its concentration has been shown to be directly related to the redox potential of the growth medium. Oxidants and reductants, used in latex systems as initiators and neutralizers, affect directly the redox potential; in addition, the reductant may also bind directly to oxygen, causing a secondary effect by lowering the concentration of dissolved oxygen.

The reducing activities of aerobes, microaerophiles and anaerobes at different redox levels are schematically shown in Figure 1. The range of the aerobe is from $+400$ mV to $+100$ mV; the microaerophile will range in its reducing activity from $+100$ mV to -200 mV, and the anaerobe will show a reducing range from -200 mV to -500 mV. In examining the redox potential of latex samples that showed no evidence of microbial growth, the following results have been found. The redox potential of latex paints ranged from $+120$ mV to -50 mV (microaerophilic), the vinyl acrylic–vinyl acetate emulsion ranged from $+550$ mV to -30 mV (aerobic to microaerophilic), and the acrylic emulsions ranged from $+40$ mV to -50 mV (microaerophilic). The redox potential of latex emulsions depends on the presence of oxidant and reductant. The presence of an oxidant, such as hydrogen peroxide or potassium persulphate, increases the redox potential thus favouring growth of aerobes. It also, however, is either bactericidal or bacteriostatic in its oxidizing effect on the cell surface. Reductants used to form redox couples and to neutralize the presence of oxidant lower the redox potential, thus favouring growth of micro-aerophiles and anaerobes. It is reasonable to conclude that the presence of an oxidant and a high redox potential is detrimental to growth of micro-aerophilic and anaerobic micro-organisms. The susceptibility of latex paint to bacterial growth may also be controlled by the presence of either an oxidant or a reductant, which in large part comes from the latex emulsion.

In evaluating the morphological and biochemical characteristics of bacterial isolates from contaminated latex paints and emulsions, Winters (1979) found that these organisms are mainly pseudomonads (85%). Common characteristics of these organisms are shown in Table 1. Many are heterotrophic organisms that could grow on and be transferred through an autotrophic medium. These organisms could derive their energy by oxidation of an inorganic substrate (thiosulphate and bisulphite) and derive their carbon from carbon dioxide. The majority of these organisms could not oxidize cellulosic ethers or the latex itself, suggesting that these raw materials could not serve as the carbon or energy source for cellular growth.

Table 1

Characteristics of isolates from contaminated latex paints and emulsions

Characteristics	Isolate
Cell morphology	Short straight rods
Motility	Single flagella; motile
Gram stain	Negative
Catalase	Positive
Indole oxidase	Positive
Indole produced	Negative
Cellulase	Negative
Hydrogen sulphide production	Negative
Nitrite from nitrate	Positive
Gelatinase	Positive
Acid produced from	Glucose
	Maltose
	Sucrose
	Galatose
Growth on autotrophic metabisulphite medium	Positive

There must be an input of biochemically useful energy in order to obtain cell growth. Without oxidizable carbon compounds, heterotrophic growth would be limited. The presence of a reductant which can be oxidized may trigger growth of a facultative autotroph population that can oxidize the reductant with reduction of carbon dioxide. In addition, these reductants may be able to bind directly with many biocides now commonly used in latex paints and latex emulsions which would lower their effectiveness.

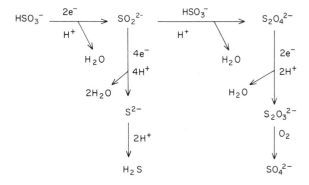

Fig. 2. Biochemical pathway for oxidation of sulphur-containing reductants.

Figure 2 outlines the general mechanism by which these organisms oxidize these reductants. If bisulphite is present, it will be reduced to thiosulphate which is then oxidized to sulphate. Sulphur and hydrogen sulphide will also accumulate. Synthesis of these reduced sulphides is probably the main cause of odours.

These reductants, while lowering the redox potential, may also serve as energy sources. It appears that certain pseudomonads will grow in the presence of reductant either because of the micro-aerophilic nature of the growth medium or because they can utilize the reductant as an energy source. One such isolate was inoculated into a latex emulsion in which the redox potential was varied by additions of sodium meta-bisulphite. An inoculum size of $1 \cdot 10^3$ viable bacterial per ml was placed in 100 ml of a vinyl acrylic latex emulsion. Maximum growth

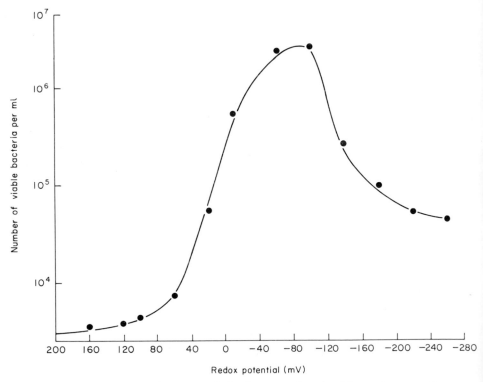

Fig. 3. Growth of a facultative autotroph *Pseudomonas* sp. inoculated into a latex emulsion at different redox potentials.

occurred at a redox range of -60 mV to -100 mV potential (Fig. 3). Whether these results are due to this being the optimum redox range for this organism or because this optimum redox range is also the optimum concentration of bisulphite for maximum growth is not certain. The data in Table 2, in which this same organism was inoculated into a hydroxyethyl cellulose-containing medium and into a nutrient broth, show the same kind of results. With the former medium, the optimum redox potential range was between $+50$ mV and -50 mV.

The time-course of growth of this isolate in latex emulsion is shown in Fig. 4. The organism reduced the latex emulsion from an initial $+50$ mV potential to a redox value of -100 mV. This organism demonstrates the reducing activity of bacteria that has been observed in many latex systems. The potential of some contaminated latex paints has been reduced to a value as low as -750 mV.

These results may also explain why many investigators have had difficulty in getting bacterial isolates from contaminated latex systems to grow in an untreated latex-containing medium. The absence of a suitable energy source may be the answer.

The problem of gas production caused by the presence of bacteria in the container is a serious problem. When this problem occurs, the containers bulge, lids of the containers fly off, and the paint is con-

Table 2

Growth of *Pseudomonas* sp. at different redox potentials. $1 \cdot 10^3$ of *Pseudomonas* sp. per ml was inoculated into 100 ml of 1.5% hydroxyethyl cellulose supplemented with 0.3% sodium nitrate, 0.3% potassium dihydrogen phosphate and 0.3% dipotassium hydrogen phosphate (pH 7.2) and nutrient broth (pH 6.9)

| | Number of pseudomonads in | |
Redox potential	Hydroxyethyl cellulose medium	Nutrient broth
$+300$	0	0
$+250$	3.1×10^3	0
$+200$	3.2×10^4	0
$+150$	4.0×10^3	0
$+100$	2.1×10^5	0
$+50$	2.4×10^6	2.1×10^2
0	2.8×10^7	4.8×10^5
-50	2.6×10^6	1.3×10^7
-100	4.5×10^5	1.3×10^7
-150	3.8×10^3	1.1×10^5

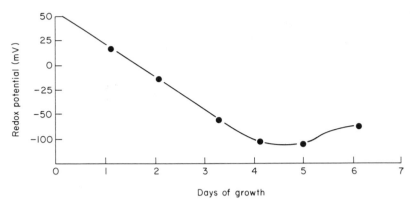

Fig. 4. Effect on redox potential due to growth of a facultative autotroph *Pseudomonas* sp. in a latex emulsion.

sidered an economic loss. Winters and Goll (1974) have isolated facultative anaerobes such as species of *Pseudomonas*, *Proteus* and *Enterobacter* from latex paints that have undergone gassing. These bacteria lower the redox potential of a latex paint, and convert nitrates into nitrogen gas and formaldehyde, and glyoxal into carbon dioxide. Production of these gases change the vapour pressure within the can of paint.

III. BIODETERIORATION OF EXTERIOR PAINT FILMS ON WOOD SUBSTRATES

Biodeterioration of paint films is a problem that has plagued the paint industry for many years. Attachment of bacteria and fungi is referred to as defacement, and the economic loss attributed to this microbiological attack has been estimated in excess of one million U.S. dollars per year. Identification studies of fungi presented on paint films have shown the presence of species of *Phoma*, *Cladosporium*, *Alternaria*, *Penicillium*, *Dematium*, *Aspergillus*, *Trichoderma*, *Mucor*, *Rhizopus* and *Aureobasidium*. In addition, species of *Pseudomonas*, *Flavobacterium*, *Micrococcus*, *Desulfovibrio*, *Pichia* and *Rhodotorula* have also been identified. *Aureobasidium pullulans*, because of the black pigment that it synthesizes, has long been implicated as the main defacer of paint films (Goll, 1958; Ross, 1964; Duncan, 1968). It is believed that the nutritional needs and other capabilities of *A. pullulans* allow for its successful colonization of paint films. It can metabolize many sugars and some strains may even utilize cellulose and hemicellulose.

The defacement syndrome is an accumulative process in which a succession of species become attached and grow over an extended period of time. The data of Winters *et al.* (1978), as well as that of Schmitt (1978), indicated many different micro-organisms involved in this mildew syndrome. Although Schmitt (1978) did not observe any succession pattern, Winters *et al.* (1978) did suggest a succession pattern of micro-organisms leading to the ultimate dominance of *A. pullulans*.

The successful dominance of *A. pullulans* is believed to be due to its adaptation to physical environments similar to those surrounding paint films, to its resistance of high temperature, ultraviolet radiation and periodic dessication. These factors, when combined with the fact that *A. pullulans* does release an antibiotic substance (Van Den Heuvel, 1970), probably account for its considerable dominance on paint films. Schmitt (1978) and Lingappa *et al.* (1963) have shown that *A. pullulans* will show different growth characteristics depending on the nature of the growth medium. Not only were the growth characteristics modified, but synthesis of the black pigment was also affected by growth medium. Catley (1971) and Merdinger (1964) observed that *A. pullulans* produced this black pigment only when grown on a suitable carbohydrate source. Metabolism of sodium acetate and glycerol did not lead to synthesis of this black pigment. If phenolic or quininoid compounds present in the atmosphere or leached from the wood substrata are toxic to fungi, then *A. pullulans* may synthesize this black pigment in order to detoxify these compounds. Plants are known to be protected from microbial colonization by secretion of toxic phenolic and quininoid compounds. Horvath and others (1976) have shown that *A. pullulans* can utilize polyphenolics and aromatic extracts from wood as carbon and energy sources. McGlown and Old (1974) have stressed that, in the biodeterioration of surface coatings, a complex ecosystem was involved. This included a delicate balance between the varieties of micro-organisms grown on the film and the environmental conditions at the surface.

Schmitt and Padgett (1973), in trying to elucidate the mechanisms of *Aureobasidium pullulans* degradation of organic coatings, have hypothesized that a symbiotic relationship exists between *A. pullulans* and bacteria of genus *Pseudomonas* which may allow *A. pullulans* to grow on paint films. The findings of Schmitt and Padgett (1973) suggest that an ecosystem may exist on paint film surfaces. The suggestion that a natural ecosystem may exist during biodeterioration of paint films is analogous to the fouling of marine surfaces (Cope, 1973; Marshall, 1973). A complex ecosystem on paint film surfaces, composed of micro-organisms, organic and inorganic materials, would suggest that a fungicide directed at a single organism may have limited use. It therefore seems important that, in order to understand fully the biodeterioration process on paint film surfaces, it will be necessary to understand both the type of microbial succession that takes place as the biodeterioration of paint film proceeds and how the micro-organisms relate enzymically to one another on the paint films. In this regard, the total ecology of the micro-organisms and their physiology will be necessary to understand the interactions between organisms that occur on paint film surfaces. Once we know what initiates microbial growth and the physiology of the succession process, a meaningful attempt can be made to inhibit this biodeterioration process on paint films.

REFERENCES

Buono, F., Stewart, W. J. and Freifeld, M. (1973). *Journal of Paint Technology* **45**, 43.
Catley, B. J. (1971). *Applied Microbiology* **22**, 650.
Cope, W. A. (1973). *Developments in Industrial Microbiology* **15**, 67.
Duncan, C. O. (1968). *Official Digest* **38**, 1003.
Floyd, J. D., Gill, J. W. and Wirick, M. G. (1966). *Journal of Paint Technology* **38**, 398.
Goll, M. (1958). *Official Digest* **30**, 399.
Goll, M. (1972). *Project Cope, Federation of Societies for Paint Technology* 56.
Goll, M. and Winters, H. (1973). *Proceedings of the 13th Annual Coatings Symposium, North Dakota State University.*
Holmes, W. F. (1969). *Journal of Paint Technology* **41**, 688.
Horvath, R. S., Brent, M. M. and Cropper, D. G. (1976). *Applied and Environmental Microbiology* **32**, 505.
Lalk, R. H. (1964). *American Paint Journal* **67**, March 2.
Lingappa, Y., Sussman, A. S. and Bernstein, I. A. (1963). *Mycopatholia et Mycologia Applicata* **20**, 109.
Marshall, K. C. (1973). *Proceedings of the 3rd International Congress on Marine Corrosion and Fouling* **3**, 62.
McGlown, D. J. and Old, G. (1974). *Journal of the Oil and Colour Chemists' Association* **57**, 13.
Merdinger, E. (1964). *Transactions of the Illinois State Academy of Science* **57**, 28.
Miller, W. G. (1973). *Journal of the Oil and Colour Chemists' Association* **56**, 307.
Rosenberg, P. (1978). *Journal of Coatings Technology* **50**, 83.
Ross, R. T. (1964). *Developments in Industrial Microbiology* **6**, 149.

Ross, R. T. (1972). *Project Cope, Federation of Societies for Paint Technology* 53.

Schmitt, J. A. (1978). *Journal of Coatings Technology* **50**, 35.

Schmitt, J. A. and Padgett, D. (1973). *Journal of Paint Technology* **45**, 31.

Van Den Heuvel, J. (1970). *Netherland Journal of Plant Pathology* **76**(3), 192.

Wienert, L. A. and Vanderstraeten, W. W. (1973). *Journal of the Oil and Colour Chemists' Association* **56**, 292.

Wimpenny, J. W. T. and Nicklen, D. K. (1971). *Biochimica et Biophysica Acta* **253**, 352.

Winters, H. (1972). *Journal of Paint Technology* **44**, 39.

Winters, H. (1979). *Proceedings of the 1st Symposium on Paint Microbiology*. Paint Research Association, London, England.

Winters, H. (1980). *Journal of Coatings Technology* **52**, 71.

Winters, H. and Goll, M. (1974). *Journal of Paint Technology* **46**, 49.

Winters, H., Isquith, I. R. and Goll, M. (1978). *Developments in Industrial Microbiology* **17**, 167.

11. Rubber

BRONISLAW J. ZYSKA

Central Mining Institute, 40–951 Katowice, Poland

I. Introduction 323
II. Interaction of Rubber and Micro-Organisms 325
   A. Micro-Organisms of *Hevea* Latex 325
   B. Micro-Organisms of Raw Natural Rubber 327
   C. Microbial Resistance of Synthetic Rubbers 329
   D. Microbial Resistance of Compounding Ingredients 333
   E. Microbiological Resistance of Vulcanizates 337
   F. Microbial Deterioration of Certain Rubber Goods 347
III. Colonization of Rubber Products by Pathogenic Micro-Organisms . . . 367
IV. Protection Against Micro-Organisms 368
   A. Preservation of *Hevea* Latex 368
   B. Preservation of Natural Rubber 370
   C. Protection of Rubber Products 371
V. Acknowledgements 379
References 380

I. INTRODUCTION

A rising trend in level of rubber consumption is one of the indices characterizing technical development in every industrialized country. Economic analyses show that during the last 30 years the demand for rubber in various countries is proportional to the Gross National Product. Figure 1 shows the increase in World production of natural rubber and synthetic rubbers in the period 1950–1978. In the year 1979, the World production of rubbers was $10.84 \cdot 10^6$ tonnes, that is $3.77 \cdot 10^6$ tonnes of natural rubber and $7.07 \cdot 10^6$ tonnes of synthetic rubbers. In many branches of industry, development is largely governed by achievements and advances made in the rubber industry. One of the properties often required in rubbers is high resistance to micro-organisms, especially for products to be used in moist air, communal and sea water, sewages, soil, deep mines and in tropical climates. In view of progress made in investigations on microbiological deterioration of raw rubbers and their vulcanizates, the development of certain rubber goods should envisage a more accurate choice of microbially resistant compounding ingredients and/or incorporation into the mix of a suitable microbiocide.

Much of the literature on rubber technology and manufacture neglects the problem of micro-organisms affecting resistance of rubber goods and their reliable life (Blow, 1971; Westphal, 1964; Dogadkin, 1972; Eirich, 1978). Extensive references on microbiological deterioration of elastomers and rubbers are given by Williams (1958), Gillespie

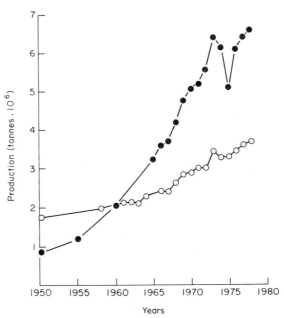

Fig. 1. World production of natural rubber ($\bigcirc$) and synthetic rubber ($\bullet$).

(1963), Becker and Gross (1974), Zyska (1977) and Biodeterioration Information Centre (1979).

In approximate calculations of economic losses due to microbial deterioration the assumption is made that 2% of all technical materials are lost because of attack of micro-organisms (Hueck-van-der-Plas, 1965). In the case of rubber, on the basis of values for the year 1977, this means that 200 thousand tonnes of rubber were lost due to deterioration, the equivalent of 400 thousand tonnes of rubber goods. The value of economic losses cannot be calculated precisely due to differentiated prices of rubber in the manufactured rubber goods. Example of rubber pipe joints, used in water or sewage pipelines, shows that microbial deterioration shortens the life of a certain proportion of the joints from 50 to 4–8 years. According to Hutchinson *et al.* (1975), the price of installing one metre of pipeline 25 mm in diameter, is one pound sterling. Every leakage in a pipeline caused by damage to rubber joints before the elapse of 50 years will be the source of economic losses.

II. INTERACTION OF RUBBER WITH MICRO-ORGANISMS

A. Micro-Organisms of *Hevea* Latex

Latex is collected from the tree *Hevea brasiliensis* by daily or periodical tapping of the cambium to a depth of 1.0–1.5 mm. The plantations are situated in the tropical zone, in the countries of south east Asia and in Africa. In new plantations, owing to cultivation of selected clones, the use of fertilizers, improved tapping systems and stimulation of latex outflow, the annual yield has been increased more than five-fold in comparison with older plantations, approaching 2 mg per hectare per year.

Latex is a complex dispersion of rubber particles in the water phase, containing about 35% rubber, 60% water and 5% of compounds, such as proteins, amino acids, lipids, quebrachitol, certain other carbo-hydrates and inorganic and organic constituents.

Latex while in the *Hevea* tree is sterile, as has been demonstrated in the investigations of Beeley (1940), McMullen (1951) and Taysum (1957a, 1966). After tapping, and as the latex flows along the tapping cut into the cup and is subsequently collected and bulked in the factory,

Table 2

Microbiological resistance of raw, unvulcanized synthetic rubbers

Type of rubbers	28-Day test (Lazar and Ioachimescu, 1972; Lazar, 1973)		14-Day test (Dubok et al., 1971)		5–12 Month test (Schwartz, 1963)		6-Month test (Schwartz, 1963)	
	Trade name of rubber	Microbial resistance	Trade name of rubber	Microbial resistance	Trade name of rubber	Microbial resistance	Trade name of rubber	Microbial resistance
Styrene-butadiene	Carom 1500	0	SKS-30	3	Buna S	2	Buna S	3
	Carom 1712 with oil	3	SKMS-30-ARKM 15	3				
			SKMS-10	3				
Polyisoprene	Cariflex I.R. 305	2	SKI-3	3	—	—	—	—
Polybutadiene	Europrene cis	1	SKD	3	—	—	—	—
Isobutene-isoprene (butyl)	—	—	not given	1	not given	4	not given	3
Ethylene-propylene	Keltan 712	0	SKEP	3	—	—	—	—
Chloroprene	Denka S-40	0	Nairit A	0	Neopren DWN	0	Neopren DWN	3
			Nairit B	0	Neopren W	1	Neopren W	0
Acrylonitrile-butadiene (nitrile)	SKN-26	0	SKN-26	0	Perbunan	3	Perbunan	2
Fluorocarbon	—	—	SKF-26	1	—	—	—	—

to deteriorate different types of synthetic rubbers and to provide a basis for more general conclusions.

D. Microbiological Resistance of Compounding Ingredients

All rubber products contain many individual constituent materials, which are brought together by a physical mixing process and cross-linked in the vulcanization process at a given pressure and temperature. A mix includes such constituents as base polymer or blend of polymers, cross-linking agents, accelerators of cross-linking and their modifiers, anti-oxidants and anti-ozonants, re-inforcing fillers, processing aids, diluents, colouring materials, blowing agents, fungicides and fibrous materials.

In manufacture of vulcanized rubbers, which are resistant to micro-organisms, compounding ingredients which have a satisfactory level of microbiological resistance must be selected. Finding ingredients resistant to micro-organisms presents fewer difficulties than finding an adequate fungicide for a rubber. Hundreds of compounding ingredients are known in rubber technology and manufacture, but their microbiological resistance was extensively investigated only by Dubok *et al.* (1971), Lazar (1973) and Lazar and Ioachimescu (1973).

Table 3 presents evidence on the fungal resistance of some compounding ingredients in 14- and 28-day tests on agar-containing media both without and with sugar as the organic source of carbon. Resistance is classified according to the scale given in Table 1. Differences found in the evaluation of microbiological resistance may be due to variations in test methods and test periods as well as differences in chemical composition of the trade products.

Carbon blacks, non-black fillers and colouring materials are resistant, but when in contact with edible substances, they are sensitive or very susceptible to fungal growth. Sulphur, as the vulcanizing agent, behaves as a semiresistant material or a slightly fungitoxic material. The fungicidal and bactericidal properties of accelerators are known since the investigations of Dimond and Horsfall (1943). The anti-oxidants and anti-ozonants include both fungitoxic and very sensitive compounds. Only a few of the plasticisers, softeners and extenders listed in Table 3 are fungicidal or semiresistant, most being very sensitive to fungal growth. The blowing agent is sensitive to attack by fungi, but it has

Table 3

Fungal resistance of certain compounding ingredients

Ingredient	Trade name, country	Fungal resistance in 28-day test[a] (Lazar, 1973; Lazar and Ioachimescu, 1973)	in 14-day test[b] (Dubok et al., 1971)
Carbon blacks			
Channel, from natural gas	Dg-100, U.S.S.R.	—	—/3
Channel, from anthracene oil	DgM-80, U.S.S.R.	—	—/3
Furnace, from hydrocarbon oils	PM-70, U.S.S.R.	—	—/3
Furnace, type HAF	Rebonex H, Romania	0/2[c]	—
Furnace, type FEF	Carbondis 50, Romania	0/3	—
	PM-50, U.S.S.R.	—	—/3
Furnace	PM-30, U.S.S.R.	—	—/3
Furnace	PM-15, U.S.S.R.	—	—/3
Furnace, type SRF	Methanex D, Romania	1/1	—
Furnace, type GPF	Furnal R-300, Romania	0/2	—
Non-black fillers and colouring materials			
Calcium silicate	U.S.S.R.	—	—/3
Calcium fluoride	U.S.S.R.	—	—/3
Magnesium oxide	U.S.S.R.	1/2	—/1
Calcium carbonate	U.S.S.R.	—	—/0
Asbestos	U.S.S.R.	—	—/3
Synthetic silica	Aerosil, G.F.R.	—	—/3
China clay	Romania	1/2	—
Lithoprone	U.S.S.R.	—	—/3
Talc	U.S.S.R.	—	—/3

Table 3—*cont.*

Ingredient	Trade name, country	Fungal resistance	
		in 28-day test[a] (Lazar, 1973; Lazar and Ioachimescu, 1973)	in 14-day test[b] (Dubok et al., 1971)
Vulcanizing agents			
Sulphur	U.S.S.R., Romania	1/1	—/0
Accelerators			
2-Mercaptobenzothiazole	Kaptaks, U.S.S.R.	—	—/00
	Vulkacit M, G.F.R.	0/00	—
Dibenzothiazyl disulphide	Altax, U.S.S.R.	—	—/0
	Vulkacit DM, G.F.R.	0/00	—
N-Cyclohexylbenzothiazylsulphenamide	Vulkacit CZ, G.F.R.	0/00	—/00
	Santocure, U.S.A.	—	—/00
Zinc diethyl dithiocarbamate	Vulkacit ZDK, G.F.R.	1/00	—
NN'-Diphenyl guanidine	Vulkacit D, G.F.R.	0/00	—
	DFG, U.S.S.R.	—	—/3
Tetramethyl tiuram disulphide	Vulkacit Th, G.F.R.	0/00	—
	Tiuram, U.S.S.R.	—	—/00
Anti-oxidants and anti-ozonants			
2,2'-Methylene di-4-methyl-6-t-butylphenol	NG-2246, U.S.S.R.	—	—/1
2-Mercaptobenzimidazole	U.S.S.R.	—	—/3
	Antioxidant MB, Japan	0/0	—
Phenyl-β-naphthylamine	Antioxidant PBN	0/0	—
	Neozon D, U.S.S.R.	—	—/1
N-Isopropyl-N'-phenyl-p-phenylenediamine	Santoflex, England	0/0	—/00
2,2,4-Trimethyl-1,2-dihydroquinoline polymers	Flectol H, England	1/2	—
6-Ethoxy-1,2-dihydro-2,2,4-trimethylquinoline	Santoflex AW, England	2/2	—

Table 3—cont.

| Ingredient | Trade name, country | Fungal resistance | |
		in 28-day test[a] (Lazar, 1973; Lazar and Ioachimescu, 1973)	in 14-day test[b] (Dubok et al., 1971)
Anti-ozonant waxes			
Ceresine	Wosk AF-1, Wosk ZV-1, U.S.S.R.	—	—/3
Paraffin wax	Antiflux 654, G.F.R.	4/4	—
Plasticizers, softeners and extenders			
Paraffin	U.S.S.R., Romania	4/4	—/3
Paraffin oil	410 oil, Romania	3/4	—
Aromatic oil	Medium T, Romania	2/3	—
Ceresine	type 57, 85, 100, U.S.S.R.	—	—/1
Ozocerite	type 60, U.S.S.R.	—	—/1
Mineral rubber	Rubrake, U.S.S.R.	—	—/3
Colophony	U.S.S.R., Romania	0/0	—/3
Linseed oil	U.S.S.R.	—	—/3
Factice	U.S.S.R., Romania	2/3	—/3
Stearic acid	Romania	4/4	—
Polypropylene glycol adipate	Icechim plasticizer	4/4	—
Aromatic polyether	Plastikator FH	2/3	—/00
Phthalic anhydride	U.S.S.R.	—	—/0
Dibutyl phthalate	U.S.S.R.	—	—/0
Dibutyl sebacate	U.S.S.R.	—	—/3
Blowing agents			
Dinitrosopentamethylenetetraamine with transformer oil-type TC2004	Micropor, Romania	2/0	—

[a,b] The resistance scale is indicated in Table 1

[c] Without carbon source in medium compared with the presence of sugar in medium

a pronounced fungicidal action when in contact with organic substrates.

The bacteriostatic action of compounding ingredients has only been tested in relation to six accelerators and three anti-oxidants (Winner, 1957). The only bacteriostatic compounds, active to *Staphylococcus aureus*, *Streptococcus pyrogenes*, *Escherichia coli*, *Proteus vulgaris* and *Pseudomonas aeruginosa* were the accelerator *NN'*-diphenylguanidine and the anti-degradant β-naphthol. The bacteriostatic activity of compounding ingredients is important when designing a mixing formulation for rubber goods used in medicine and pharmacy.

E. Microbiological Resistance of Vulcanizates

1. Vulcanizates of Natural Rubber

The action of such degradative agents as light, heat and atmospheric exposure, on rubber vulcanizates has been the subject of investigations since the early 1920s, and much work has been done to search for effective antidegradants. The role of micro-organisms in degradation of rubber goods manufactured from natural rubber was first reported in the years 1942–1949, when World production of natural rubber and synthetic rubbers exceeded two million tonnes per annum. This type of biodegradation was first observed by ZoBell and Grant (1942) on natural rubber stoppers used for determining the biochemical oxygen demand of water samples, and also by Blake and Kitchin (1949) on underground electric cables with natural rubber vulcanizate insulation. Another instance of microbial degradation of natural rubber products was reported in Holland in 1949, where severe deterioration of rubber rings was found in asbestos–cement pipelines after 10 years of service (Rook, 1955; Leeflang, 1963). The only earlier report on the attack of natural rubber vulcanizate by the fungus *Stemphylium macrosporoideum* in India, was given by Scott (1920), when he isolated this organism from wrinkled and distorted natural rubber. An overall review of this problem was made by Cundell and Mulcock (1972) and Zyska (1976, 1977).

According to Cundell and Mulcock (1972), before an organic material such as rubber can be degraded by micro-organisms, certain conditions must be met, i.e. micro-organisms capable of attacking the rubber must be present in the environment and the metabolic require-

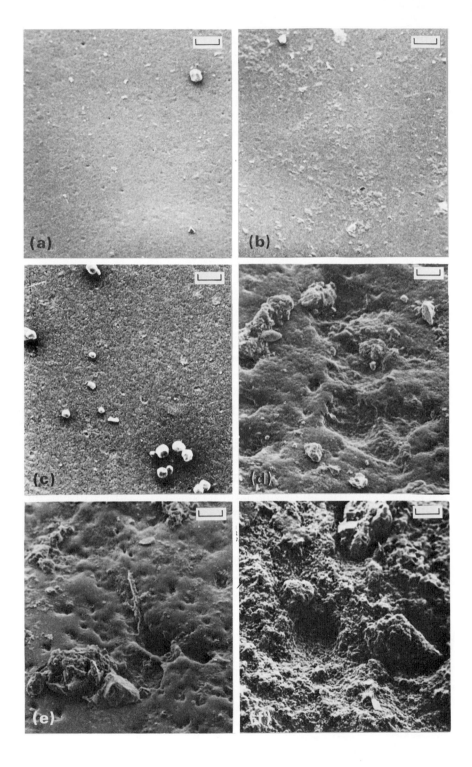

ments of these organisms must be met, environmental conditions must be conducive to growth of these organisms, liquid water must be present and compounds inhibiting microbial growth absent.

Criteria for investigation of microbial deterioration of rubber vulcanizates may be based on changes in measured rubber properties found from samples tested in controlled conditions and on microbiological tests (Zyska et al., 1971; Cundell and Mulcock, 1975; Reszka et al., 1975; Fudalej et al., 1976; Kwiatkowska et al., 1979). The measurement of rubber properties may include scanning electron microscope examination of surface appearance, swelling in organic liquids, estimation of sol fraction of rubber in solvent, network chain density, infrared spectrum, diffusivity of gases, calorific value, weight loss, strength and electrical properties and hardness. In microbiological tests, microbial activity may be measured by estimation of the number of organisms, oxygen uptake, carbon dioxide evolution, nitrogen flux and enzyme content, always observing Koch's rules.

When exposed to coil micro-organisms, the surface of natural rubber vulcanizates shows marked craters and cavities (Fig. 3). In unpublished tests by D. Kwiatkowska and B. Zyska at the Central Mining Institute, Poland, it was shown that carbon-black filler content up to 45 parts per hundred parts of rubber (p.p.h.r.) in natural rubber (0.07 mm thick) has no influence on the geometry of cavities during 45 days soil exposure. However, Reszka et al. (1975), examining natural rubber (1.0 mm thick) exposed to soil micro-organisms in the soil-burial test up to 360 days, found that the higher the content of carbon-black filler in natural rubber, the fewer surface structural changes are visible in a comparable time interval. Evidence of deteriorating action on natural rubber containing a small quantity of filler by *Streptomyces* sp. in pure culture was given earlier by Rook (1955) and Leeflang (1963) in samples exposed to these micro-organisms for two years at 25°C. Changes in the properties of natural rubber vulcanizates due to microbial degradation are given in Table 4, which shows that micro-

---

Fig. 3. Effect of soil burial for 45 days on the surface appearance of specimens of natural rubber vulcanizate containing different proportions of carbon black (expressed as parts per hundred parts of rubber). Micrographs (a), (b) and (c) are controls which were not buried. Specimen (a) lacked carbon black, (b) contained 5 and (c) 45 parts of carbon black per hundred parts of rubber. Micrographs (d), (e) and (f) are of buried specimens. Specimen (d) lacked carbon black, (e) contained 5 and (f) 45 parts of carbon black per hundred parts of rubber. Specimens were examined in the scanning electron microscope. Bars indicate 50 μm.

Table 4

Changes in natural rubber vulcanizates due to microbial degradation

Property	Test micro-organism	Sample thickness (mm)	Testing period (days)	Value of the property in		References
				Control samples	Deteriorated samples	
Equilibrium degree of swelling in toluene (g solvent per g sample)	Soil micro-organisms	0.07	7 14 28	2.81[a] 2.65[b] 2.81[a] 2.65[b] 2.81[a] 3.65[b]	3.49[a] 3.32[b] 3.39[a] 3.18[b] 3.13[a] 3.09[b]	D. Kwiatkowska and B. F. Zyska (unpublished observations)
	Fusarium solani	0.07	7 28 56	2.69[a] 2.43[b] 2.69[a] 2.43[b] 2.69[a] 2.43[b]	3.04[a] 2.82[b] 3.20[a] 3.02[b] 2.95[a] 2.59[b]	
Network chain density ($v \cdot 10^4$, mol/cm³)	Soil micro-organisms	0.07	7 14 28	1.672[a] 1.667[b] 1.672[a] 1.667[b] 1.672[a] 1.667[b]	1.242[a] 1.205[b] 1.311[a] 1.302[b] 1.462[a] 1.375[b]	
	Fusarium solani	0.07	7 28 56	1.626[a] 1.696[b] 1.626[a] 1.696[b] 1.626[a] 1.696[b]	1.483[a] 1.359[b] 1.431[a] 1.253[b] 1.483[a] 1.483[b]	Kwiatkowska et al. (1979)
Infrared spectrum	Micro-organisms in the Leeflang test bath	1.78	360	Normal	Decline in prominence $R{>}C{=}CH_2$—peak at 890 cm^{-1}; appearance of a broad peak in 1,000 cm^{-1}	Cundell and Mulcock (1975)

Table 4—*cont.*

			7–28 years					
Calorific value (cal/g)	Sewage micro-organisms	—		7510	5420			Cundell and Mulcock (1973c)
Loss in weight (%)	Soil micro-organisms	0.07	7 / 28 / 91	0	18.2[a] / 21.7[a] / 40.2[a]	5.8[b] / 19.1[b] / 34.3[b]	5.9[c] / 10.5[c] / 15.0[c]	Kwiatkowska *et al.* (1979)
	Micro-organisms in the soil percolation unit	1.78	547	0	5.3			Cundell and Mulcock (1973a)
	Pink and grey actinomycete	1.78	730	+2.1	7.3–13.4			Cundell and Mulcock (1976b)
Tensile strength (kG/cm^2)	Streptomycetes in liquid medium	0.2	365	122	70–80			Rook (1955)
	Micro-organisms in the Leeflang test bath	1.78	730	272 / 220	79 / 117			Cundell and Mulcock (1975)
Elongation (%)	Streptomycetes in liquid medium	0.2 / 1.78	365 / 730	480 / 560	300–390 / 470			Rook (1955); Cundell and Mulcock (1973b)

[a] carbon black content 0 p.p.h.r.
[b] carbon black content 5 p.p.h.r.
[c] carbon black content 45 p.p.h.r.

organisms cause changes in the chemical, physicochemical and strength properties of natural rubber.

The mechanisms by which properties of rubber vulcanizates are improved and controlled by the use of fillers are often studied using a carbon black suspension in paraffin oil. The kinetics of the structures formation and the thixotropic phenomena are examined using electrical conductivity techniques. Zyska *et al.* (1971, 1973) found that the microbial degradation of natural rubber may cause not only degradation of chains responsible for transmission of stresses but also influence the ability for interaction between filler and rubber. The higher the filler content in the paraffin oil, the greater the destructive influence of the micro-organisms.

Cundell and Mulcock (1973a) measured both the oxygen uptake by micro-organisms growing at the exposure of natural rubber in a Leeflang test bath or in soil percolation units, and carbon dioxide evolu-

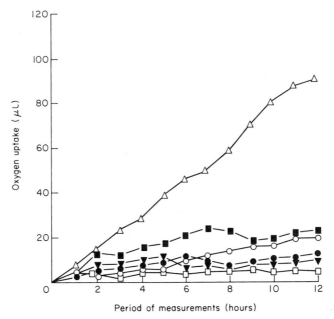

Fig. 4. Time-course of oxygen uptake by micro-organisms growing at the expense of vulcanized rubber after 12 months in a Leeflang test bath, and measured by Warburg microrespirometry. △ indicates the response of natural rubber, ▼ chloroprene rubber, ● butadiene rubber, ■ isobutene-isoprene rubber, □ acrylonitrile-butadiene rubber, and ○ styrene-butadiene rubber. From Cundell and Mulcock (1973a). Reproduced by permission of Verlag Duncker and Humblot and the authors.

tion from surfaces of natural rubber strips removed from a Leeflang
test bath. In Figure 4 are given the oxygen uptake data for 12.7·6.35
·1.77 mm sample by soil micro-organisms growing in natural rubber,
chloroprene, polybutadiene, isobutene-isoprene (butyl), acrylonitrile-
butadiene and styrene-butadiene vulcanizates after 12 months in a
Leeflang test bath (Cundell and Mulcock, 1973a). The natural rubber
included in the mix 30 parts of hard clay per hundred parts of rubber.
Figure 5 shows the oxygen uptake per hour per g of rubber by soil
micro-organisms colonizing natural rubber (0.07 mm thick sheets) with
a carbon black loading up to 45 p.p.h.r. (Kwiatkowska et al., 1979).

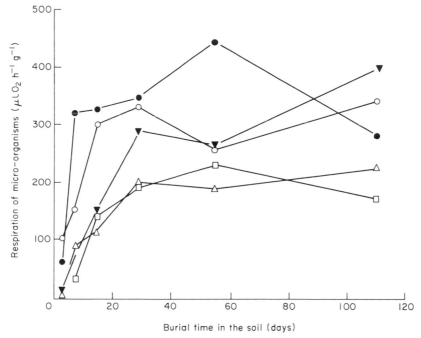

Fig. 5. Time-course of respiratory activity of micro-organisms colonizing vulcanizates of natural
rubber with carbon black contents (expressed as parts per hundred parts of rubber) of 0 (O),
5 (●), 15 (▼), 30 (△) and 45 (□), after exposure in the soil-burial test. Reproduced by
permission of the Biodeterioration Information Centre, University of Aston in Birmingham.

As the susceptibility of vulcanized natural rubber to degradation by
soil and water-borne micro-organisms has been sufficiently confirmed
by research data, Dickenson's (1965a, b, 1968, 1969) conflicting
hypotheses may be rejected. Cundell and Mulcock (1975) postulate

gartikelen KIWA N.V., 1961) and New Zealand (Cundell and Mulcock, 1973a) investigations after contact with potable water or sewage
for up to 18 years.

The microscopic symptoms of microbial deterioration of pipe-joint
rings, seen in scanning electron microscope are growth of a *Streptomyces*
sp. on the surface of ring exposed to the top of the sewage flow line
(Cundell and Mulcock, 1973e) and cavities and craters in the surface
layer of solid rubber rings of natural, styrene-butadiene and isobutene
rubbers as well as of cellular rubber rings of natural, styrene-butadiene
and compounded rubber (Kerner-Gang, 1973, 1977). The *Streptomyces*
species as deteriogens of pipe-joint rings have been confirmed on pure
cultures by Rook (1955) and Leeflang (1963) in Holland, by Cundell
and Mulcock (1973e, 1975, 1976) in New Zealand and by Kerner-Gang
(1973) in the German Federal Republic. Actinomycetes were isolated
from sewage and water-mains joint rings in Great Britain by Hutchinson *et al.* (1975). Table 7 shows some more important characteristics
of these actinomycetes.

Certain other micro-organisms are often isolated from deteriorated
pipe-joint rings. Rook (1955) found that *Fusarium* species could be
readily isolated from such rings though, after two months, no weighable
loss of rubber was detected. Recent investigations on degradation of
natural-rubber sheet by soil micro-organisms conducted at the Central
Mining Institute, Katowice in Poland gave evidence of physicochemical
changes in natural rubber vulcanizates due to attack of pure cultures
of *Fusarium solani* (Kwiatkowska *et al.*, 1979).

According to Hutchinson *et al.* (1975), deteriorated British natural
rubber rings from sewage and water mains contain (per gramme of
rubber), $4.5 \cdot 10 – 4.41 \cdot 10^6$ actinomycete propagules and $0 – 2.26 \cdot 10^4$
sulphur-oxidizing *Thiobacillus neapolitanus*, *T. thioparus* and *T. thiooxidans*. This occurrence of thiobacilli in deteriorated rings in a British
environment is the first report since Rook (1955). These findings show
that the colonization process of natural rubber in pipe-joint rings may
be considerably more complex than previously supposed and further
investigations are needed.

Ecological factors possibly influencing microbial attack of natural
rubber pipe-joints are: type of medium in the pipe-line; presence of
chlorine in potable water; microbial activity of soil; the nature of the
pipe material; the type of pipe-joint rings; the presence of certain
compounds in the rings.

Observations on biodeterioration of rings refer only to water mains, stormwater and sewage lines. Of three grades of water in Dutch pipe lines, surface water, subterranean and dune waters, the highest percentage of deteriorated rings was encountered in dune waters (Table 6). In Great Britain, a large number of actinomycetes were isolated from damaged rings of both water and sewage pipelines after only four years of service (Hutchinson *et al.*, 1975). Attack occurs more frequently in sewage than in water mains. Dickenson (1968, 1969) postulated that chlorination of water may render natural rubber more resistant to microbial deterioration and inhibit the growth of micro-organisms on its surface. Cundell and Mulcock (1973b) measured the evolution of carbon dioxide by micro-organisms respiring on the natural rubber surface in Leeflang test bath using unchlorinated and chlorinated water. Figure 7 shows the regression lines for results from both series of tests.

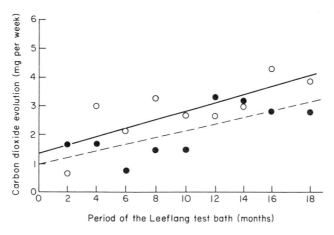

Fig. 7. Carbon dioxide evolution from micro-organisms growing on vulcanized natural rubber suspended in a Leeflang test bath in chlorinated (●) or unchlorinated (○) water. From Cundell and Mulcock (1973b).

These relations are time-dependent but, due to the scatter of results, regression analysis provides no evidence of the microbiological effect of chlorine in water. That chlorine has no effect on microbial deterioration of natural rubber in British water mains is stressed by Hutchinson *et al.* (1975). Deterioration is more evident on the internal surface of the ring than on the soil side. In New Zealand, however, observations on sewage pipelines showed 35.8% of rings deteriorated on the inside circumference, 33.4% on the outside circumference in contact with the soil in the backfilled trench, and 6.1% on both sides of the rings. In addition, 9.0% showed distinct signs of softening at the sewage flowline. The influence of the pipe material on the deterioration of natural rubber rings is difficult to assess. Cundell and Mulcock (1973e) and Hutchinson *et al.* (1975) contend that the use of concrete or asbestos cement pipes favours the development of streptomycetes in natural rubber pipe-joint rings. However, in polyvinylchloride sewage mains and water mains, actinomycetes and thiobacilli colonized the natural rubber rings causing in four years roughening of their surface with some pitting.

Comparison of solid-rubber and cellular-rubber rings gives evidence of remarkable sensitivity of the latter to deterioration (Kerner-Gang, 1973, 1977). The natural content of nitrogen in the natural rubber has no influence on the resistance of rings to deterioration (Leeflang, 1963). Including sulphur in the mix, to increase the number of cross-links in the vulcanized rubber (Cundell *et al.*, 1973a, b) must be used with caution as sulphur may favour growth of thiobacilli, possibly followed by acid breakdown of rings and corrosion of cement or asbestos cement pipes (Thaysen *et al.*, 1945; Hutchinson *et al.*, 1975). The type of rubber used in pipe-joint ring manufacturer may be chosen to give increased microbial resistance of the product. In Dutch investigations, rings of acrylonitrile-butadiene, chloroprene, styrene-butadiene and isobutene-isoprene (butyl) rubbers after 21 months of test, were found to be inert to microbial deterioration (Keuringsinstituut voor Waterleidingartikelen KIWA N.V., 1961). American formulations of chloroprene, isobutene-isoprene (butyl), acrylonitrile-butadiene and styrene-butadiene rubbers were also shown to be resistant to micro-organisms after 24 months (Leeflang, 1968). In the German Federal Republic, resistance of styrene-butadiene rings to microbial deterioration after two years observations is reported (Kerner-Gang, 1973), and similarly for chloroprene and isobutene-isoprene (butyl) rubber rings in Great Britain by Hutchinson *et al.* (1975). The findings of Cundell and Mulcock (1973a) provide evidence for excluding styrene-butadiene rubber from rubber-ring formulations.

Failure of a pipe-joint ring due to microbial deterioration could represent a grave public-health hazard and repair work would be expensive. Leakage of raw sewage from a failed joint may lead to con-tamination of ground waters and subsidence of the surrounding soil. No pipe failures in Britain and New Zealand have been attributed directly to biodeterioration, although the possibility of such a process being a contributory factor cannot be excluded (Cundell and Mulcock, 1973a; Hutchinson *et al.*, 1975). In industrial cases of failure in materials, the microbial factor is most often neglected, and the micro-biological expert will be the last to be called for consultation. Again, more co-operation in the design stage could lead to manufacture of rubber products largely resistant to micro-organisms, giving enormous economic advantages. Practical exploitation in industry of reliable microbiological research results is usually delayed for many years.

3. Elastomer-Based Electric Cables, Telephone and Electronic Equipment

Progress in design and technology demands a continual improvement in the formulation of compounding materials for cables, telephone and electronic equipment. The majority of results on formulations derived from tests of environmental influence on the reliability of cables and equipment, performed in industrial laboratories, are unpublished. The scarcity of information on microbial resistance of modern insulation and sheath polymers used in cables or equipment makes it difficult to give an adequate review of the current state of knowledge of microbial processes in elastomers.

Blake and Kitchin (1949) and Blake and his coworkers (1950, 1953, 1955) were the first to publish detailed results of tests on microbial degradation of rubber insulations, made of natural, styrene-butadiene and isobutene-isoprene (butyl) rubber vulcanizates. Their findings are now of limited value as significant progress has been made in the manufacture of synthetic rubbers and the design of mixing formulation for cables insulation and sheaths, and many new ingredients are now used. Nevertheless their results are of special interest because controls were made in sterile soil. Tests were performed on No. 14 tinned copper wire with 1.19 mm rubber insulation. Thin-walled vulcanized natural, styrene-butadiene or isobutene-isoprene (butyl) rubber latex insulation was applied by the dip process to give a 0.38–0.51 mm wall. The results are shown in Figure 8. The rubrax-softened natural rubber insulation vulcanized with benzothiazyl disulphide and zinc dimethyldithio-carbamate, showed a loss in insulation resistance in active soil of pH 8 from 10^6 to 10 $M\Omega$ during 3.5 months. In sterile soil, the insulation resistance remained unchanged for 20 months (Fig. 8, curves 1,2 and 7,8). The influence of soil micro-organisms on natural rubber is empha-

Table 8

Changes in properties of electroinsulating elastomers after burial in soil. From Connolly (1976.)

Elastomer	Order of magnitude change in DC insulating resistance of elastomer-insulated copper–steel conductors after one year of soil-burial exposure	Lowering in tensile strength in elastomeric compounds after up to eight years soil-burial exposure (%)
Natural rubber insulation	−5	—
Natural rubber sheath	−0	over 51
Styrene-butadiene rubber insulation	−0.5	—
Styrene-butadiene rubber sheath	−2.5	16–25
Chloroprene rubber sheath, type W, magnesium oxide	−2.5	—
Chloroprene rubber sheath, type W, red lead cure	−2.5	—
Chloroprene rubber sheath, type W, red lead cure and microbiodices	−3.0	—
Chloroprene rubber with clay	—	over 51
Chloroprene rubber, no clay, carbon black	—	1–15
Isobutene-isoprene rubber insulation	−3.5	26–50
Acrylonitrile-butadiene rubber, sulphur cure	—	increase 15
Acrylonitrile-butadiene rubber, heat-resistant	—	16–25
Chlorosulphonated polyethylene	+0.5	16–25
Dimethyl silicone	−6.0	—
Ethylene-propylene copolymer	—	16–25
Ethylene-propylene terpolymer	—	26–50
Polyurethane (ester)	—	over 51
Polyurethane (ether)	—	1–15

for deterioration of materials. This has been stressed by Chamberlain et al. (1965), Mikhailov et al. (1967) and Zyska (1964, 1968). The colonization of chloroprene-rubber sheaths of mining cables by micro-organisms is shown in Table 9 (Zyska et al., 1972). In the Prezydent coal mine in Poland, the cables were used for two years, but in the remaining mines for only one year. It was found that, in deep mines, the association of micro-organisms with chloroprene rubbers differs. During the isolation of the micro-organisms (Fig. 9), it was found that

Frequencies of micro-organisms in rubber cables in coal mines

Frequency of micro-organisms in each rubber cable (%)

Type	Source	Depth of isolation (mm)	Number of micro-organisms in the layer	*Penicillium* sp.	*Aspergillus* spp.	*Cephalosporium* spp.	*Trichoderma* spp.	*Scopulariopsis* spp.	*Botrytis* spp.	*Fusarium* spp.	*Trichothecium* sp.	*Verticillium* sp.	*Cladosporium* sp.	*Chaetomium* sp.	*Gliocladium* sp.	*Spicaria* sp.	Non-identified fungi	Actinomycetes	Bacteria
4 × 10 mm^2; 0.75 kV	Prezydent coal mine, level 212 m	0.0–0.5	9	19.3	34.6	26.9	7.7	7.7	—	—	—	—	—	—	—	—	—	—	3.8
		0.5–1.1	17																
1 × 400 mm^2; 1.0 kV		0.0–0.5	18	8.3	58.4	11.2	—	2.7	2.7	—	—	2.7	—	2.7	—	—	5.6	—	5.6
		0.5–1.0	18																
5 × 4 mm^2; 0.75 kV		0.0–0.4	19	14.5	35.3	3.0	—	11.8	3.0	11.8	3.0	—	5.8	—	—	—	5.8	3.0	3.0
		0.4–1.1	15																
3 × 35 mm^2; 1.0 kV	Andaluzja coal mine, level 190–303 m	0.0–0.5	13	—	11.4	28.5	11.4	—	—	11.4	—	5.7	5.7	—	—	—	22.9	—	2.9
		0.5–1.2	22																
4 × 50 mm^2; 1.0 kV		0.0–0.5	12	—	2.9	31.4	25.7	—	—	—	—	5.7	5.7	—	2.9	2.9	14.3	—	8.5
		0.5–1.2	23																
5 × 4 mm^2; 0.75 kV		0.0–0.5	12	—	3.4	27.6	34.6	—	—	—	—	3.4	—	—	—	—	27.6	—	3.4
		0.5–1.1	17																
5 × 4 mm^2; 0.75 kV	Radzionków coal mine, level 300–400 m	0.0–1.1	19	16.6	75.0	—	—	—	—	—	—	—	5.6	2.8	—	—	—	—	—
		1.1–1.8	7																
		1.8–2.9	10																
4 × 16 mm^2; 0.75 kV		0.0–1.1	16	53.8	11.6	15.5	—	—	—	3.8	—	—	3.8	—	—	—	3.8	—	7.7
		1.1–2.2	8																
		2.2–2.8	2																
5 × 4 mm^2; 0.75 kV	Kazimierz coal mine, level 300 m	0.0–0.9	14	39.1	30.3	13.4	—	—	—	4.3	—	—	4.3	4.3	—	—	4.3	—	—
		0.9–1.8	9																
5 × 4 mm^2; 0.75 kV		0.0–1.1	7	23.1	46.1	—	—	—	—	—	—	—	7.7	15.4	—	—	7.7	—	—
		1.1–1.9	5																
		1.9–2.3	1																
5 × 4 mm^2; 0.75 kV		0.0–1.3	12	31.8	31.8	9.1	—	—	—	—	—	—	—	18.2	—	—	9.1	—	—
		1.3–2.6	4																
		2.6–4.4	6																

they are able to colonize the sheath rubber up to a depth of 4.4 mm. Evaluation of electrical and mechanical properties in 4 · 2.5 mm² rubber-cable sections under the influence of mine and soil micro-organisms, is shown in Table 10. A close parallel may be observed between the influence of mine roadway environments and that of the soil-burial test. The direction and magnitude of changes seen under practical conditions are in agreement with those observed in tests. In the author's opinion, microbial degradation of rubber sheaths in mine cables is certainly common to all deep mines although cable failures are generally regarded as being of exclusively mechanical origin.

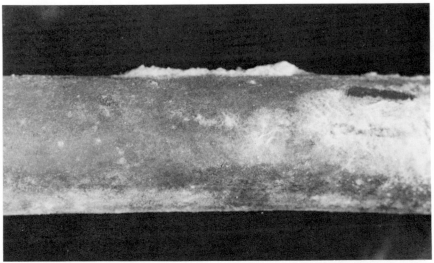

Fig. 9. Photograph of a chloroprene rubber mining cable, with a volume of 4 · 50 mm², an internal diameter of 64 mm and carrying 1 kV, from a coal mine at a depth of 350 m. The fungi colonizing the sheath rubber of the cable are species of *Aspergillus*, *Cephalosporium*, *Cladosporium*, *Fusarium*, *Spicaria*, *Trichoderma* and *Verticillium*, together with unidentified fungi and bacteria. For details, see information on the fifth cable sample in Table 9, p. 359.

Fungal growth on rubber earpad-covers of telecommunications head-phones was reported by Uniyal *et al.* (1971) from India. In earlier work on mould growth on gas masks, carried out at the Defence Research Laboratory in Kanpur in India, as many as 18 fungi were isolated, most of them belonging to the genus *Aspergillus*. Apart from the clinical aspects, fungal growth on rubber or rubberized components of tele-communications and electronic equipment has a psychological aspect, as operators doubt the reliability of this equipment.

Silicon rubber is a very versatile elastomer, no other sealant appears

Table 10

Electrical and Mechanical properties of rubber cable (4 × 2.5 mm²) influenced by mine and soil micro-organisms

Testing environment	Duration (days)	Capacitance		Leakance		Electrical strength		Mechanical fatigue strength (Contra-flexures × 10³)	
		(pF)	(%)	(tg × 10⁻³)	(%)	(kV)	(%)		(%)
Control	0	75.9 ± 0.39	100.0	32.4 ± 0.01	100.0	37.5 ± 1.13	100.0	32.4 ± 4.80	100.0
Out-take air Kazimierz mine	120	107.2 ± 0.51	141.2	79.9 ± 1.51	246.6	29.5 ± 0.41	78.6	5.8 ± 0.38	17.9
	240	182.0 ± 18.13	239.8	68.8 ± 9.64	212.3	17.7 ± 0.66	47.2	4.6 ± 0.17	14.2
Out-take air Gottwald mine	210	105.3 ± 7.15	138.7	70.0 ± 4.24	216.0	28.0 ± 1.00	74.6	6.0 ± 0.61	18.5
Ageing in oxygen	1.25	68.2 ± 1.17	89.8	25.0 ± 0.18	77.1	45.5 ± 0.05	121.3	18.6 ± 2.48	57.4
Soil-burial test	120	96.8 ± 0.53	127.7	67.5 ± 1.35	208.3	28.3 ± 0.07	75.4	6.0 ± 0.87	18.5
Ageing in oxygen and soil-burial test	1.25 120	87.6 ± 0.90	115.4	62.1 ± 6.90	191.6	34.0 ± 2.71	90.6	4.7 ± 0.20	14.5

such as *Trichophyton rubrum*, *T. mentagrophytes* or *Epidermophyton floccosum* has been reported several times (Ajello and Getz, 1954; Alteras and Ignatescu, 1952; Gentles, 1957; English, 1969; English and Gibson, 1959; Schönborn *et al.*, 1968). These three fungi are responsible for dermatophytosis of the foot, and the first two for onychomycosis in human beings. *Candida albicans* is reported as a pathogen in moist inter-digital lesions of dermatophytosis of the foot (Emmons *et al.*, 1977). Koch *et al.* (1966) gave evidence that the hyphae and conidia of dermatophytes colonize rubber products over a long period, being the source of repeated infection.

IV. PROTECTION AGAINST MICRO-ORGANISMS

A. Preservation of *Hevea* Latex

To keep *Hevea* latex fluid for a day or two, it is essential to protect the protein–lipid layer surrounding the rubber particle. This is com-monly achieved by addition of small quantities of preservatives. Such short-term preservatives are called anticoagulants. If the latex is to be kept for long periods, the normal practice is to treat the field latex with a strong preservative, which will either kill most of the acid-producing bacteria or prevent them from producing any acids. Higher concentra-tions of preservatives, commonly called long-term preservatives, are required in the manufacture of latex concentrate.

Ammonia, formaldehyde and sodium sulphite are short-term pre-servatives (Cook, 1960). The anticoagulant effect of ammonia was discovered in 1791 and is still the best general purpose compound for this use. During the past 30 years great efforts have been made to find further short-term preservatives (Turner, 1967). The most successful has been sodium pentachlorophenate but, due to medical and environ-mental hazards, its use in practice is negligible. A full list of actual and proposed short-term preservatives for *Hevea* latex, and their control of fluidity for 24 hours and volatile fatty acids build up is given by John (1978). They include formaldehyde, sodium sulphite, ammonia, sodium hypochlorite, streptomycin sulphate and tridimethylamino-methyl phenol. Further to this, Cheong and Ong (1974) developed two composite short-term preservatives to overcome some defects of the pure ammonia system, one comprising 0.15% hydroxylamine neutral sulphate and 0.05% ammonia, the other 0.4–0.5% boric acid and 0.07% ammonia. In production of latex concentrate, increase in volatile fatty acids content of field latex has to be prevented or arrested by the use of long-term preservatives. By adding 0.5% ammonia, it is possible to arrest the build-up of volatile fatty acids for a short while, but bacteria often become resistant and, consequently, the concentration of ammonia has to be progressively raised to undesirable levels. Build-up of volatile fatty acids can be effectively controlled for an indefinite period by treating the latex with 0.3% ammonia and storing at 45°C. This system is still being investigated (John, 1978).

Sodium hypochloride, streptomycin sulphate and tridimethylamino-methyl phenol were tested for their suitability as long-term preservatives in field and concentrate latices (John, 1978). Sodium hypochlorite at 0.1% effectively controlled build-up of volatile fatty acids in field latex containing 0.3% ammonia. It is, however, unsuitable due to its adverse effect on the latex. Streptomycin sulphate, effective at 0.05%, is commercially too expensive. Using 0.2% ammonia and 0.3% tridi-methylaminomethyl phenol can be considered suitable as a long-term preservative. In further investigations, 1-3-chlorallyl-3,5,7-triazy-1-azoniaadamante chloride (Dowicil 75) at 0.4%, tetramethyltiuram disulphide at 0·05%, 1,2-benzoisothiazoline-3-one (Proxel CRL) at 0.03%, 2,2'-dihydroxy-5,5'-dichlorodiphenylmethane at 0.2% and hydrazine hydrate and hydroxylamine neutral sulphate each at 0.1% were effective preservatives in the presence of 0.2% ammonia (John et al., 1976; John, 1978). However, the combination of tetramethyl-thiuram disulphide and zinc oxide was found to be more effective than all the other preservatives mentioned (John, 1978; John and Verstraete, 1979). The best concentration of the tetramethylthiuram disulphide–zinc system in the field latex treated with ammonia to 0.2% is 0.025%, with a 50/50 ratio of the two chemicals. The combination satisfactorily keeps the latex fluid for up to 20 days. The resulting latex concentrate has good mechanical and chemical stability, good preservation character-istics, easy handling, good processability and light colour, which is evident in the finished product. The advantage of the system is a significant saving in cost compared with latex treated with ammonia up to 1.0%. It can be used in the estate sector and by smallholders producing *Hevea* latex.

Table 11

Microbiocides for rubber products

Active microbicidal ingredient	Trade name	Field of application	References
Organometallic compounds			
Copper-8-quinolinate	—	In chloroprene rubber gaskets, Perbunan 26, acrylonitrile-butadiene (nitrile) rubber (Hycar OR-15), Thiokol	Phillips (1947a, b)
	Milmer-1	In rubber products	Bakanauskas and Prince (1955)
		In chloroprene vulcanizates including rubber grade waxes	Ritzinger (1959)
Copper-8-quinolinate with dithiobisbenzothiazole	—	In styrene-butadiene, chloroprene and polysulphide rubbers fungistatic and antiseptic properties, not for natural rubber	Monsanto Chemical Company (1952)
Phenylmercuric acetate	Aseptised H 31	In rubber products	Hueck-van der Plas (1966)
Diarylmercuric compounds	—	In rubber products fungistatic and bacteriostatic properties	Sowa (1952)
Tributyltin derivative	Metatin 29–10	In rubber products	Hueck-van der Plas (1966)
Sulphur-containing compounds			
Tetramethylthiuram disulphide	Tuads	In vulcanizates fungicidal	Dimond and Horsfall (1943)
	—	In natural rubber ring-joints 3 p.p.h.r. gives protection; 0.2 p.p.h.r. no biocidal effect with actinomycetes	Rook (1955)
	—	In isobutene-isoprene rubber	Bieri (1959)
	Tiuram	In natural rubber vulcanizates inhibits superficial growth of fungi	Dubok *et al.* (1971)

373

Table 11 —*cont.*

Active microbicidal ingredient	Trade name	Field of application	References
Tetramethylthiuram disulphide with 2-mercaptobenzothiazole and sulphur	Tiuram with Kaptaks	In isobutene-isoprene rubber vulcanizates inhibits superficial growth of fungi	Dubok *et al.* (1971)
Zinc dimethyldithiocarbamate	—	In natural rubber ring-joints 0.2 p.p.h.r., no biocidal effect with actinomycetes	Rook (1955)
Zinc dimethyldithiocarbamate with 2-mercaptobenzothiazole	—	In rubber products	Bakanauskas and Prince (1955)
Zinc dimethyldithiocarbamate with zinc-mercaptobenzothiazole	—	In vulcanized natural rubber protects the surface from microbial deterioration after twelve months Leeflang test bath	Yeager (1954); Hills (1967)
Copper dimethyldithiocarbamate	—	In vulcanizates fungicidal	Fajfr and Jirsak (1959)
Potassium methyldithiocarbamate	—	In rubber products	Ross (1971)
Dithiocarbamate esters	—	In rubber bacteriostatic	Société des Usines Chimiques Rhone-Poulenc (1951)
2-Mercaptobenzothiazole	Captax	In vulcanizates moderately good fungicide	Dimond and Horsfall (1943)
	—	In natural rubber ring-joints 0.5 p.p.h.r., no biocidal effect to actinomycetes	Rook (1955)
Sodium salt of 2-mercaptobenzothiazole and dimethyldithiocarbamic acid	Vancide 51	In rubber products	Vanderbilt R.Z. Co. (1952)
Halogen derivatives of 2-mercaptobenzothiazole and their salts	—	In rubber products	Monsanto Chemical Company Ltd. (1961)
Lauryl pyridinium salt of 5-chloro-2-mercaptobenzothiazole	—	In vulcanized natural rubber protects the surface from microbial deterioration after twelve months Leeflang test bath	Hills (1967)

Table 11—cont.

Active microbicidal ingredient	Trade name	Field of application	References
N-Trichloromethylthio-4-cyclohexene-1,2-dicarboximide	Captan	In rubber products	Hueck-van der Plas (1966)
	Captan	In vulcanized natural rubber ineffective	Hills (1967)
N-Trichloromethylthiophthalimide	Folpet	In rubber products	Hueck-van der Plas (1966)
Azomethane derivatives of thiophene	—	In natural-butadiene, chloroprene and butadiene rubber vulcanizates for electric cables	Dubrovin (1959)
Thiopene with hydrocarbons	Albichtol	In rubber formulations for cables, 5% gives good protection for 10 years	Rudakova and Popova (1973)
2,3,5,6-Tetrachloro-4-methyl-sulphonyl-piridine	SA 1013	In natural rubber specimens, 2 p.p.h. rubber failed to prevent deterioration	Cundell and Mulcock (1973b)
Aromatic compounds Benzylbenzoate	—	In natural rubber ring-joints ineffective	Hills (1967)
Para-Dichlorobenzene	Chlobol	In natural rubber ring-joints, 3% failed to prevent deterioration	Leeflang (1963); Hills (1967)
Cresylic acids produced in Roumania	—	In natural rubber and synthetic rubber vulcanizates, 2% prevents superficial growth of fungi	Pitis (1972); Lazar and Ioachimescu (1974)
Ortho-Phenylphenol	—	In rubber vulcanizates	Kost *et al.* (1959)
	—	In rubber products contacting with milk it is bactericidal	Brown D.S. Company (1961)
Thymol	—	In natural rubber ring-joints, 2% failed to prevent deterioration	Leeflang (1963); Hills (1967)
4-Chlorothymol	—	In rubber vulcanizates	Kost *et al.* (1959)
Ortho-Benzyl-*p*-chlorophenol	—	In natural rubber ring-joints, ineffective	Hills (1967)
		In natural rubber ring-joints	Hills (1967)

Table 11 —*cont.*

Active microbicidal ingredient	Trade name	Field of application	References
Sodium 2,4,5-trichlorophenate	—	In rubbery products	Ross (1971)
Pentachlorophenol	—	In polysulphide-type Thiokol latex MX linings at 5% it is effective to fungi and bacteria for 4 years	Allen and Fore (1953)
	—	In rubber vulcanizates	Kost *et al.* (1959)
		In chloroprene rubber sheaths of mining cables at 2.5%	Zyska *et al.* (1972)
Pentachlorophenol with substituted *p*-phenylenediamines	PCP with Santoflex IP	In chloroprene rubber sheaths of mining cables at 3.0%	Zyska *et al.* (1972)
Sodium pentachlorophenate	—	In rubbery products	Ross (1971)
2,4-Dinitrophenol	—	In rubber vulcanizates effective, but in natural rubber ring-joints, ineffective	Kost *et al.* (1959); Hills (1967)
Salicylanilide	—	In rubber products	British Cotton Industry Research Association (1940)
Zinc, nickel and antimony salts of salicylanilide	—	In natural-butadiene, chloroprene and butadiene rubbers vulcanizates for electric cables	Dubrovin (1959)
Sodium salt of salicyl-*o*-(or-*p*-)-toluidine	—	In rubber products	British Cotton Industry Research Association (1940)
Salicyl-*o*-anisidine	—	In rubber products	British Cotton Industry Research Association (1940)
Para-Hydroxy-diphenylmethane	—	In bathing mats, 4% active against dermatophytes	Schönborn *et al.* (1968)
2,2'-Dihydroxy-5,5'-dichlorodiphenylmethane	—	Shaped rubber products	Dunlop Rubber Company (1950)
	—	In rubber products	Bakanauskas and Prince (1955)
	Dichlorophene, Preventol GD	In natural rubber specimens, 3% or 2 p.p.h. rubber failed to prevent deterioration	Leeflang (1963); Cundell and Mulcock (1973b)

Table 11—*cont.*

Active microbicidal ingredient	Trade name	Field of application	References
3-Alkyl-2,2'-dihydroxy-3',5,5',6'-tetrachlorodiphenyl-methane	—	In rubber products, bactericidal and fungicidal	Dow Chemical Company (1950)
N-Cyclohexylmethyl-N'-phenyl-p-phenylenediamine	—	In natural rubber products	Monsanto Chemical Company (1970)
2,4-Trimethyl-1,2 dihydroquinoline polymer	—	In natural rubber vulcanizate specimens, effective after 12 months of the Leeflang test bath	Cundell *et al.* (1973b)
NN'-Di((1,4-dimethylpentil)-phenyl-p-phenylenediamine Nitrosodiphenylamine	—		
Ortho- and Para-Benzyl phenol in mixture with salicylanilide and tetramethylthiuram disulphide	Antimykotikum A	In bathing mats, rubber footwear, rubber mattresses and aprons, sterile medical supplies	Hofmann (1962)
		In rubber products, 0.5–2.0% of ready mixture	Hueck-van der Plas (1966)
		In natural rubber and synthetic rubber vulcanizates, at 2% prevents superficial growth of fungi	Pitis (1972); Stanescu and Chican (1974); Lazar and Ioachimescu (1974)
		In natural rubber specimens at 2 p.p.h.r., failed to prevent deterioration	Cundell and Mulcock (1973b)
Quaternary ammonium compounds			
Alkyl dimethyl ammonium chloride in mixture with dialkyl dimethyl ammonium chloride	Arquad 12	In rubbery products	Ross (1971)
Quaternary ammonium compounds of higher alkyl dimethyl benzyl-ammonium chloride type	Germ-I-Tol	In rubbery products	Ross (1971)
Quaternary ammonium salts, oil-soluble	—	In rubbery products	Nippon Oil Company (1976)

Table 11 —*cont.*

Active microbicidal ingredient	Trade name	Field of application	References
Miscellaneous microbiocides			
Antibiotic acidophilin	—	In rubber suppresses clouding caused by growth of pseudomonads	University of Nebraska (1972)
Butylcarbitol formal	TP90B	In natural rubber specimens at 2 p.p.h.r., failed to prevent deterioration	Cundell and Mulcock (1973b)
Silvered anion-active guanidine–nitrate–urea–formaldehyde resin	—	In natural rubber, styrene-butadiene rubber (Buna S) and acrylonitrile-butadiene (nitrile) rubber, bactericidal	American Cyanamid Company (1952)
Rosin amines with metal salts	—	In rubber compounding	Hercules Powder Company (1952)
Zinc oxide	—	In earpads for telecommunications sets and mine detectors, 10% prevents superficial growth of fungi	Uniyal *et al.* (1971)

relatively non-hazardous in handling in the rubber industry; (g) resistant to leaching by water; (h) durable for the service life of the rubber product; (i) non-corrosive to metals; (j) no effect on the physical properties of the vulcanized rubber; (k) non-hazardous in the final rubber product; and (l) price justifying its economic effectiveness.

Table 12

Fungal resistance of certain vulcanizates with fungicides in the mix. From Lazar and Ioachimescu (1974).

Type of rubber	Code formulation	Fungal resistance of vulcanizate in a 28-day test[a]		
		Without fungicide	Anti-mykoticum A (Bayer)	Cresylic acids
Natural-styrene-butadiene	CN-C 1502	1	00	00
Chloroprene rubber	7 NE 544	0	0	0
Natural-chloroprene-styrene-butadiene rubber	$DS_{40}C_{1507}$	3	2	00
Natural-chloroprene rubber	CN-NGNA	2	1	0
Styrene-butadiene-ethylene-propylene rubber	K_{720}	1	0	00
Styrene-butadiene rubber oil extended	C_{1712}-RA	3	2	1
Natural-acrylonitrile-butadiene rubber	SKN_{26}	4	3	1
Natural-styrene-butadiene	$CN-C_{1500}$	2	1	00
Natural-styrene-butadiene oil extended	$CN-C_{1500}$-H	1	0	0
Chloroprene rubber with blowing agent	GBM	3	2	1

[a] Scale of evaluation according to Table 1.

As the processing technology of rubber products requires vulcanization temperature of 135–175°C, only a small group of microbiocides can be considered. So far an ideal compound, which would fulfill all of the requirements, has not been found. Every compound used in the past as a bactericide or fungicide has both advantages and disadvantages. Since the research of the late 1930s and the early 1940s, a number of compounds have been found to be microbiostatic or microbicidal in rubber products. The nearest possible attempt at a full list is given in Table 11, which includes compounds tested in rubber products for their microbicidal effect or claimed in patents for such effect. Compounds suggested or tested for rubberized textile materials are not included, as these materials are not discussed in this chapter. Forecasting the microbial resistance of any rubber formulation with a given microbicide in the mix is impossible. Lazar and Ioachimescu (1974) tested 10 types of vulcanizates which incorporated as biocides Antimykotikum A or cresylic acids. As can be seen in Table 12, the chosen microbicides improve the fungal resistance of tested rubber specimens to varying degrees and more investigations are needed for some formulations to achieve good fungal resistance. Pitis (1972), testing 12 other rubber formulations, showed that these two fungicides can give better results in protecting rubber vulcanizates against superficial growth of fungi. Effectiveness is evidently dependent on rubber formulation. Evaluation techniques for effectiveness of microbiocides against micro-organisms growing in the rubber product at the expense of rubber or mix ingredients is described by Hills (1967), Zyska et al. (1972), Cundell and Mulcock (1973b), Cundell et al. (1973b) and recently by Kerner-Gang (1977). The effectiveness of microbiocides against dermatophytes should be tested according to the technique by Schönborn et al. (1968).

V. ACKNOWLEDGEMENTS

I am grateful to Mrs. M. Kuczera and Mrs. M. Mierzyńska for valued technical assistance, to Mrs. J. Reszka and Mr. K. R. Reszka at the Engineering College in Koszalin for preparing scanning electron photomicrographs. I wish to thank all my colleagues who have helped in work on microbial deterioration of rubber at the Central Mining Institute in Katowice, especially Mrs. Dr. K. Bohdanowicz-Murek, Mrs. D. Kwiatkowska M.Sc. and Mrs. Dr. B. J. Rytych. To Miss Dr. P. M. Stockdale and Miss S. Daniels at Commonwealth Mycological Institute in Kew, Dr. H. O. W. Eggins and Dr. M. Allsopp at Biodeterioration Information Centre in Birmingham, Dr. C. K. John at Rubber Research Institute of Malaysia in Kuala Lumpur and Dr. I. Ionita at the Research Institute for the Electrotechnical Industry in Bucharest my thanks are due for generous assistance with publications. Finally, I should like to thank my wife, Maria Monika, for helping in my research work. This paper is published by permission of the Central Mining Institute in Katowice, Poland.

REFERENCES

Ajello, L. and Getz, M. E. (1954). *Journal of Investigative Dermatology* **22**, 17.
Allen, F. H. and Fore, D. (1953). *Industrial and Engineering Chemistry* **45**, 374.
Alteras, I. and Ignatescu, M. (1952). *Derm-Vener. Rev. Soc. Med. Roumania* **7**, 433.
American Cyanamid Company (1952). United States Patent 2,578,186.
Arens, P. (1912). *Zentralblat für Bakteriologie, Parasitenkunde Infectionskrankheiten und Hygeine Abbeilung II.* **35**, 465.
Backer, H. and Gross, H. (1974). *Material und Organismen* **9**, 81.
Bakanauskas, A. and Prince, A. E. (1955). *Applied Microbiology* **3**, 86.
Beeley, F. (1940) *Annual Report of the Rubber Research Institute of Malaya*, p. 13.
Bell, A. A. and Muse, R. R. (1965). *Plant Disease Reporter* **49**, 323.
Bieri, B. (1959). *Schweizer Archiv für Wissenschaft und Technik* **25**, 210.
Biodeterioration Information Centre (1979). Rubber and Plastics Biodeterioration Document SB8.
Blake, J. T. and Kitchin, D. W. (1949). *Industrial and Engineering Chemistry* **41**, 1633.
Blake, J. T., Kitchin, D. W. and Pratt, O. S. (1950). *American Institution of Electric Engineers* **69** (II), 748.
Blake, J. T., Kitchin, D. W. and Pratt, O. S. (1953). *Transactions of the American Institution of Electric Engineers* **72** (III), 321.
Blake, J. T., Kitchin, D. W. and Pratt, O. S. (1955). *Applied Microbiology* **3**, 35.
Blow, C. M., ed. (1971). 'Rubber Technology and Manufacture'. Butterworths, London.
Bobilioff, W. (1929). *Archief voor de Rubber Cultuur* **13**, 118.
British Cotton Industry Research Association (1940). Mildew-proofing of textile materials, etc. British Patent 323,579.
British Standard 2011 (1963). Basic Methods for the Climatic and Durability Testing of Components for Telecommunication and Allied Electronic Equipment. Part 2, Test 7, Mould Growth.
Brown, D. S. Co. (1961). British Patent 883,448; United States Patent 2,947,282.
Calderon, O. H. and Staffeldt, E. E. (1965). *International Biodeterioration Bulletin* **1**, 33.
Cantwell, S. G., Lau, E. P., Watt, D. S. and Fall, R. R. (1978). *Journal of Bacteriology* **135**, 324.
Chamberlain, E. A. C., Meek, I. R., Kewell, C. B. and Finlay, A. J. (1965). *Transactions of the Institution of the Rubber Industry* **41**, T24.
Cheong, S. F. and Ong, C. O. (1974). *Journal of the Rubber Research Institute of Malaysia* **24**, 118.
Connolly, R. A. (1962). *Bell Laboratories Record*, No. 4, 124.
Connolly, R. A. (1976). *In* 'Biodeterioration of Materials' (A. H. Walters and E. H. Hueck-van der Plas, eds.), vol. 2, pp. 168–178. Applied Science Publishers Ltd., London.
Cook, A. S. (1960). *Journal of the Rubber Research Institute of Malaya* **16**, 65.
Cook, A. S. and Sekhar, K. C. (1953). *India Rubber Journal* **125**, 132.
Crum, M. G., Reynolds, R. J. and Hedrick, H. G. (1967). *Applied Microbiology* **15**, 1352.
Cundell, A. M. and Mulcock, A. P. (1972). *International Biodeterioration Bulletin* **8**, 119.
Cundell, A. M. and Mulcock, A. P. (1973a). *Material und Organismen* **8**, 1.
Cundell, A. M. and Mulcock, A. P. (1973b). *Developments in Industrial Microbiology* **14**, 253.

Cundell, A. M. and Mulcock, A. P. (1973c). *International Biodeterioration Bulletin* **9**, 17.
Cundell, A. M. and Mulcock, A. P. (1973d). *International Biodeterioration Bulletin* **9**, 91.
Cundell, A. M. and Mulcock, A. P. (1973e). *Material und Organismen* **8**, 165.
Cundell, A. M. and Mulcock, A. P. (1973f). *International Biodeterioration Bulletin* **9**, 91.
Cundell, A. M. and Mulcock, A. P. (1975). *Developments in Industrial Microbiology* **16**, 88.
Cundell, A. M. and Mulcock, A. P. (1976a). *New Zealand Journal of Science* **19**, 291.
Cundell, A. M. and Mulcock, A. P. (1976b). In 'Proceedings of the Third International Biodegradation Symposium' (J. M. Sharpley and A. M. Kaplan, eds.), pp. 659–664. Applied Science Publishers, London.
Cundell, A. M., Hills, D. A. and Mulcock, A. P. (1973a). *European Rubber Journal* **155**, 22, 25, 28, 35.
Cundell, A. M., Mulcock, A. P. and Hills, D. A. (1973b). *Rubber Journal* **35**, 22.
Dachselt, E. (1971). Thioplaste. VEB Deutscher Verlag für Grundstoffindustrie, Leipzig.
De Vries, O. (1927). *Archief voor de Rubber Cultuur* **11**, 279.
Dickenson, P. B. (1965a). *Rubber Development* **18**, 85.
Dickenson, P. B. (1965b). *Rubber Journal* **147**, 54.
Dickenson, P. B. (1968). In 'Proceedings of the Natural Rubber Research Conference, Kuala Lumpur', Preprint 1–10.
Dickenson, P. B. (1969). *Journal of the Research Institute of Malaya* **22**, 165.
Dimond, A. E. and Horsfall, J. G. (1943). *Science* **97**, 144.
DIN 40046. Blatt 10. Entwurf (1970). Klimatische und mechanische Prüfung für elektrische Bauelemente und Geräte der Nachrichtentechnik. Prüfung J: Schimmelwachstum.
Dogadkin, B. A. (1972). Elastomer chemistry. Izd. Khimiya, Moscow.
Dolezel, B. (1964). Korroziya plasticheskikh materialov i rezin Izd. Khimiya, Moskva.
Dow Chemical Company (1950). British Patent 676,859.
Dubok, N. N., Angert, L. G. and Ruban, G. I. (1971). *Soviet Rubber Technology*, March, 17.
Dubrovin, G. I. (1959). *Zhurnal Prikladnoi Khimii* **32**, 2547.
Dunkley, W. E. (1964). *Rubber Development* **35**, 22.
Dunlop Rubber Company (1950). British Patent 675,200.
Eaton, B. J. and Fullerton, R. G. (1929). *Rubber Research Institute of Malaya*, Bulletin No. 2, pp. 1–36.
Eaton, B. J. and Grantham, J. (1915). *Agricultural Bulletin F.S.M.* **4**(2), 26.
Eaton, B. J. and Grantham, P. (1918). *Rubber Research Institute of Malaya Incorporated*, Bulletin No. 27, pp. 163–169.
Eirich, F. R., ed. (1978). *Science and Rubber Technology*. Academic Press, New York.
Elphick, J. J. (1965). In 'Microbiological Deterioration in the Tropics' (N. J. Butler and H. O. W. Eggins, eds.), pp. 253–259. Monograph No. 23, Society of Chemical Industry, London.
Elphick, J. J. (1971). In 'Microbial Aspects of Metallurgy (J. D. A. Miller, ed.), pp. 157–172. Medical and Technical Publishing Co. Ltd., Aylesbury.
Emmons, C. W., Binford, C. H., Utz, J. P. and Kwon-Chung, K. J. (1977). 'Medical Mycology'. Lea and Febiger, Philadelphia.
English, M. P. (1969). *British Journal of Dermatology* **81**, 705.
English, M. P. and Gibson, M. D. (1959). *British Medical Journal* **i**, 1446.
Esuruoso, O. F. (1970). *Mycopathologia et Mycologia Applicata* **42**, 187.
Fajfr, M. and Jirsak, K. (1959). Czechoslovak Patent 92,727.
Fudalej, P. S., Zyska, B., Fudalej, D. S. and Kuczera, M. H. (1976). In 'Proceedings

of the Third International Biodegradation Symposium' (J. M. Sharpley and A. M. Kaplan, eds.), pp. 347–355. Applied Science Publishers, London.

Gentles, J. C. (1957). *British Medical Journal* **1**, 746.

Gillespie, J. M. (1963). Rubber and Plastic Research Association of Great Britain, Shawbury, Information Report 5825.

Gorter, K. and Swart, N. L. (1916). *Meded v.h. Rubber-proefstation W. Java* **6**, 1.

Groenewege, J. (1921). *Meded. Alg. Proefst. Landb., Buiteng. No. 9*, 1.

Hedrick, H. G. and Gilmartin, J. N. (1964). *Developments in Industrial Microbiology* **6**, 124.

Heinisch, K. F. and Kuhr, P. (1957). *Archief voor Rubber Cultuur* **34**, 1.

Heinisch, K. F. and Nadarajah, M. (1960). In 'Proceedings of the Natural Rubber Research Conference—Kuala Lumpur', pp. 893–902.

Heinisch, K. F., Nadarajah, M. and Muthukuda, D. S. (1962). *The Rubber Research Institute of Ceylon* **38**, 40.

Hercules Powder Company (1952). United States Patent 2,492,939.

Hills, D. A. (1967). *Rubber Journal* **77**, 14.

Hofmann, W. (1962). *Kautschuk und Gummi* **15**, 501.

Hueck, H. J. (1965). *Material und Organismen* **1**, 5.

Hueck, H. J. (1968). In 'Biodeterioration of Materials' (A. H. Walters and J. J. Elphick, eds.), pp. 6–12. Elsevier Publishing Co. Ltd., Amsterdam, London and New York.

Hueck-van der Plas, E. (1965). *International Biodeterioration Bulletin* **1**, 1.

Hueck-van der Plas, E. (1966). *International Biodeterioration Bulletin* **2**, 69.

Hutchinson, M., Ridgway, J. W. and Cross, T. (1975). In 'Microbial Aspects of the Deterioration of Materials' (R. J. Gilbert and D. W. Lovelock, eds.), pp. 187–202, Technical series No. 9. Academic Press, London, New York, San Francisco.

International Electrotechnical Commission. Publication 68-J (1960). Recommended Basic Climatic and Mechanical Robustness Testing Procedures for Components of Electronic Equipment, Test J, Mould Growth, 2nd edition.

International Organization for Standardization Recommendation R 846 (1968). Recommended practice for the evaluation of the resistance of plastics to fungi by visual examination.

John, C. K. (1966). *Journal of the Rubber Research Institute of Malaya* **19** (4), 219.

John, C. K. (1968). *Journal of the Rubber Research Institute of Malaya* **20** (4), 173.

John. C. K. (1978). D.Sc. Thesis: University of Ghent, Belgium.

John, C. K. and O'Connell, J. (1967). *Journal of the Rubber Research Institute of Malaya* **20**, 112.

John, C. K. and Verstraete, W. (1979). In 'Proceedings of the 4th International Biodeterioration Symposium'. In press.

John, C. K., Nadarajah, M. and Lau, C. M. (1976). *Journal of Rubber Research Institute of Malaysia* **24**, 261.

Kalinenko, V. O. (1938). *Mikrobiologiya (Moscow)* **17**, 119.

Kennedy, H. (1971). *Journal American Water Works Association* **63**, 189.

Kereluk, K. and Baxter, R. M. (1963). *Developments in Industrial Microbiology* **4**, 235.

Kerner-Gang, W. (1973). *Material und Organismen* **8**, 17.

Kerner-Gang, W. (1977). In 'Biodeterioration Investigation Techniques' (A. H. Walters, ed.), pp. 41–49. Applied Science Publishers Ltd., London.

Keuringinstituut voor Waterleidingartikelen KIWA N.V. (1961). Aantasting van rubberingen voor waterleidingbuizen. Rapport van de Commissie Rubberingen. Rijswyk, Holland.

Koch, H. A., Breitbarth, W. and Jung, R. (1966). *Zeitschrift für ärtliche Fortbildung* **14**, 869.

Kost, A. N., Nette, I. T., Pomorsteva, N. V. (1959). *Vestnik Moskovskogo Universiteta* **14**, 213.

Kulman, F. E. (1958). *Corrosion* **14**, 213.

Kwiatkowska, D., Zyska, B. and Zankowicz, L. P. (1980). *In* 'Proceedings of the Fourth International Biodeterioration Symposium, Berlin' (T. A. Oxley, J. Becker and D. Allsopp, editors), pp. 135–141. Pitman Publishing Ltd. and The Biodeterioration Society.

Lazar, V. (1973). *In* 'Modelling of environmental effects on electrical and mechanical equipment. 4th International Symposium', pp. 261–265. Czechoslovak Academy of Sciences, Praha.

Lazar, V. and Ioachimescu, M. (1972). *Studii si cercetari de Biologie, Seria botanica (Bucuresti)* **24**, 43.

Lazar, V. and Ioachimescu, M. (1973). *Revue Roumaine de Biologie, Serie de Botanique (Bucuresti)* **18**, 227.

Lazar, V. and Ioachimescu, M. (1974). *In* 'Al Partulea Simpozion de Biodeteriorare', pp. 18–23. Institutul de Cercetare si Proiectare pentru Industria Electrotehnica, Bucharest.

Leeflang, K. W. H. (1963). *Journal American Water Works Association* **55**, 1523.

Leeflang, K. W. H. (1968). *Journal American Water Works Association* **60**, 1070.

Lowe, J. S. (1959). *Transactions of the Institution of the Rubber Industry* **35**, 10.

Lowe, J. S. (1960). *In* 'Proceedings of the Natural Rubber Research Conference, Kuala Lumpur', pp. 822–825.

MacLachlan, J., Heap, W. M. and Pacitti, J. (1966). *In* 'Microbiological Deterioration in the Tropics' (N. J. Butler and H. O. W. Eggins, eds.), pp. 185–200. Monograph No. 23, Society of Chemical Industry, London.

Macura, A. and Laskownicka, Z. (1979). *Postepy Microbiologii* **18**, 73.

May, M. E. and Neihof, R. A. (1977). *Naval Research Laboratory Memorandum Report 3666, Washington, D.C.*

McMullen, A. I. (1951). *Journal of the Rubber Research Institute of Malaya* **13**, 29.

Mikhailov, W. A., Davidenko, W. N., Logvin, V. V. and Shnyrev, I. M. (1967). Nadezhnost apparatury gornykh avtomaticheskikh ustroistv. Tekhnika, Kiev.

Miller, R. N., Herron, W. C. and Krigens, A. G. (1964). *Materials Protection* **3**, 60.

Monsanto Chemical Company (1952). United States Patent 2,608,551.

Monsanto Chemical Company Ltd. (1961). British Patent 873,602.

Monsanto Chemical Company Ltd. (1970). United States Patent 3,510,432; Australian Patent 414,712, 1971.

Muraoka, J. S. (1966). *Materials Protection* **5**, 35.

Nette, I. T., Pomorsteva, N. V. and Kozlova, E. I. (1959). *Mikrobiologiya (Moscow)* **28**, 881.

Nippon Oil Company (1976). British Patent 1,440,238.

O'Brien, T. E. H. (1926). *Rubber Research Station Ceylon Bulletin*, No. 42.

Pepper, M. (1960). *In* 'Proceedings of the Natural Rubber Research Conference, Kuala Lumpur', pp. 887–892.

Phillips, L. C. (1947a). U.S. Air Material Command, Engineering Division, Materials Laboratory, Dayton, Ohio, PB 98 831, pp. 16.

Phillips, L. C. (1947b). U.S. Air Material Command, Engineering Division, Materials Laboratory, Dayton, Ohio, PB 98 848, p. 20.

Philpott, M. W. and Sekar, K. C. (1953). *Journal of the Rubber Research Institute of Malaya* **14**, 93.

Pitis, I. (1972). *In* 'Biodeterioration of Materials'. Vol. 2. (A. H. Walters and E. Hueck-van der Plas, eds.), pp. 294–300. Applied Science Publishers, London.

Reszka, J., Zyska, B., Fudalej, P. S. and Reszka, K. R. (1975). *International Biodeterioration Bulletin* **11**, 71.

Reynolds, R. J., Crum, M. G. and Hedrick, H. G. (1967). *Developments in Industrial Microbiology* **8**, 260.

Ritzinger, G. B. (1959). *Rubber and Plastics Age* **40**, 1067.

Rook, J. J. (1955). *Applied Microbiology* **3**, 302.

Ross, E. O. (1971). United States Patent 3,591,410.

Rubber and Plastic Biodeterioration (1979). SB 8, Biodeterioration Information Centre, Birmingham, U.K.

Rudakova, A. K. and Popova, T. A. (1973). *Proceedings of Symposium on Biodeterioration of Buildings and Industrial Materials*, Moscow, pp. 72–79, IBBRIS 75/4-HX-2680.

Rychtera, M. and Bartáková, B. (1963). Tropicproofing electrical equipment. Leonard Hill Books Ltd., London SNTL, Praha.

Rytych, B. J. (1969). *International Biodeterioration Bulletein* **5**, 3.

Schade, A. L. (1937). *Mycologia*, **29**, 295.

Schönborn, Ch., Jung, H. D. and Kaben, U. (1968). *Deutsches Gesundheits-Wesen* **23**, 2429.

Schwartz, A. (1963). Abhandlungen der deutschen Akademie der Wissenschaften zu Berlin. Klasse für Chemie, Geologie und Biologie, Nr. 5, Akademie-Verlag, Berlin, DDR, 138 pp.

Scott, J. (1920). *India Rubber Journal* **60**, 410.

Scott, J. A. and Hill, E. C. (1971). Microbiological Aspects of Subsonic and Supersonic Aircraft Symposium on Microbiology, London, 27–28 January, 14 pp. Preprint, IBBRIS 71/2-30-4491.

Seubert, W. (1960). *Journal of Bacteriology* **79**, 426.

Seubert, W. and Fass, E. (1964). *Biochemische Zeitschrift* **341**, 35.

Seubert, W. and Remberger, U. (1963). *Biochemische Zeitschrift* **338**, 245.

Seubert, W., Fass, E. and Remberger, U. (1963). *Biochemische Zeitschrift* **338**, 265.

Shaposhnikov, V. N., Rabotnova, I. L., Yarmola, G. A., Kuznetsova, V. M. and Mozokhina-Porshnyakova, N. N. (1952a). *Mikrobiologiya* (*Moscow*) **21**, 146.

Shaposhnikov, V. N., Rabotnova, I. L., Yarmola, G. A. and Kuznetsova, V. M. (1952b). *Mikrobiologiya* (*Moscow*) **21**, 280.

Snoke, L. R. (1957a). *The Bell System Technical Journal* **36**, 1098.

Snoke, L. R. (1957b). *Bell Laboratories Record* No. 8, 287.

Sowa, F. J. (1952). British Patent 605,442.

Société des Usines Chemiques Rhone-Poulenc (1951). Esters of dithiocarbamic acid. British Patent 684,379.

Söhngen, N. L. and Fol, J. G. (1914). *Zentralblatt für Bakteriologie, Parasitenkunde, Infektionskrankheiten und Hygeine, Abteilung II* **40**, 87.

Spence, D. and van Niel, C. B. (1936). *Industrial and Engineering Chemistry* **28**, 847.

Stanescu, E. and Chican, M. (1974). *In* 'Al Partulea Simpozion de Biodeteriorare'. Institutul de Cercetare si Proiectare pentru Industria Electrotehnica, Bucharest, pp. 122–125.

Taysum, D. H. (1957a). *Journal of Applied Bacteriology* **20**, 188.

Taysum, D. H. (1957b). *Applied Microbiology* **56**, 349.

Taysum, D. H. (1960). *In* 'Proceedings of the Natural Rubber Research Conference, Kuala Lumpur', pp. 834–840.

Taysum, D. H. (1966). *In* 'Microbiological Deterioration in the Tropics' (N. J. Butler and H. O. W. Eggins, eds.), pp. 105–120. Monograph No. 23, Society of Chemical Industry, London.

Thaysen, A. C., Bunker, H. J. and Adams, M. E. (1945). *Nature, London* **155**, 323.

Treichler, R. (1951). *Rubber Age, New York* **69**, 579.

Tsuchii, A., Suzukii, T. and Takahara Y. (1978). *Agricultural and Biological Chemistry* **42**, 1217.

Turner, J. N. (1967). 'The Microbiology of Fabricated Materials'. Churchill Ltd., London.

Turner, J. N. and Kennedy, M. E. (1954). *Nature, London* **173**, 506.

Turner, J. N., Limpel, K. L. E., Bathershall, R. D., Bluestone, H. and Lamont, D. (1964). *Contributions of the Boyce Thompson Institute* **22**, 303.

University of Nebraska (1972). United States Patent 3,689,640.

Uniyal, B. P., Dayal, H. M. and Tandan, R. N. (1971). *Defence Science Journal* **21**, 75.

Vanderbilt, R. C. Co., Inc. (1952). *Vanderbilt News* **18**, 10.

Voegeli, H. E. and Cousminer, J. J. (1978). *International Biodeterioration Bulletin* **14**, 119.

Wastie, R. L. (1971). *International Biodeterioration Bulletin* **7**, 121.

Westphal, H. (1964). Handbuch der Fördergurte. VEB Deutscher Verlag für Grundstoffindustrie, Leipzig.

Williams, K. E. (1958). The Research Association of British Rubber Manufacturers, Shawbury, Information Bureau Circular No. 444.

Winner, H. I. (1957). *Journal of Applied Bacteriology* **20**, 88.

Yeager, C. C. (1954). *Canadian Chemical Process* **38**, 58.

ZoBell, C. A. and Beckwith, J. D. (1944). *Journal American Water Works Association* **36**, 439.

ZoBell, C. E. and Grant, C. W. (1942). *Science* **96**, 379.

Zyska, B. (1964). *Biuletyn Głównego Instytutu Górnictwa* **14**, 28; *In 'Przeglad Górniczy'* vol. 20, No. 10.

Zyska, B. (1968). *In* 'Biodeterioration of Materials' (A. H. Walters and J. J. Elphick, eds.), pp. 111–119. Elsevier, Amsterdam, London and New York.

Zyska, B. (1976). *In* 'Proceedings of the Third International Biodegradation Symposium' (J. M. Sharpley and A. M. Kaplan, eds.), pp. 331–341. Applied Science Publishers, London.

Zyska, B. (1977). Microbial corrosion of materials. Wydawnictwa Naukowo-Techniczne, Warsaw. (In Polish.)

Zyska, B., Fudalej, P. S. and Rytych, B. J. (1971). *International Biodeterioration Bulletin* **7**, 155.

Zyska, B., Fudalej, P. S., Rytych, B. J. and Fudalej, D. S. (1973). *International Biodeterioration Bulletin* **9**, 23.

Zyska, B., Rytych, B. J., Zankowicz, L. P. and Fudalej, D. S. (1972). *In* 'Biodeterioration of Materials' (A. H. Walters and E. H. Hueck-van der Plas, eds.), vol. 2, pp. 256–267. Applied Science Publishers, London.

12. Drugs and Cosmetics

ROSAMUND M. BAIRD

Department of Medical Microbiology, St. Bartholomew's Hospital, London, U.K.

I.	Introduction	387
II.	Spoilage	388
	A. Biodegradation of the Active Ingredient	390
	B. Changes in Physical Properties	393
	C. Aesthetic Deterioration	394
III.	Health Hazard	395
	A. Sterile Pharmaceuticals	395
	B. Non-Sterile Pharmaceuticals and Cosmetics	397
IV.	Assessment of the Risk	401
	A. Incidence of Contamination	401
	B. Hazard to the User	406
V.	Sources of Contamination	408
	A. Contamination in Manufacture	408
	B. Contamination During Use	414
VI.	Factors Influencing Microbial Survival	418
VII.	Monitoring and Control	420
	A. International Standards	421
	B. Control During Manufacture	423
	C. Control During Use	424
VIII.	Economic Aspects	424
References.		426

I. INTRODUCTION

For centuries, man has had an implicit faith in the mystical powers of lotions and potions. An almost impenetrable aura has surrounded not only the compounding and ministering of materia medica, but also the healing of the blemish or disease in question. When the cure was unsuccessful, it was assumed that this was judgement from an external source. It was inconceivable that use of these galenicals might worsen the existing condition. During the Twentieth Century, however, there has been a gradual understanding of the mechanisms by which drugs and cosmetics act, and an accompanying decline in the belief of the occult. Of many factors now known to influence their action, one of the most fascinating studies has been the discovery that, concealed within these products, there may be a population of living micro-organisms. The study of microbial contamination of pharmaceutical and cosmetic products has had, and continues to have, far-reaching implications from the point of view of both spoilage and economics, and as a potential hazard to health.

At first sight, pharmaceuticals and cosmetics may seem to have little in common, the one being used internally or externally on the human body for therapeutic purposes, the other being applied externally for cleansing or decorative purposes. In practice, however, similar problems, including the microbiological aspects, are encountered during the formulation, manufacture, packaging and use of both cosmetics and pharmaceuticals, and particularly of non-sterile products (creams, ointments, lotions). Useful comparisons between these two groups may therefore be made, although the emphasis may fall on different aspects. Thus, contamination studies in pharmaceutical products have drawn attention to the potential risk to health, based on hospital experience. Here the effects have been observed on groups of susceptible patients, by nature of some pre-existing disease or else compromised by antibiotic or immunosuppressive therapy. In general practice, opportunities for detecting medicament-borne infections in individual patients have been much less. There is even less information on the effect of using contaminated cosmetics on healthy people in the community. Manufacturers of cosmetics have traditionally been more concerned with the spoilage aspects. Certain cosmetics are, however, used in hospitals and there their microbiological flora may have serious health implications.

II. SPOILAGE

A wide variety of cosmetic and pharmaceutical products are susceptible

Table 1

Some genera of micro-organisms isolated from pharmaceutical and cosmetic products

Bacteria	Fungi	Yeasts
Acinetobacter	*Alternaria*	*Candida*
Alkaligenes	*Aspergillus*	*Monilia*
Bacillus	*Cladosporium*	*Torula*
Citrobacter	*Fusarium*	
Clostridium	*Monospora*	
Corynebacterium	*Mucor*	
Enterobacter	*Penicillium*	
Erwinia	*Rhizopus*	
Escherichia	*Torulopsis*	
Flavobacterium	*Trichoderma*	
Klebsiella		
Micrococcus		
Proteus		
Pseudomonas		
Salmonella		
Serratia		
Staphylococcus		
Streptococcus		

to bacterial and fungal invasion, resulting in a spoiled product that is unfit for use (Table 1). The formulations of these products are often complex. In addition to an active ingredient, a range of natural gums, suspending and thickening agents, carbohydrates, animal or plant oils, protein hydrolysates, amino acids, surfactants, flavourings, stabilizers and appreciable quantities of water may be present, many of which can support microbial growth. Other ingredients, such as beer, milk and egg extracts, vitamins and herbs, may be incorporated for promotional reasons. In formulations where anionic emulsifying agents have been replaced by non-ionic surfactants to give a more stable product, the risk of microbial contamination has been shown to increase (Bryce and Smart, 1965). Growth may be further affected by variations in pH value, oxygen concentration and osmotic pressure within the product, and by conditions of storage including humidity and temperature. Spoilage may present itself in three different ways, namely biodegradation of the active ingredients, changes in the physical properties of the product, and aesthetic deterioration.

isolation of *Ps. aeruginosa* from an antibiotic- and steroid-containing ointment. Subsequent investigations showed that, during production, a film of condensed moisture formed on the surface of the ointment, so enabling contaminants to multiply.

Contamination has also been reported in cosmetics used for skin care, such as handcreams and lotions. These may be used extensively in hospitals to prevent chapping of hands and in the control of cross-infection. An outbreak of *Klebsiella pneumoniae* septicaemia in patients with venous catheters in an intensive-care unit was traced to use of a contaminated lanolin handcream (Morse *et al.*, 1967). Further investigations showed that several brands of opened and unopened hand-creams were contaminated with *Serratia marcescens*, *Ps. aeruginosa*, *Escherichia intermedia*, *Alkaligenes faecalis* and *K. pneumoniae* (Morse and Schonbeck, 1968). Lotions used in the skin care of neonates may also give rise to concern. A detergent application used for cleaning the napkin area was found to be contaminated with the same strain of *Ps. aeruginosa* as that found in the faeces of newborn infants (Cooke *et al.*, 1970).

Simple aqueous lotions, such as detergents or shampoos, are highly prone to contamination by Gram-negative bacteria, and a large inoculum may be present with no visual signs of deterioration. Medi-cated shampoos also suffer from the same problem, since their active ingredients may have a narrow spectrum of antimicrobial activity. A potential health hazard may be presented to patients with injured eyes who use contaminated shampoos, for it is almost impossible to keep shampoo away from the eyes. Victorin (1967) reported an outbreak of *Pseudomonas* otitis in neonates which was traced to a contaminated liquid detergent used for cleaning the babies' baths.

4. Disinfectants and Antiseptics

Disinfectants are widely used in hospitals in the control of cross-infection. Proprietary products with a wide spectrum of activity may be used for disinfecting surfaces, equipment or the skin. Through misuse, disinfectants may become contaminated with Gram-negative bacteria and, ironically, may even contribute to the spread of nosoco-mial infections in hospital. Much evidence has accumulated over the past 20 years on how contamination arises, yet reports of infection continue to punctuate the literature. Factors affecting the behaviour of disinfectants have been extensively reviewed by Maurer (1978) and include not only antimicrobial activity but also pH value, concentration and temperature of the solution, contact time, deterioration on storage, and the presence of organic matter, man-made fibres and detergents, all of which may result in inactivation.

Broadly speaking, reports of contaminated disinfectants fall into three categories, namely those where the need for disinfection was unappreciated, those where disinfection was unsuccessful and those where the disinfectant itself was responsible for the spread of infection.

Two outbreaks of respiratory-tract infections caused by *Ps. aeruginosa* in patients treated by mechanical ventilation were described by Phillips and Spencer (1965) and Phillips (1967). In the first case, it had not been appreciated that the equipment could pump the organism directly into the patients' lungs and, in the second case, the method of disinfection was shown to be inadequate. Disinfectants inactivated during the manufacturing process, or in use, may subsequently become contaminated. A strain of *Ps. multivorans*, responsible for nine cases of wound infection, was shown to survive in dilutions of Savlon (Bassett *et al.*, 1970). A difference in pH value between solutions prepared with tap or distilled water was later shown to affect the antimicrobial activity of the chlorhexidine and cetrimide components, thereby enabling the strain to survive only in Savlon prepared with distilled water (Bassett, 1971).

The use of cork stoppers has frequently been associated with inactivation of disinfectant solutions (Nelson, 1942; Lowbury, 1951; Boud, 1967). Linton and George (1966) studied survival of *Pseudomonas* spp. in chlorhexidine solutions and suggested that solutions in contact with cork were inactivated by release of a potent tanin. Similarly, cotton and other cellulose fibres may inactivate quaternary ammonium compounds during prolonged contact (Plotkin and Austrian, 1958; Malizia *et al.*, 1960; Lee and Fialkow, 1961).

IV. ASSESSMENT OF THE RISK

A. Incidence of Contamination

The foregoing reports have largely been concerned with specific contaminated medicines and cosmetics and with ill-effects caused by their

B. Hazard to the User

The likelihood of a contaminated product giving rise to untoward effects during its use on patients or consumers is not easily assessed. It is possible, however, to predict with some accuracy the outcome of certain events based on past experience. The effect is dependent on the combination of several factors, amongst which the type and degree of microbial contamination, the route of administration and the resistance of the host are considered to be of prime importance. These cannot be considered in isolation; the outcome is determined by the interaction of these factors. Inevitably, most of the relevant information has come from hospital studies of contaminated pharmaceutical products. Opportunities for recognizing this type of infection have generally been greater here as a larger but contained population has been at risk.

1. Type and Degree of Contamination

The advent of the antibiotic era has resulted in changes in the number and types of infections seen in hospitals today and, to a certain extent, in the general community. The emergence of the Gram-negative 'opportunist pathogens' over the past 20 years has resulted in the spread of nosocomial infections, causing particular problems in the elderly, newborn, patients with pre-existing diseases and those whose normal defence mechanisms are suppressed. Opportunist pathogens are widely distributed in hospital environments. They have frequently been isolated from cosmetics and medicaments and, in some cases, infections have developed following the use of products contaminated with these organisms. It is significant that, in four outbreaks of septicaemia in the United States and Great Britain, opportunist pathogens were isolated from contaminated intravenous fluids which caused the outbreak. Topical medicaments contaminated with *Ps. aeruginosa* may cause infections when applied to broken skin, but not to intact skin. Free-living bacteria, such as *Flavobacterium meningosepticum*, can cause severe or fatal infections when applied to the skin of newborn infants but may cause only short episodes of fever when given intravenously to adults in large doses (Olsen, 1967). Opportunist pathogens may therefore cause a variety of unforeseen complications, depending on the route of administration and the condition of the patient. On the other hand, the presence of pathogens in products may be associated with more predictable results. Thus, use of powders contaminated with *Cl. tetani* in internal cavities or as an umbilical cord dressing in neonates must be considered highly dangerous.

There is very little information on the dose of organisms required to initiate an infection, although some work has been done with animal models and human volunteers. Crompton *et al.* (1962) showed that, in a rabbit model, a dose of 60 cells of *Ps. aeruginosa* injected into the anterior chamber of the eye was sufficient to cause panophthalamitis and loss of sight. Even smaller doses may be required if the eye has been damaged. Elek and Conen (1957) reported that 1×10^6 cells of *Staph. aureus* were needed to form pus after intradermal injection. A much smaller inoculum was required in the presence of a foreign body, such as a suture. Buck and Cooke (1969) found that at least 10^6 cells of *Ps. aeruginosa* had to be ingested before intestinal colonization took place in normal adults. In contrast, Shooter *et al.* (1969) found that, in a hospital patient who had inadvertently swallowed contaminated peppermint water, a dose of 10^3 organisms per ml was required for colonization. In an outbreak of salmonellosis from infected thyroid tablets, the infecting dose appeared to have been about 10^5 organisms (Kallings *et al.*, 1966). A very much smaller dose caused salmonellosis in babies with cystic fibrosis who had been treated with contaminated pancreatin powder. Glencross (1972) suggested that the intestine of babies with a maladsorption syndrome may be more easily colonized than that of healthy children or adults. The fact that newborn babies rapidly acquire a normal intestinal flora suggests that they can be colonized with very low numbers of organisms (Lennox-King, 1976).

Some strains of organisms are known to be more pathogenic than others. For example, the significance of environmental strains of *Ps. aeruginosa* in the development of patient infections is controversial. Certain strains may be better equipped to survive in different hosts, and this may explain why some strains of *Ps. aeruginosa* can apparently pass from environment to patients via the medicaments as vectors and others do not (Baird *et al.*, 1980).

2. *Route of Administration*

Contaminated solutions injected into the body may result in serious consequences, particularly when the contaminants are Gram-negative bacteria and endotoxins have been formed. Intraspinal injection of contaminated drugs has potentially the most harmful effect, since the resistance of the meninges is believed to be significantly lower than that of the blood stream. In the presence of certain drugs, normal body-defence mechanisms may be impaired by those having a vaso-constrictive or necrotic action.

Contaminated products instilled into or used around the eye may also have damaging effects. *Pseudomonas aeruginosa* has invariably been held responsible. Minute abrasions caused by fingernails or the im-proper use of tweezers, cosmetic applicators or contact lenses may initiate infection.

Intact skin provides an excellent barrier to invading micro-organisms. When the surface is breached during surgery or damaged in patients with burns or pressure areas, the skin may become colonized and infected with opportunist organisms not normally part of the skin flora. These organisms may also be found on intact skin which is occluded (such as under a wet dressing or plaster) or diseased, and in patients whose normal skin flora has been suppressed by the use of certain therapeutic agents. Use of topical steroids may encourage development of local infections by suppressing phagocytosis and by a local vasoconstrictor action. When contaminated steroid-containing products are applied, the effect is likely to be compounded.

3. Resistance of the Host

The resistance of the host to infection plays a crucial part in determining the outcome of a contaminated product episode. This may range from a normal healthy individual in the community to a gravely ill patient in hospital. Normal defence mechanisms may be impaired in the case of leukaemics or diabetics, or deficient in the case of newborn and young children.

V. SOURCES OF CONTAMINATION

A. Contamination in Manufacture

1. Raw Materials

The quality of raw materials is of prime importance in determining the quality of the finished product. Their microbiological content varies according to the origin of the product. The material may be natural or synthetic, aqueous or oily, organic or inorganic. Unless adequately processed, naturally occurring raw materials may be heavily contaminated, usually with faecal organisms. *Salmonella muenchen, Sal. bareilly* and *Sal. agona* from thyroid and pancreatin extracts have been held responsible for outbreaks of salmonellosis in Sweden and Britain. Gums, such as tragacanth and acacia, and powders, such as starch, are often heavily contaminated with coliforms. Potato starch has been implicated as the source of coliforms in tranquillizer tablets and subsequent improvements in the processing and extraction of starches has resulted in a higher quality raw material. Natural earths, such as Fuller's earth, French chalk, kaolin or bentonite, are extracted from open-cast mines, and may be heavily contaminated with *Cl. tetani* and other spore bearers. Today, ethylene oxide is widely used during processing to lower the size of the microbial populations of these products.

Synthetic, inorganic and oily raw materials are inherently less susceptible to microbial contamination, and only small numbers of Gram-positive spore bearers are generally found. Accidental ingress of water into an oily product or the presence of a humid atmosphere may be sufficient to allow growth of spoilage organisms. Liquid paraffin and arachis oil may be metabolized by moulds, resulting in obnoxious tastes, odours and slimy deposits. Irrespective of their original microbial flora, all raw materials may be subject to spoilage, particularly mould growth, during improper storage in damp or fluctuating temperature conditions.

Aqueous raw materials, including water, dyes and detergents, constitute the greatest microbiological hazard. In the manufacture of cosmetics and pharmaceuticals, large volumes of water may be incorporated into the formulation or used for cleaning equipment during processing. Mains water in this country is generally of high quality and free from coliforms; however, it invariably contains small numbers of micro-organisms that can survive and multiply in susceptible products. Before use, water may be purified by de-ionization, distillation or reverse osmosis. Treated water may nevertheless support microbial growth since traces of organic matter may still be present (Eisman *et al.*, 1949). In hospitals, 'tank water' has frequently been associated with outbreaks of infection and its use is no longer recommended (Kelsey and Beeby, 1964; Last *et al.*, 1966; Thomas *et al.*, 1972). Collection, storage and delivery of water to the point of use also present micro-biological problems and, unless the water is kept at an elevated temperature, multiplication of contaminants will occur. *Pseudomonas* spp. and other Gram-negative organisms, which thrive in minimal media, present the greatest problem. At this stage, formation of endo-toxin may occur which can have serious repercussions in the manu-facture of sterile fluids. *Pseudomonas thomasii* has been isolated from hospital supplies of distilled water used for production of sterile fluids (Phillips *et al.*, 1972; Baird *et al.*, 1976b). Beveridge and Hope (1971) isolated *Ps. aeruginosa* in high numbers from many stock solutions, including anise, cinnamon, camphor, dill and peppermint waters.

2. Equipment

Plant design is of prime importance in assuring product cleanliness. Equipment with smooth impervious surfaces and a minimum number of 'dead legs', and which is easily dismantled for cleaning, may be better maintained than more complex designs. Pumps, joints, pipelines and valves may not be readily accessible, and small quantities of product or stagnant water may become trapped enabling microbial growth to take place between batches. Sokolski *et al.* (1962) reported degradation of a product containing clay, pectin, and a hydroxybenzoate preserva-tive by *Cladosporium resinae*, following an accumulation of this organism in a trap of the production pipeline.

3. Environment

Micro-organisms may be introduced into the production environment via raw materials, packaging materials, such as sacks, drums and cardboard, the air and water supplies, cleaning and other equipment, and from the staff who work in the unit. Unless adequately processed and preserved, the microbial flora of the final product may reflect that found in the production environment. Furthermore, badly designed units may result in cross-contamination of products. Appreciable quantities of water may be used in production units for manufacture and cleaning of equipment and surfaces. Here and in other wet sites, *Ps. aeruginosa* and other Gram-negative bacteria may grow profusely. In a study of environmental contamination in hospital pharmacies, *Ps.*

Table 7

Pharmaceutical products contaminated with *Pseudomonas aeruginosa* during manufacture
and possible environmental origin of strains

Pharmacy	Product	Proportion of samples contaminated	Possible environmental origin of strain
A	Steroid-containing cream	3/3	Puddle of water
B	Oily cream	1/1	Not found
	Cetomacrogol cream	1/1 (NT)	(NT) Mop, rubber taps, sink, draining board
	Chlorhexidine disinfectant	1/1 (NT)	(NT) Mop, rubber taps, sink, draining board
C	Oily cream	6/6	Sink, surround, drain
D	Chlorxylenol disinfectant	7/7	Water supply, label damper
	Pine disinfectant	7/8	Water supply
	Hand cream	5/12	Sink
	Zinc oxide cream	1/1	Sink, label damper
E	Calamine cream with steroid	2/6	Tap, sink
	Calamine lotion	1/1	Tap, sink
	Steroid cream	1/6	Tap, nailbrush

NT indicates that the strain was non-typable.

aeruginosa was isolated from a variety of wet sites, including sinks, drains, taps, draining boards, tank-water supplies, label dampers and cleaning equipment (Baird *et al.*, 1976a). The same strains of *Ps. aeruginosa* were invariably isolated from contaminated products made in these pharmacies (Table 7) and subsequently from patients who used these products (Baird and Shooter, 1976).

The development of clean-room technology in the past few years has revolutionized manufacture of sterile products in industry and, more recently, in hospitals. Here, air supplied to a clean room is passed through a 'high-efficiency particulate air' filter and emerges in laminar flow patterns at a slight positive pressure. Products made under these conditions are thus protected from operator and other environmental contamination. It is only a matter of time before some version of clean-room technology is applied to the production of non-sterile pharmaceuticals, and possibly cosmetics.

4. Staff

Undoubtedly, staff present the greatest microbiological hazard, because they are the least predictable source of contamination. Much is now known of the ways in which micro-organisms are disseminated from the human body. Skin scales covering the surfaces are shed continuously and a complete layer of 10^8 cells is shed every four days. About 10% of these scales carry micro-organisms, chiefly micrococci and aerobic and anaerobic diphtheroids. The contribution of staff to contamination in the production environment may be lowered by the use of protective clothing. The wearing of close-weave clothing with head and foot coverings, mask and gloves has greatly decreased skin dispersal from operators in clean rooms compared with those wearing conventional laboratory coats. The value of using antibacterial agents (such as hexachlorophane, chlorhexidine and povidone-iodine) in handwashing procedures in production units has been based on their proved effectiveness in operating theatres.

Clearly, the health of personnel is an important consideration in engaging production staff, although the value of routine microbiological checks is controversial. Emphasis on personal and production hygiene may be considered more worthwhile. Studies of wound infections caused by staphylococci from operating-theatre staff suggests that people with open-skin disease should not be employed in pharmaceutical production units (Payne, 1967).

5. Containers and Closures

Packaging materials are responsible for maintaining the integrity of the finished product during storage and subsequent use. In the field of cosmetics and toiletries, packaging is considered to be an essential part of the overall presentation of a product and must meet with consumer approval. Packaging designed to minimize the opportunities for contamination is not always compatible with that having the greatest marketing appeal; a compromise between the two may be necessary.

Plastic materials have been used extensively for packaging in recent years. Unlike paper and card, these are not biodegradable and hence should decrease the potential for microbial spoilage in the product. Compared with glass, however, they are porous and may allow diffusion of carbon dioxide and oxygen, thereby encouraging microbial growth within the container. Plastic closures have now replaced paper and card liners and cork closures, which were often heavily contaminated and, being adsorbent, served as a further substrate for mould growth.

The filling of containers with products requires care. Large air spaces may be left which provide sufficient atmospheric moisture to enable growth of moulds in otherwise unsuitable environments, such as oils (Smart and Spooner, 1972). A film of condensed moisture on the surface of a product may permit multiplication of contaminants in an otherwise hostile environment (Kallings *et al.*, 1966).

Re-use of containers and closures in hospitals presents its own problems. Containers returned to the pharmacy for washing may be contaminated, and this continuous process may be responsible for introduction of new strains of *Ps. aeruginosa* into pharmacy environments. A residual inoculum remaining in the dregs of containers after washing may be sufficient to contaminate fresh stocks. The practice of 'topping up' containers has been widely deprecated. Organisms present in the dead space behind plastic liners in bottles have been known to survive hot-water washing procedures and later have contaminated phenolic disinfectants (Simmons and Gardner, 1969; Cragg and Andrews, 1969).

6. *Cleaning Equipment*

The use of contaminated cleaning equipment has contributed to the spread of cross-infection in hospitals, and has been extensively reviewed by Maurer (1978). Mops, buckets, cloths, scrubbing machines and other items, particularly when stored wet, may provide a suitable breeding ground for Gram-negative bacteria and especially for *Pseudomonas* spp. In a study of the distribution of *Ps. aeruginosa* in hospital pharmacies, wet cleaning equipment was frequently cited as a source of this organism (Baird *et al.*, 1976a) and, subsequently, the same strains were isolated from contaminated products made in these pharmacies. There is no reason to suppose that equipment used for cleaning manufacturing areas in industry is exempt from contamination problems.

7. *Processing*

Inadequacies in the manufacturing process may result in contamination of the product and, furthermore, in a product that is unsuitable for use. Without doubt, a failure in the sterilization process is the most serious oversight, owing to the hazardous consequences that result from using improperly sterilized materials. In 1972, commercially made dextrose infusions were found to be the cause of an outbreak of septicaemia in which six people died (Meers *et al.*, 1973). At a subsequent inquiry it was revealed that, during sterilization of the batch, the process had been controlled by faulty instruments; thus one-third of the batch was inadequately sterilized and later found to be contaminated with *Ps. thomasii*, *K. aerogenes*, *E. cloacae* and coryneform bacteria (Clothier Report, 1972).

Generally speaking, the more a product is handled during processing, the greater the chances of contamination. Many pharmaceutical products are supplied in bulk and dispensed at a later date into smaller containers. Medicines dispensed in pharmacy containers have been shown to be contaminated twice as often as those packed in the manufacturer's original container (Public Health Laboratory Service Report, 1971).

B. Contamination During Use

There have been few published studies on in-use contamination, although the problem has been appreciated for many years. Such contamination is accepted as inevitable and to a large extent is outside the manufacturer's control, especially when products are not used for their intended purpose. A prime example of this is eye cosmetics which are frequently moistened with saliva before use, so that isolation of micro-organisms from the buccal cavity and portions of food in these products is hardly surprising.

Many pharmaceuticals and cosmetics that are not packaged as single-dose units are therefore formulated to contain a preservative system specifically designed for this purpose. A review of preservative studies is outside the scope of this work, but the subject has been discussed extensively by Bean (1972) and Croshaw (1977). Attempts to control in-use contamination have largely centred on the preservation aspects of lowering microbial numbers once in the product, rather than limiting their introduction into the product, through an understanding of the sources of contamination and their control. A review of these sources may, however, be considered a worthwhile exercise.

1. Human Sources

The user, the patient, other patients and medical or nursing staff may all be responsible for introducing micro-organisms into the product. The type and degree of contamination largely depend on the way in which the product is used. Thus, oral medicines that are handled minimally and generally given by spoon or in a small measure have a low incidence of contamination. On the other hand, products for rectal use, such as barium enemas, may become heavily contaminated with faecal organisms (Steinbach *et al.*, 1960). Studies on in-use contamination of intravenous fluids have shown that micro-organisms from human sources may be introduced at two stages. Administration sets may become contaminated with skin flora from the patient when the set is put up or changed, and contamination from the staff may occur when drug additives are introduced into the set. The incidence of contamination has been shown to increase with the number of additives (Kundsin *et al.*, 1973). In recent years, increasing use has been made of this route for multiple drug administrations. Clearly the problem of microbial contamination has important implications here and in the developing field of total parenteral nutrition.

Cosmetics and pharmaceutical products applied to the skin present the greatest problem in terms of in-use contamination. During use they are continuously challenged by the resident flora of the skin (staphylococci, micrococci, aerobic and anaerobic diphtheroids). Myers and Pasutto (1973) reported that these organisms were frequently isolated from used cosmetics. In a study of eye cosmetics, Wilson *et al.* (1971) noted that the predominant contaminants in used products represented the indigenous flora of the skin around the eye. The ease with which these bacteria can survive in certain types of product suggests that pathogenic organisms with similar growth requirements may likewise contaminate these products, when the opportunity arises. Topical products used in hospitals may additionally be challenged by transient organisms, such as *Ps. aeruginosa*, which do not normally survive on the skin, but may be isolated from the skin of certain types of patients or recovered intermittently from the hands of nursing staff. Several factors are known to increase the survival of *Ps. aeruginosa* on the skin, including suppression of the flora of Gram-positive bacteria by certain therapeutic agents (such as antibiotics, steroids and immunosuppressants), skin disease, skin damage or occlusion, and the presence of high-humidity atmospheres. Medicaments belonging to patients in these categories may thus become heavily contaminated with *Ps. aeruginosa* and other Gram-negative organisms (Baird *et al.*, 1979b, 1980).

The problem of in-use contamination in hospitals is compounded by the practice of issuing topical medicaments in multidose stock pots for the convenience of nursing staff. These medicaments may be used on several patients in a ward until the stock pot is empty. Once contaminated, these medicaments may serve as a vehicle for cross-contamination between other patients and the task of pinpointing the source of contamination may be extremely difficult. Creams and oint-

Table 8

Contamination in topical medicaments: the effect of packaging

| Medicaments | Hospital | | |
	A Stock pots	B Stock pots	C Individual tubes
Number examined	163	44	250
Number contaminated	63	41	22
Number contaminated with *Pseudomonas aeruginosa*	32	11	0
Number contaminated with *Staphylococcus aureus*	34	26	0

ments used in the treatment and prevention of pressure sores are an example of products frequently packed in stock pots. In one study it was found that these medicaments often became contaminated with *Ps. aeruginosa* and *Staph. aureus* (Table 8) and, furthermore, that the same contaminating strains could be isolated from skin sites, including pressure sores of these patients (Baird *et al.*, 1979b). In contrast, when the same products were packed in tubes for individual patient's use, they became contaminated much less often.

Handcreams and handlotions may also become contaminated during their use by nursing staff in the hospital. Ayliffe *et al.* (1969) drew attention to this problem in finding that 29 out of 70 hexachlorophane-containing handcreams were contaminated with Gram-negative bacilli, most of which were *Klebsiella* spp., *Ps. aeruginosa* or *E. coli*. Handcreams are widely applied after handwashing procedures, in itself considered to be the most important precautionary measure in the control of hospital cross-infection. Ironically, the use of contaminated handcreams may result in the hands becoming more contaminated than before washing. Contaminants may subsequently be transmitted from the hands to patients during nursing procedures. Morse *et al.* (1967) described an outbreak of *Klebsiella pneumoniae* septicaemia in patients with venous catheters in an intensive-care unit which was traced to a contaminated lanolin handcream. The contaminating organism, it was suggested, was inadvertently transferred from the handcream to the catheters by the nurses' hands.

2. *Applicators*

Cosmetic and pharmaceutical products, particularly those for external use, may sometimes be applied with spatulas, pads, sponges or brushes. If re-used, these may become heavily contaminated and serve as a continuing source of contamination for the remaining product and potential infection for the user. In one study of cosmetic applicators, all samples were found to be heavily contaminated and many contained *Staphylococcus* spp. (Myers and Pasutto, 1973). In contrast, the use of spatulas on a single occasion may lower appreciably the incidence of contamination in topical products, even when stock pots are involved (Baird *et al.*, 1979b). Svanberg (1978) reported that subjects heavily infected with salivary *Streptococcus mutans* could contaminate their toothbrushes and toothpastes with the same strains, and that use of these items could account for transmission of *S. mutans* between individuals.

3. *Equipment*

In hospitals, a variety of equipment is utilized during the course of patient therapy. Humidifiers, ventilators, aspirators, resuscitators, nebulizers and other apparatus may become contaminated during use, particularly with *Ps. aeruginosa* and other opportunist pathogens. These intricate and sometimes delicate instruments may be difficult or impossible to sterilize by heat and chemical disinfection may be attempted less successfully. Disinfectants with a narrow antibacterial spectrum or those deteriorating with storage, or inactivated in the presence of organic matter, may become heavily contaminated during use. Equipment treated in this way provides a source of cross-infection for further

patients and undoubtedly contributes to the overall level of contamination in the environment.

4. The Environment

Large numbers of opportunist pathogens may be found in hospital
environments and, under suitable conditions, these may find their way
into products. *Pseudomonas aeruginosa* is widely distributed in the hospital
environment and particularly in moist sites, such as drains, taps and
cloths. It has been postulated that splash-back from drains of hand-
basins may be responsible for introduction of *Ps. aeruginosa* into pots
of handcream kept at the side of basins. Baths may also be a source
of cross-infection if they are not cleaned properly after use. Patients
with diseases of the skin frequently use special ointments or creams
as a soap substitute in the bath. Clearly these are exposed to microbial
contamination during use from bath water which repeatedly dilutes
the product. Furthermore, a film of ointment on the sides of the bath
may impair cleaning. Cross-infection problems are compounded when
these medicaments are supplied in multidose pots for use by several
patients.

Products used in the home have been found to be less frequently
contaminated than those used in hospitals (Baird *et al.*, 1979a) and,
in particular, *Ps. aeruginosa* has been found much less often in these
products (0.1% compared with 2.7%). The fact that domicillary
products are generally issued in reasonably small quantities and often
for use by one person must be largely responsible for this fact. Indeed,
it has been contended that *Ps. aeruginosa* is rarely found in the domestic
environment, in contrast to its incidence in hospitals (Whitby and
Rampling, 1972).

VI. FACTORS INFLUENCING MICROBIAL SURVIVAL

Microbial contaminants may be described as dynamic populations in
which the organisms are actively multiplying, or static populations in
which no growth occurs. In the latter situation, this may be due to
inherently adverse conditions within the product, to the presence of
antimicrobial agents in sublethal concentrations, or to the presence of
spores. Both populations should be regarded as changeable and both

may exist within the same product. Thus, aerobic organisms may grow within a closed system until the oxygen tension has been lowered to a level to permit growth of anaerobes. Similarly, metabolic byproducts formed during multiplication of one organism may serve as a substrate for growth of another.

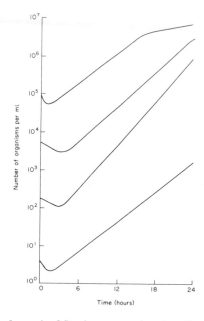

Fig. 5. Time-course of growth of *Pseudomonas aeruginosa* in sodium lactate solutions.

Several key factors are known to influence microbial survival within pharmaceutical and cosmetic products. The importance of highly nutritious ingredients and available water to aid survival has already been discussed. Of particular clinical significance, products for intravenous injection, ranging from simple salt solutions to the complex lipid emulsions used in total parenteral nutrition, may support growth of a variety of bacteria, including opportunist pathogens (Fig. 5). The presence and efficacy of antibacterial agents has also been mentioned as a critical factor. The activity of certain compounds has been shown experimentally to vary with changes in pH value and to increase with higher temperatures (Cowen and Steiger, 1976). On the other hand, higher ambient temperatures may favour microbial growth. In practice, products may be subjected to varying temperatures during storage and use, ranging from warm drug cupboards in the hospital ward and warm humid bathroom cupboards in the home to tropical climates overseas.

Generally speaking, the longer a product remains in use, the higher the incidence of microbial contamination. In practice, it is often impossible to know for how long a product has been in use. In a study of 1,977 pharmaceutical products returned from use in the home in the year 1977, 82% of samples were of indeterminate age (Baird *et al.*, 1979a), and a small number were over 40 years old. Similarly, in a survey of hospital medicines, 60% of samples were of unknown age (Public Health Laboratory Service Working Party Report, 1971). Pharmaceutical and cosmetic products are commonly hoarded for some future use. The labelling of all preparations with an expiry date, after which time the product should not be used, would to some extent alert consumers and patients to the potential hazards of this practice.

Variations in the inoculum size of the invading organism and the number of times a product is challenged during use can also affect survival patterns. These variations may be substantially affected by the type of packaging. Products packed in multidose containers may be subjected to multiple and continuous challenges. Large stock pots present the greatest problem, since the contents are often removed by hands which are never sterile. Products packed in small tubes or pots and used in individual patients become contaminated much less often (Baird *et al.*, 1979b). In practice, the inoculum size varies according to how and where the product is used and, occasionally, a multiple inoculum may be introduced.

VII. MONITORING AND CONTROL

There are several approaches to control of microbial contamination in pharmaceutical and cosmetic products. Through the establishment of standards, products are required by law to comply with such regulations. Through the observance of accepted guidelines, products will meet a minimum common standard. By following good manufacturing practices, product quality can be maintained, regularly inspected and subsequently sustained during use by means of effective preservation, suitable packaging and through education of product users.

A. International Standards

Although first recognized as an international problem by Kallings *et al.* (1966), the control of microbial contamination in products has been tackled in different ways in the United States, Britain and in other European countries. To date, there has been no attempt to establish universally accepted standards. A distinction has been made, however, between pharmaceuticals that must be made as sterile products, and pharmaceutical and cosmetic preparations made as non-sterile products. Sterile products have traditionally included injectable and ophthalmic solutions, and the most recent edition of the Nordic Pharmacopoeia also requires all ophthalmic products, eardrops, preparations used in certain body cavities and preparations containing corticosteroids to be sterile. In Britain, contact-lens solutions will shortly have to be made as sterile products, and the licensing of new contact-lens fluids in Great Britain began on January 1st, 1980. Standards will continue to change as more information becomes available on the microbiological hazards of product contamination.

The establishment of standards for non-sterile products has presented even greater problems. There have been two schools of thought. European standards have tended to follow the Swedish suggestion that a limitation on the total microbial count should be imposed on all products and that this should not exceed 100 micro-organisms per g or per ml. Standards adopted by Britain and the United States have required the absence of viable harmful organisms such as *Salmonella* spp., *E. coli*, *Ps. aeruginosa* and *Staph. aureus* in certain products. In accordance with this, the United States Pharmacopoeia has established microbial limits for ten product monographs, including acacia, digitalis, gelatin, starch and thyroid. Both sets of regulations have their limitations. Standards based on a total viable count may be practicable at the time of manufacture, but may be difficult to maintain during storage. On the other hand, it is becoming increasingly difficult to define what is meant by a harmful or pathogenic organism. The importance of opportunist pathogens in contaminated products has already been discussed, and often their presence must be assessed in relation to the route by which they are to be given.

As a result, many pharmaceutical manufacturers have set up their own in-house standards, based on past experience with their own products. These may include a limitation on the number of spoilage organisms, such as a 100 or 1,000 organisms per g or per ml, and a requirement for the absence of certain organisms considered to be injurious to health.

Except for the United States, there is no specific legislation governing the microbial content of cosmetic products. The amended Federal Food, Drug and Cosmetic Act of 1967 states that a product shall be deemed adulterated if it consists in whole or in part of any filthy, putrid or decomposed substance. Thus, the presence of pathogens or opportunist pathogens constitutes adulteration within the Act by reason of potential

Table 9

Comparison of international microbiological guidelines for cosmetics

Association	Limits
Cosmetic, Toiletry and Fragrance Association, United States (1973)	Less than 1,000 organisms per g or per ml. Less than 500 organisms per g or per ml in eye or baby products. No product should have a microbial content recognized as harmful to the user.
Cosmetic, Toiletry Manufacturers Association of Australia (1974)	Less than 100 organisms per g or per ml. Less than 10 organisms per g or per ml for eye products. Less than 1,000 organisms per g or per ml for those with a mineral earth component and certain organisms (*Staphylococcus aureus*, *Escherichia coli*, and species of *Salmonella*, *Pseudomonas*, *Candida*, *Clostridium*) should be excluded.
Cosmetic, Toiletry and Perfumery Association of Great Britain (1976)	Less than 1,000 aerobic organisms per g or per ml. Less than 1,000 yeasts or moulds per g or per ml. Less than 100 organisms per g or per ml in eye or baby products and significant numbers of harmful organisms should be excluded (test methods for clostridia, *Pseudomonas aeruginosa* and *Staphylococcus aureus*).
Amended Cosmetic regulations in Yugoslavia (1977)	Less than 1,000 aerobic mesophilic bacteria per g or per ml. Less than 100 yeasts or mould spores per g or per ml, and *Staphylococcus aureus*, *Streptococcus pyogenes*, *Enterococcus* sp., *Pseudomonas aeruginosa*, *Escherichia coli*, *Proteus* sp., *Salmonella* sp. and *Shigella* sp. should be excluded in 0.1 g or 0.1 ml.

danger to health. Manufacturers of cosmetic products have traditionally been more concerned with spoilage organisms, and it is only
recently that the importance of organisms harmful to health has been
appreciated. The Cosmetic, Toiletry and Fragrance Association in the
United States took the initiative by publishing guidelines in 1973 which
suggested that all products should contain less than 1000 organisms
per g or per ml, and that those for use in the eye or on babies should
contain less than 500 organisms per g or per ml. The vexed question
of harmful organisms was met by a requirement that no product should
have a microbial content recognized as harmful to the user as determined by a standard method. Cosmetic manufacturers in Great Britain,
Yugoslavia and Australia have subsequently published their own guidelines and codes of practice (Table 9).

B. Control During Manufacture

In Britain and many other countries, control of pharmaceutical preparations and medicated cosmetics begins with the registration of new
products. New products are unlikely to be accepted for registration
if their standards are markedly lower than those of related published
monographs. Manufacture of these products in industry and more
recently in hospitals is controlled under the terms of the Medicines Act,
1968. Products are required to be made under conditions described
in the Guide to Good Pharmaceutical Manufacturing Practice (1977)
and come under the surveillance of the Medicines Inspectorate.

Factors affecting the microbial quality of the final product are well
established and must be controlled during manufacture. These include
quality control of raw materials, including water, formulation of components into a stable preparation and the inclusion of preservatives
if required, provision of a suitable manufacturing environment for the
process to be undertaken, scrupulous hygienic maintenance of environment and plant, control of all processes including sterilization, super-
vision and training of operators, and control and suitability of
packaging components. The end product is then inspected for quality.
For sterile products, a sterility test is performed on a representative
sample of the batch to demonstrate the absence of viable micro-
organisms under the test conditions, as described in the appropriate
pharmacopoeia. Non-sterile products should be tested for their overall

Pseudomonas spp. (Harbord, 1976). In a later survey of nine recalls initiated by the Food and Drug Administration between 1970 and 1974, it was suggested that there had been a marked improvement in product quality (Sanders, 1976). Until more conclusive evidence has been published it is difficult to state with conviction whether recent improvements in manufacturing processes, the establishment of guidelines, and comprehensive research and quality-control testing programmes have been rewarded by an overall improvement in the microbiological quality of the end product. In all developments, a cost–benefit equation must be closely examined. For, although it is possible to produce all products in a unit-dose form and free from microbial contamination, the benefit to the user must be carefully balanced against the inevitable cost of such a process.

REFERENCES

Allen, H. F. (1959). *Transactions of the American Ophthalmological Society* **57**, 377.

Association of British Pharmaceutical Industry (1978). 'The Pharmaceutical Industry and the Nation's Health'. 9th edition, London.

Awad, Z. A. (1977). Ph.D. Thesis: University of London.

Ayliffe, G. A. J., Barry, D. R., Lowbury, E. J. L., Roper-Hall, M. J. and Martin Walker, W. (1966). *Lancet* **i**, 1113.

Ayliffe, G. A. J., Barrowcliff, D. F. and Lowbury, E. J. L. (1969). *British Medical Journal* **1**, 505.

Baird, R. M. (1977). *Journal of the Society of Cosmetic Chemists* **28**, 17.

Baird, R. M. and Shooter, R. A. (1976). *British Medical Journal* **2**, 349.

Baird, R. M., Brown, W. R. L. and Shooter, R. A. (1976a). *British Medical Journal* **1**, 511.

Baird, R. M., Elhag, K. M. and Shaw, E. J. (1976b). *Journal of Medical Microbiology* **9**, 493.

Baird, R. M., Crowden, C. A., O'Farrell, S. M. and Shooter, R. A. (1979a). *Journal of Hygiene* **83**, 277.

Baird, R. M., Farwell, J. A., Sturgiss, M., Awad, Z. A. and Shooter, R. A. (1979b). *Journal of Hygiene* **83**, 445.

Baird, R. M., Awad, Z. A., Shooter, R. A. and Noble, W. C. (1980). *Journal of Hygiene* **84**, 103.

Bassett, D. C. J. (1971). *Journal of Clinical Pathology* **24**, 708.

Bassett, D. C. J., Stokes, K. J. and Thomas, W. R. G. (1970). *Lancet* **i**, 1188.

Bean, H. S. (1967). *Pharmaceutical Journal* **199**, 289.

Bean, H. S. (1972). *Journal of the Society of Cosmetic Chemists* **23**, 703.

Beveridge, E. G. (1975). *In* 'Microbial Aspects of the Deterioration of Materials' (R. J. Gilbert and D. W. Lovelock, eds.), p. 213. Academic Press, London and New York.

Beveridge, E. G. and Hope, I. A. (1971). *Pharmaceutical Journal* **207**, 102.

Beveridge, E. G. and Hugo, W. B. (1964). *Journal of Applied Bacteriology* **27**, 304.

Bignell, J. L. (1951). *British Journal of Ophthalmology* **35**, 419.

Blackwell, B. (1972). *Clinical Pharmacology and Therapeutics* **13**, 841.

Boud, M. R. (1967). *Medical Journal Australia* **1**, 1258.

Bryce, D. M. and Smart, R. (1965). *Journal of the Society of Cosmetic Chemists* **16**, 187.

Bucherer, H. (1965). *Zentralblatt für Bakteriologie, Parasitenkunde, Infektionskrankheiten und Hygiene, Abt II*, **119**, 232.

Buck, A. C. and Cooke, E. M. (1969). *Journal of Medical Microbiology* **2**, 521.

Butler, N. J. (1968). *In* 'Biodeterioration of Materials' (A. H. Walters and J. J. Elphick, eds.), vol. 1, p. 269. Elsevier, Amsterdam.

Chong, Y. H. and Beng, C. G. (1965). *Medical Journal of Malaya* **20**, 49.

Clothier, C. M. (Chairman) (1972). *In* 'Report of the Committee appointed to enquire into the circumstances, including the production, which led to the use of contaminated infusion fluids in the Devonport section of Plymouth General Hospital', Cmnd 5035, H.M.S.O., London.

Cooke, E. M., Shooter, R. A., O'Farrell, S. M. and Martin, D. R. (1970). *Lancet* **ii**, 1045.

Cosmetic, Toiletry and Fragrance Association Report (1975). *Cosmetic Journal* **7**, 3.

Cowen, R. A. and Steiger, B. (1976). *Journal of the Society of Cosmetic Chemists* **27**, 467.

Cragg, J. and Andrews, A. V. (1969). *British Medical Journal* **3**, 57.

Crompton, D. O. (1962). *Australasian Journal of Pharmacy* **43**, 1020.

Crompton, D. O., Anderson, K. F. and Kennare, M. A. (1962). *Transactions of the Ophthalmological Society of Australia* **22**, 81.

Croshaw, B. (1977). *Journal of the Society of Cosmetic Chemists* **28**, 3.

Dale, J. K., Nook, M. A. and Barbiers, A. R. (1959). *Journal of the American Pharmaceutical Association, Practical Pharmacy Edition* **20**, 32.

Dunnigan, A. P. and Evans, J. R. (1970). *Toilet Goods Association Cosmetic Journal* **2**, 39.

Editorial (1908). *British Medical Journal* **1**, 892.

Eisman, P. C., Kull, F. C. and Mayer, R. L. (1949). *Journal of the Pharmaceutical Association, Scientific Edition* **38**, 88.

Elek, S. D. and Conen, P. E. (1957). *British Journal of Experimental Pathology* **38**, 575.

Falkow, S. (1975). *In* 'Infectious Multiple Drug Resistance' (J. R. Lagnado, ed.), pp. 205–224. Pion Ltd., London.

Felts, S. K., Schaffner, W., Melly, M. A. and Koenig, M. G. (1972). *Annals of Internal Medicine* **77**, 881.

Forkner, C. E. (1960). *In* '*Pseudomonas aeruginosa* Infections', (I. S. Wright, ed.), p. 59. Grune and Stratton, New York.

Gale, E. F., Cundliffe, E., Reynolds, P. E., Richmond, M. H. and Waring, M. J. (1972). *In* 'The Molecular Basis of Antibiotic Action', pp. 386–391. John Wiley and Sons, London and New York.

Glencross, E. J. (1972). *British Medical Journal* **2**, 376.

Grant, D. J. W., De Szocs, J. and Wilson, J. V. (1970). *Journal of Pharmacy and Pharmacology* **22**, 461.

Guide to Good Pharmaceutical Manufacturing Practice (1977). H.M.S.O., London.

Harbord, P. E. (1976). *Symposium of the Society of Cosmetic Chemists of Great Britain, Birmingham*, 11.

Hardy, P. C., Ederer, G. M. and Matsen, J. M. (1970). *New England Journal of Medicine* **282**, 33.

Hart, A. and Moore, K. E. (1977). *Pharmaceutical Journal* **218**, 245.

Heiss, F. (1967). *Fette Seifen Anstrichmittel* **69**, 365.

Hills, S. (1946). *New Zealand Medical Journal* **45**, 219.

Hugo, W. B. and Beveridge, E. G. (1962). *Journal of Applied Bacteriology* **25**, 72.

Kallings, L. O., Ringertz, O., Silverstolpe, L. and Ernerfeldt, F. (1966). *Acta Pharmaceutica Suecica* **3**, 219.

Kedzia, W., Lewon, J. and Wismienski, T. (1961). *Journal of Pharmacy and Pharmacology* **13**, 614.

Kellaway, C. H., MacCallum, P. and Terbutt, A. H. (1928). *Report of the Commission to the Governor of the Commonwealth of Australia, Melbourne.*

Kelsey, J. C. and Beeby, M. M. (1964). *Lancet* **ii**, 82.

Kohn, S. R., Gershenfeld, L. and Barr, M. (1963). *Journal of Pharmaceutical Sciences* **52**, 967.

Krupka, L. R. and Racle, F. A. (1967). *Nature, London* **216**, 486.

Kuehne, J. W., Ahearn, D. G. and Wilson, L. A. (1971). *Developments in Industrial Microbiology* **12**, p. 173.

Kundsin, R. B., Walter, C. W. and Scott, J. A. (1973). *Surgery* **73**, 778.

Kunz, L. J. and Ouchterlony, O. T. (1955). *New England Journal of Medicine* **253**, 761.

Last, P. M., Harbison, P. A. and Marsh, J. A. (1966). *Lancet* **i**, 74.

Lee, J. C. and Fialkow, P. J. (1961). *Journal of the American Medical Association* **177**, 144.

Lennox-King, S. M. J. (1976). Ph.D Thesis, University of London.

Lepard, C. W. (1941). *Transactions of the American Academy of Ophthalmology and Otolaryngology* **46**, 55.

Linton, K. B. and George, E. (1966). *Lancet* **i**, 1353.

Lowbury, E. J. L. (1951). *British Journal of Industrial Medicine* **8**, 22.

McCall, C. E., Collins, R. N. and Jones, D. B. (1966). *American Journal of Epidemiology* **84**, 32.

McCullough, J. C. (1943). *Archives of Ophthalmology* **29**, 924.

Maki, D. G., Rhame, F. S., Mackel, D. C. and Bennett, J. V. (1976). *American Journal of Medicine* **60**, 471.

Malizia, W. F., Gangarosa, E. J. and Goley, A. G. (1960). *New England Journal of Medicine* **263**, 800.

Maurer, I. M. (1978). 'Hospital Hygiene', 2nd edn., pp. 42–57. Edward Arnold, London.

Meers, P. D., Calder, M. W., Mazhar, M. M. and Lawrie, G. M. (1973). *Lancet* **ii**, 1189.

Midtvedt, T. and Linstedt, G. (1970). *Acta Pathologica et Microbiologica Scandinavica, Section B* **78**, 488.

Mitchell, R. G. and Hayward, A. C. (1966). *Lancet* **i**, 793.

Morse, L. J. and Schonbeck, L. E. (1968). *New England Journal of Medicine* **278**, 376.

Morse, L. J., Williams, H. I., Grenn, F. P., Eldridge, E. F. and Rotta, J. R. (1967). *New England Journal of Medicine* **277**, 472.

Myers, G. E. and Pasutto, F. M. (1973). *Canadian Journal of Pharmaceutical Sciences* **8**, 19.

Nelson, J. H. (1942). *Journal of Pathology and Bacteriology* **54**, 449.

Noble, W. C. and Savin, J. A. (1966). *Lancet* **i**, 347.

Norton, D. A., Davies, D. J. G., Richardson, N. E., Meakin, B. J. and Keall, A. (1974). *Journal of Pharmacy and Pharmacology* **26**, 841.

Olin, G. and Lithander, A. (1948). *Acta Pathologica et Microbiologica Scandinavica* **25**, 152.

Olsen, H. (1967). *Danish Medical Bulletin* **14**, 6.

Payne, R. W. (1967). *British Medical Journal* **4**, 17.

Phillips, I. (1967). *Journal of Hygiene* **65**, 229.

Phillips, I. and Spencer, G. (1965). *Lancet* **ii**, 1325.

Phillips, I., Eykyn, S. and Laker, M. (1972). *Lancet* **i**, 1258.

Plotkin, S. A. and Austrian, R. (1958). *American Journal of Medical Sciences* **235**, 621.

Public Health Laboratory Service Working Party Report. (1971). *Pharmaceutical Journal* **207**, 96.

Robinson, E. P. (1971). *Journal of Pharmaceutical Sciences* **60**, 604.

Sanders, A. C. (1976). *Symposium of the Society of Cosmetic Chemists of Great Britain, Birmingham*, 10.

Shooter, R. A., Gaya, H., Cooke, E. M., Kumar, P., Patel, N., Parker, M. T., Thom, B. T. and France, D. R. (1969). *Lancet* **i**, 1227.

Simmons, N. A. and Gardner, D. A. (1969). *British Medical Journal* **2**, 668.

Smart, R. and Spooner, D. F. (1972). *Journal of the Society of Cosmetic Chemists* **23**, 721.

Sokolski, W. T., Chidester, C. G. and Honeywell, G. E. (1962). *Developments in Industrial Microbiology* **3**, 179.

Steinbach, H. L., Rousseau, R., McCormack, K. R. and Jawetz, E. (1960). *Journal of the American Medical Association* **174**, 1207.

Stewart, R. B. and Cluff, L. E. (1972). *Clinical Pharmacology and Therapeutics* **13**, 463.

Svanberg, M. (1978). *Scandinavian Journal of Dental Research* **86**, 412.

Theodore, F. H. and Feinstein, R. R. (1952). *American Journal of Ophthalmology* **35**, 656.

Thomas, M. E., Piper, E. and Maurer, I. M. (1972). *Journal of Hygiene* **70**, 63.

Victorin, L. (1967). *Acta Paediatrica Scandinavica* **56**, 344.

Walker, H. W. and Ayres, J. C. (1969). *In* 'The Yeasts' (A. H. Rose and J. S. Harrison, eds.), vol. 2, p. 463. Academic Press, London and New York.

Wargo, E. J. (1973). *American Journal of Hospital Pharmacy* **30**, 332.

Whitby, J. L. and Rampling, A. (1972). *Lancet* **i**, 15.

Wilson, L. A. and Ahearn, D. G. (1977). *American Journal of Ophthalmology* **84**, 112.

Wilson, L. A., Kuehne, J. W., Hall, S. W. and Ahearn, D. G. (1971). *American Journal of Ophthalmology* **71**, 1298.

Wilson, L. A., Julian, A. J. and Ahearn, D. G. (1975). *American Journal of Ophthalmology* **79**, 596.

Wolven, A. and Levenstein, I. (1969). *Toilet Goods Association Cosmetic Journal* **1**, 34.

Wolven, A. and Levenstein, I. (1972). *American Cosmetics and Perfumery* **87**, 63.

Yasufuku, M., Hashimoto, K., Hamai, J. and Uesug, I. (1968). *Fifth Congress of the International Federation of the Society of Cosmetic Chemists, Tokyo*.

ductivity characteristics) as well as development of varied odours or appearance of unwanted pigmentation.

What may be considered as biodeterioration in one application may not be considered so in another. Sparse mycelial growth, which may have serious consequences in a fine-tuned electronic system, may be absolutely inconsequential on the surface of a plastic cover used to protect a piece of equipment in outdoor storage. Further, what may be considered biodeterioration during the useful life of a plastic may become beneficial waste disposal once that plastic is discarded.

During the past 35 years, plastics have found an almost infinite number of uses. In many of these uses, there are reports of bio-deterioration. It would be fruitless to list the entire spectrum of situations in which microbial degradation was either proved or suspected. A few brief examples are listed to exemplify the range of problems presented. These include degradation of pressure-sensitive tapes used in coating gas pipes, polyethylene lids or milk churns, coatings for buried pipes and cables, foam placed in the fuel tanks of military aircraft for explosion prevention, raincoats, liners of gasoline storage tanks, electronic component insulation, and plastic upholstery materials.

Most of these problems can be overcome with current knowledge. Solutions usually include the selection of proper polymers or additives, or protection with preservative treatments. However, because of the versatility of plastics, they are continually being used in new situations, and by manufacturers unfamiliar with the role of micro-organisms. Test laboratories and manufacturer's technical representatives continue to see biologically damaged materials.

II. PLASTICS—AN INTRODUCTION

It is useful, in discussing microbial degradation of plastics, to presume that the reader is familiar with plastics technology. Unfortunately, most microbiologists, and indeed most other readers not employed in the plastics industry, are not. It then becomes necessary to provide a brief indoctrination into the field of plastics so that the reader can follow this discourse. If one is prompted to initiate studies dealing with the interaction of microbes and plastics, a thorough background into the complete system studied is beneficial. In fact, consultation with a polymer chemist or plastics engineer may be necessary.

One of the basic terms in the plastics industry is *polymer*. This term is often used interchangeably with *resin*, and the terms will be used synonymously in this article. A polymer is a compound of high molecular weight with a structure composed of multiple interacted simpler chemical entities. For example, cellulose is a polyglucose. Names such as polyethylene (polythene), polypropylene and polystyrene are given to polymers to reflect the nature of the repeating unit (monomer). A polymer may be composed of more than one kind of monomer, in which case it may be called a copolymer. Protein is a natural polymer composed of several amino-acid monomers. Acrylonitrile-butadiene-styrene polymers are copolymers of the three monomers.

There are more than 40 generic types of polymers (including acrylics, epoxies, nylons, vinyls and urethanes) on the market. Most of the generic types of polymers represent a variety of resins having similar, but not identical, chemical structures. For example, Nylons-6/6, -6/10, -6/12, -6, -11 and -12 differ chemically and physically.

Various polymers may differ from batch to batch in degree of branching, cross-linking and distribution of molecular weights. In addition, resins having such properties may be deliberately produced to meet various engineering requirements. Depending on the type of process, and process controls, the geometric arrangement of side groupings on a polymeric chain may vary slightly, or considerably, thus altering the final product characteristics. Most polymers used in the plastics industry have usable properties in the molecular-weight range of a few thousand to a few hundred thousand.

All of the molecules of a nominal molecular weight in a polymer batch are not the same size. The nature of the polymerization reaction is such that a group of similar molecules, differing in size, are formed. The nature of the distribution of the molecular sizes is essentially gaussian. The molecular weight designated for a synthetic polymer is an average. The sample contains some fraction of lower molecular-weight polymers (oligomers) and some molecules significantly larger than the average molecular weight.

The term *plastic* is usually applied to those polymers that are neither highly elastic (elastomers, rubbers) nor highly crystalline (fibres). In the engineering sense, a plastic is a mixture containing one or more polymers formulated with a variety of additives and fabricated into a usable product having specific physical and aesthetic characteristics. This plastic formulation may contain some combination of polymers (on occasion, representatives from several generic types, or several varieties of the same type), plasticizers (to improve flexibility), antioxidants, antistatic agents, colourants, fillers (to add strength), flame retardants, heat stabilizers (to assist in processing), biocides, ultraviolet-radiation stabilizers, and any of several processing aids (viscosity depressants, mould-release agents, emulsifiers, slip agents and anti-blocking agents). A great many of these additives are organic compounds of relatively low (as compared to polymers) molecular weights. Examples of the chemical nature of some additives are listed in Table 1. Additives other than plasticizers are usually added at concentrations (by weight) ranging from less than 0.1% to several per cent.

Table1
Examples of additives used in plastic formulations

Additive type	Examples
Anti-oxidant	Distearyldithiopropionate esters, alkylated phenols
Antistatic agent	Quaternary ammonium compounds
Colourants	Organic dyes, carbon black, titanium dioxide
Flame retardants	Halogenated organic compounds
Heat stabilizers	Barium, cadmium, zinc or lead salts of organic acids (i.e. dioctyl tin maleate)
Lubricants	Metallic stearates, esters, fatty acids, hydrocarbon waxes
Biocides	Salicylanilide, copper 8-hydroxyquinolinolate
Ultraviolet stabilizers	Benzophenones, benzotriazoles, salicylates, acrylates

The additive to plastic formulations which represents the greatest bulk is plasticizer. Plasticizers are usually oils of various sorts, including natural or modified vegetable oils, esters of organic acids, polyesters and chlorinated hydrocarbons. Approximately three-quarters of all plasticizers sold are for use with vinyl formulations, although plasticizers are also used for certain cellulosic, polymethyl methacrylate and polystyrene applications. Plasticizer production, World-wide, exceeds a billion kg per year. There are approximately 500 different plasticizers manufactured by some 80 different international suppliers. About two-thirds of the plasticizer market is occupied by various phthalate esters, the principal materials being the di-2-ethylhexyl and the di-isooctyl esters.

A plasticizer may be defined as a material that is mechanically mixed into a plastic formulation to increase flexibility, workability or extensibility. The resins that require plasticization generally are hard and brittle. Plasticizer addition makes the formulation softer, more flexible and easier to process. In general, plasticizers maintain their chemical identity, and are compatible with the polymer through hydrogen bonding and/or van der Waals forces.

The problem of selecting the proper plasticizer(s) to meet specific end-use requirements is quite complicated. Although phthalate esters are essentially the universal plasticizers, many applications require formulation properties that cannot be met with phthalates alone. For example, phthalate-plasticized vinyls are brittle at below-freezing temperatures. Phosphate esters improve flame resistance. Polyesters resist extraction by oils and solvents. However, phosphate esters tend to cause instability of the formulation to heat and light; polyesters may affect low-temperature characteristics. The final selection of plasticizers to obtain the desired properties requires a sophisticated balancing and compromising. Plasticizer content in a final formulation depends on the end-use of the plastic, but in some cases may approach 40% in flexible vinyl films.

As late as 1957, polyvinyl resins led the plastics market in usage in the United States and continue to be among the most popular plastics. At that time, only about one per cent of the total production was utilized in rigid, non-plasticized formulations. Prior to that time, vinyls were primarily utilized in a wide variety of 'soft' applications, including many military applications throughout the World (for example, raincoats, insulation on wires and cables and waterproof coverings). Even today, plasticized vinyls enjoy wide application.

In view of the widespread use of plasticized vinyls and the nature of some materials usable as plasticizers (such as natural oils and esters of fatty acids), it is not surprising that many biodeterioration problems developed with the soft vinyl formulations. Indeed, most of the literature dealing with biodeterioration of plastics is associated with plasticized vinyl systems.

There are, however, data about the susceptibility of other types of formulations to microbial deterioration. Some of these data are contradictory, which is not surprising when the wide variety of microbially susceptible components which can be formulated into a plastic are considered.

III. MICROBIOLOGY OF PLASTIC FORMULATIONS

The study of the microbiology of plastic formulations has centred mostly on the involvement of the microbe with the additive plasticizers. Plasticizers are added to vinyl resin complexes to provide flexibility. They are also added to some of the less-used polymer formulations, notably the modified cellulosics.

Because of their molecular structure and chemical properties, most synthetic resins in common usage do not support growth. Microbial degradation of but a few polymers has been reported in the literature. These include low molecular-weight polyolefins, polyurethanes derived from esters, and a few polyesters, all of which have functional groups and arrangements susceptible to microbial attack. The microbiology of polymer degradation, and that of plasticizer deterioration will be addressed separately.

A. Microbiology of Polymers

Most of the polymers used in commercial plastic formulations do not support microbial growth. The various chemical configurations and bonds found in these polymers are also found in biodegradable organic compounds that occur in nature. The resistance, therefore, of plastic polymers to biodegradation does not appear to lie entirely in 'exotic' chemical configurations. Neither does the resistance of these compounds lie solely on the basis of their high molecular weights, since naturally occurring biodegradable polymers (i.e. proteins, polysaccharides, nucleic acids) may have molecular weights at least as great as synthetic polymers. It has been suggested by some researchers that, in some cases, the nature of the polymers may be such that chain ends are unavailable for initial oxidation by microbial enzymes. Others have speculated that, since most plastic formulations are hydrophobic in nature, the attainment of the conditions (water absorption, swelling and proper pH value) favourable for the proper functioning of microbial enzymes, which are proteinaceous, cannot be met. The cause of the microbial inertness of these synthetic polymers is probably due to a combination of conditions including chemical bonding, molecular size and configuration, ability to absorb water, and other factors yet to be considered.

It is possible to synthesize biodegradable polymers. During the first half of this decade several laboratories undertook studies to prepare such compounds. Biodegradable polymers, if they meet all other physical characteristics necessary for plastic packaging, could solve much of the litter and waste problems associated with disposable plastic packages. Bailey *et al.* (1976) described regularly alternating copolyamides with an α-amino acid comonomer as being biodegradable. These polymers are broken down in several weeks by both fungi and bacteria.

Huang *et al.* (1976) synthesized a variety of polymers which were found to be biodegradable as determined by conventional exposure to microbes, or by measurement of changes that occur when the polymers were exposed to a variety of enzymes (papain, subtilisin, urease, trypsin, chymotrypsin). They found that phenyl-substituted polyamides, polyesters, poly(amide-urethane) and poly(ester-urea) polymers derived from mandelic acid, phenylalanine and benzyl-malonic acid were biodegradable.

However, the interest in development of biodegradable polymers does not seem to be continuing. Discussions with laboratories previously active in this area indicate, in general, that their interests have turned to other endeavours.

1. Polyethylene

Polyethylene belongs to a class of synthetic polymers known as polyolefins which have the greatest usage of all available plastics. Some commercial polyethylene formulations (which may contain non-polymeric additives) will support growth. For example, Potts *et al.* (1972) noted that commercial polyethylene household wrap supported medium growth (30 to 60% coverage) of fungi from a mixed-spore test inoculum. However, this same material, after extraction with toluene, supported only sparse growth under identical conditions.

Some polyethylene resins also support microbial growth. Klemme and Watkins (1950), in an obscure military study, using extreme care in cleanliness, noted that the ability of polyethylenes to support growth when inoculated with spores of a mixture of fungi (*Aspergillus niger, Aspergillus flavus, Penicillium luteum* and *Trichoderma* T-1) was related to the molecular weight of the samples. Samples having average molecular weights below 10,000 supported small amounts of growth, whereas those with higher average molecular weights did not support growth.

Similar results were reported by Jen-Hao and Schwartz (1961). Their inoculum, however, consisted of a bacterial mixture containing four strains of *Pseudomonas aeruginosa*, a *Nocardia* species and a *Brevibacterium* species. They noted that with low average molecular-weight polymers (17,000 and below) bacterial growth increased almost a 100-fold during the first week, then gradually declined. Growth in this situation is almost certainly due to the presence of low molecular-weight oligomers present in the plastic materials. Water-soluble nutrients could not be extracted from the polyethylene samples. If, after termination of growth, the polyethylene materials were recovered, cleaned and re-used as the carbon source in fresh medium in further experiments, only the lowest molecular-weight material (average molecular weight, 4,800) stimulated significant growth. On the other hand, if the 'spent' media were filtered, re-inoculated and supplemented with fresh plastic, growth was essentially equivalent to that in the original studies.

Potts *et al.* (1972) have broadened the studies on polyethylene susceptibility to include both high-density and low-density polyethylenes. High-density polyethylene is essentially an unbranched more crystalline polymer, whereas the low-density polymer is more branched, and so less likely to fit into a crystalline pattern. Using a mixed-fungal inoculum (*Aspergillus niger*, *A. flavus*, *Chaetomium globosum* and *Penicillium funiculosum*) they noted that high-density polyethylene at lower molecular weights (approximately 11,000 and 14,000) supported light growth. Low-density polymers supported growth at essentially the same molecular weights as the high-density materials. It should be noted that of two low-density samples (molecular weights of 1,350 and 2,600) the higher molecular-weight sample supported the greater growth. The sample with a molecular weight of 1,350, however, was an amorphous, highly branched grease whereas that with a molecular weight of 2,600 was a wax with a much higher degree of crystallinity.

Potts *et al.* (1972) also demonstrated that anaerobic pyrolysis of either high- or low-density polyethylene yielded low molecular-weight polymers that will support abundant growth. It should be noted, however, that the pyrolyzate of the branched, low-density material, supported less growth at a molecular weight of 2,100 than a similar material from the high-density polyolefin with a molecular weight of 3,200.

Colin *et al.* (1976) and Potts *et al.* (1972) indicated that linear alkanes up to dotriacontane ($C_{32}H_{66}$, mol. wt. 451) can support abundant growth of micro-organisms. Haines and Alexander (1974) have noted that some micro-organisms could destroy alkanes up to tetratetracontane ($C_{44}H_{90}$, mol. wt. 620). Potts *et al.* (1972) also illustrated that branching can seriously interfere with the ability of a micro-organism to degrade a polymer chain. Although dodecane, hexadecane and tetracosane all support abundant growth, their methylated counterparts (2,6,11-trimethyldodecane, 2,6,11,15-tetramethylhexadecane and 2,6,-10,15,19,23-hexamethyltetracosane) will not.

On the basis of these data, one might suspect that any polyolefin which supports growth must contain short-chain oligomers with few branches and with molecular weights of less than 500.

Some insight into the biochemical mechanism of microbial degradation of polyolefins may be derived from studies intended to make these polymers more amenable to biodegradation. Mills and Eggins (1970) have, by extensive oxidation with boiling nitric acid, converted polyethylene (mol. wt. 10,000 to 50,000) into a waxy material (purportedly consisting of dicarboxylic acids of mean molecular weight 250) which will support growth of thermophilic fungi.

On the other hand, Potts *et al.* (1972) synthesized low molecular-weight polyethylenes in which the chains were terminated with either a carboxyl or a methylketone group. The molecular weight of the carboxy-terminated polymer was 3,000, that of the carbonyl-containing macromolecules, 1,500. Neither of these modified polymers supported growth of the test fungi, although polymers having molecular weights in these ranges do.

Photodegradation (ultraviolet radiation) of polymers leads to a wide variety of physical and chemical changes, including formation of carbonyl groups.

Colin *et al.* (1976) noted that addition of a single carbonyl grouping on one of the 3 through 18 positions on the chains of linear alkanes (C_{19} through C_{35}) had no effect on the ability of that alkane to support growth. This suggested that internal oxidation of polyolefins was not related to the susceptibility of the polymer to biodegradation.

Jones *et al.* (1974) studied thermally degraded air-oxidized low-density polyethylene (average mol. wt. 2,200), with an infrared spectrum indicating the presence of carbonyl groups. This treatment supposedly results in a product similar to ultraviolet- and photo-degraded polyethylene. Addition of the degraded polymer to various soils, and to sewage sludge, resulted in increased oxygen consumption.

The rate of oxygen consumption for enriched garden soil was approximately three times that of the control. Over 70 days almost 9% of the theoretical amount of oxygen had been consumed. In sewage sludge, oxygen uptake with the degraded polymer was quite rapid at first (0.7 ml net oxygen consumed during the first 70 hours), then it slowed dramatically. The total consumption was nearly 2% of theoretical values. Enrichment culture techniques were used to isolate soil cultures capable of oxidizing the thermally degraded polymer (Spencer et al., 1976). After ten serial transfers the same number and type of organisms (as judged solely by Gram stain and colony morphology) persisted and were considered to be capable of existing on the polymer fragments supplemented with a mineral salts solution. The organisms isolated from the degraded polyethylene included species of *Acinetobacter*, *Flavobacterium*, *Pseudomonas*, *Alcaligenes*, *Gamella*, *Arthrobacter*, *Aerococcus*, and *Cellulomonas*. Of these, only *Aerococcus* did not grow on the polymer on retesting, although not all isolates of each genus grew successfully.

Colin et al. (1976) extensively irradiated a variety of low-density polyethylene, and compared the growth obtained in mineral-salts medium supplemented with the irradiated plastics with that obtained from the unirradiated samples. The inoculum consisted of a mixed culture of organisms isolated from soil enriched with high molecular-weight ketones. They were unable to detect any significant growth either on intact polymers or on irradiated plastics.

Albertsson and Ranby (1975) quantitatively evaluated degradation of low-density polyethylene after irradiation, by measuring $^{14}CO_2$ production from samples labelled with radioactive carbon. This method allows quantitative measurements of oxidation of as little as 0.001% of the polymer. Their studies have continued for more than two years. Unirradiated samples, after more than 700 days incubation in composted garbage, had been degraded no more than 0.1%. Those samples irradiated with ultraviolet radiation for 42 days suffered slightly more than a 1% degradation when incubated under the same conditions. Under optimum conditions, without irradiation, the polyethylene samples degraded at an initial rate of 0.005 to 0.1% per month. It is not really surprising that irradiation does not significantly enhance the susceptibility of polyethylene to biodeterioration. Wiles (1973) noted that the ultraviolet chromophores (carbonyl and hydroperoxide groups) of low-density polyethylene predominantly exist on the surface of the film, having been produced by thermo-oxidation during processing. No more than approximately 0.2 μm on each surface of film undergoes photochemical changes resulting in cleavage into short chains which may be amenable to microbial metabolism.

Another type of problem affiliated with the effects of microbes on polyethylene was reported by Leesment (1958). A red discolouration of polyethylene lids for milk churns was found to be due to the presence of a species of the genus *Phoma*. It is not clear whether the organism grew on the polyethylene or on milk residue not completely removed by cleaning and disinfection processes. Other such 'pink staining' of plastics (plasticized vinyls) has been reported by Girard and Koda (1959) and Yeager (1962). Girard and Koda (1959) were able to identify one causative organism as a *Penicillium* sp. which excreted a pigment (red to pink under acidic conditions) into the plastic formulation. No colouration was noted under alkaline conditions. Yeager (1962) found that *Streptomyces rubrireticuli* produced a pigment with similar characteristics. In both cases, the microbes appeared to grow on organic materials (debris) on the vinyls rather than on the vinyl itself. The stain then migrated into the plastic formulation.

Inasmuch as polyolefins are used successfully in environments conducive to microbial growth, the polyethylenes that comprise the bulk of the World's production apparently have formulations and molecular weights well above those levels susceptible to microbial damage. Their continued use as protective packaging seems assured. Biodegradation as a means of disposal would not seem the method of choice.

2. Polyesters

Polyesters with a variety of molecular weights and physical characteristics have been manufactured. Some very low molecular-weight polyester polymers are oils that have been used as plasticizers in polyvinyl chloride compositions. This will be discussed in more detail below. Several of these polyester plasticizers have been reported to serve as an excellent nutrient for micro-organisms (Berk *et al.*, 1957), and also to undergo hydrolysis mediated by organisms growing on extraneous nutrients (Klausmeier, 1966). Other polyesters used as prepolymers for preparation of polyurethanes, and having average molecular weights ranging from 2,000 to 5,000, have been shown to support excellent fungal growth (Darby and Kaplan, 1968).

Another report illustrated that a commercial polyester (polycaprolactone, molecular weight not listed) supported heavy microbial growth (Potts *et al.*, 1972). The authors of this report also examined a variety of other polyesters for their abilities to support growth of a mixed group of fungi (*Aspergillus niger, A. flavus, Chaetomium globosum* and *Penicillium funiculosum*). They demonstrated that a branched low molecular-weight pivalactone polyester and three high molecular-weight terephthalate esters failed to support any growth. A high molecular-weight polycaprolactone, a high molecular-weight poly-hexamethylene succinate, a medium molecular-weight polyethylene succinate and a low molecular-weight polyethylene adipate all supported heavy growth. Growth in these studies does not seem to be absolutely related to either molecular size or structure. Both high and low molecular-weight polyhexamethylene adipates support low levels of growth. On the other hand, a low molecular-weight polytetra-methylene succinate supports heavy growth and a high molecular-weight sample only traces of growth.

The same authors noted that polycaprolactone samples (average mol. wt. 40,000), moulded into tensile test bars with an ultimate tensile strength of 2,610 p.s.i., had, after a year of soil burial, lost 42% of their original weight (presumably due to microbial activity) and were too fragile for strength properties to be measured. Injection-moulded containers of the same material, buried in soil for 12 months, lost 95% of their weight. The greater percentage weight loss was obviously due to the larger surface area exposed per unit weight of the containers.

Fields *et al.* (1974) provided a more in-depth study of polycaprolac-tones, using *Pullularia pullulans* as the test organism. They layered a polycaprolactone ester in the bottom of culture dishes, covered the polymer with a layer of culture medium (yeast extract–glucose–suc-cinate–mineral salts–agar) and inoculated the surface of the culture medium with the test organism. Polymers of various average molecular weights (1,250, 2,000, 17,000 and 30,000) were evaluated. Weight losses (reported in mg per cm² of polymer surface) were used for quantitative assessment of deterioration. After 42 days the lowest molecular-weight polymer had lost 16 mg per cm²; the highest, approximately 4 mg per cm².

By fractional precipitation from acetone by addition of increasing amounts of water, the authors collected nine fractions having average molecular weights ranging from 5,700 to greater than 61,000. Three fractions were evaluated for degradation by *P. pullulans*. The fraction with a molecular weight of 17,000 was partially degraded, whereas those with molecular weights of 38,000 and 61,000 were not.

In further evaluating the effect of molecular weight on polycaprolactone degradation, the authors prepared polymer mixtures of a fraction with a molecular weight of 60,000, which was not biologically degraded, and a polymer of molecular weight 2,000 which was degraded. On evaluation of the results, it appeared that the weight losses were a linear function of the amount of 2,000 molecular-weight polymer available.

The data on *P. pullulans* do not correlate well with the data on burial of polycaprolactones already described. Almost 40% of the polylactone with a molecular weight 30,000 is comprised of fractions that are not degraded by *P. pullulans* (mol. wt. 38,000 and above), yet studies showed that containers of molecular weight 40,000 (whole polymer) lost 95% of their initial weight after one year of soil burial. The differences perhaps can be ascribed to different micro-organisms (the pure culture, *P. pullulans*, compared with the entire spectrum of soil microflora), or perhaps to a more adverse environment in the soil, which could lead to some non-biological hydrolysis of the polyesters. This, in conjunction with microbial action, could perhaps lead to a more complete destruction.

Diamond *et al.* (1975) also noticed differences in results between laboratory culture exposure and soil burial tests. Films of 10 different polyesters were either inoculated with a mixture of five aspergilli (*A. flavus*, two species; *A. terreus*; *A. niger*; *A. versicolor*) and incubated on mineral salts–agar, or were buried in soil. As can be seen (Table 2), differences occur in the degree of degradation not only with polycaprolactone but with other polyesters as well. There is no correlation between either fungal growth or weight loss and molecular weight in these studies. It should also be noted that two polycaprolactones (one commercial, one laboratory synthesized) of essentially identical molecular weights both supported heavy fungal growth in plate tests. However, the laboratory sample lost more than 2.5 times the weight lost by the commercial sample in soil burial.

Fields and Rodriguez (1976) isolated soil organisms capable of degrading polycaprolactones and compared these with *P. pullulans*. The differences in degradation rates are variable and do not account for the differences noted in plate culture and soil-burial evaluations. These authors also examined a variety of other polyesters for susceptibility

Table 2

Comparative degradation of polyester films after four weeks incubation with fungi (growth) and after one month soil burial (weight loss). From Diamond *et al.* (1975)

Polyester	Average mol. wt.	Fungal growth	Weight loss (%)
Polyhexamethylene azelate	32,100	heavy	19.1
Polycaprolactone (commercial)	29,300	heavy	14.9
Polycaprolactone (laboratory)	27,500	heavy	38.5
Polydecamethylene succinate	21,700	medium	8.8
Polydecamethylene azelate	20,400	trace	3.4
Polyethylene succinate	20,100	trace	19.4
Polyhexamethylene sebacate	19,700	light	40.1
Polydecamethylene sebacate	14,800	trace	4.2
Polyethylene azelate	11,100	medium	17.0

to degradation by *P. pullulans*. Polyhexamethylene sebacate and poly-tetramethylene adipate inoculated with *P. pullulans* lost at least 14 mg per cm² of exposed surface. Under the same conditions, polytetramethylene succinate, polybetapropiolactone and two polycaprolactones lost considerably less weight.

These studies were broadened by examinations of polyesters made from 1,4-butanediol and mixtures of adipic and sebacic acids. Although the polymers in which the acid moiety is entirely adipic or sebacic acid lost only 5 to 10 mg per cm² of exposed surface, the copolymers containing 50 and 75 mol of adipic acid per 100 mol lost from 35 to 50 mg per cm², depending on the molecular weight. The authors ascribe the higher degree of attack of the copolymer to the higher amount of amorphous material in the copolymers. Obviously, some polyesters are susceptible to microbial degradation, and both molecular structure and size are factors in determining degradability. However, the inter-relationship of these two factors, as well as environmental factors which may mediate hydrolysis, is not clear.

Although several organizations have attempted to find a truly biodegradable plastic to help resolve plastic-waste problems, the only biodegradable practical plastic items on the market, to the authors' knowledge, are the containers used for transplanting tree seedlings for reaforestation. These containers, made of polycaprolactone, contain the planted seedling and eventually disintegrate after the seedlings are transplanted. The above data suggest that microbial action may play some role in this degradation.

3. Polyurethanes

Polyurethanes are a complex group of macromolecules produced by addition reactions between organic polyfunctional isocyanates and hydroxyl-rich compounds (at least two hydroxyl groups per molecule). Some of the more common isocyanates include toluene diisocyanate, diphenylmethane diisocyanate, hexamethylene diisocyanate and naphthalene diisocyanate. The hydroxylated compounds may be, among others, glycols, polyethers, polyesters or castor oil. By increasing the functionality of the reactants, or by altering the chemical groupings between the urethane linkages, a wide variety of polymers with various characteristics can be prepared to meet specific engineering requirements.

Since the early 1950s, there have been occasional reports about the susceptibility of polyurethanes to microbial attack. As early as 1958, Farbenfabriken Bayer A. G. was awarded a patent for protection of polyurethane foams from microbial attack. Not all early reporters found urethane to be susceptible to biodeterioration. However, Crum et al. (1967) studied microbial susceptibility of polyurethane lining materials used in aircraft fuel tanks. In these studies, two organisms selected for their ability to grow on jet aircraft fuel (Pseudomonas aeruginosa and Cladosporium resinae) served singly or in combination as an inoculum in an environment containing a mineral salts solution overlayed with jet aircraft fuel. Quantitative measurements were made of changes in electrical resistivity of the liner, and of microbial growth and oxygen consumption. The polyurethane linings lost integrity, as measured by changes of resistivity, within 3–4 weeks. Bacterial counts increased to 10^{10} cells per ml after one week. Mycelial weights of the C. resinae approached 15 mg per 100 ml medium. Oxygen consumption by C. resinae after seven days was no more than 1.4 times the controls with any of the linings tested. For the bacteria, oxygen utilization was as

luteum, Monilia geophila). Their findings indicated that plasticizers of fatty-acid esters, and long-chain dicarboxylic acid esters, could serve as nutrients for the fungi. Phthalate and phosphate esters with alkyl alcohol substituents were generally resistant. Also, glycol and glycollate had little nutritive value unless fatty-acid substituents were present.

In the same year, a brief report from the United States National Bureau of Standards on studies with 117 different plasticizers was published. These plasticizers were inoculated with spores of *Chaetomium globosum, Aspergillus niger* or *Penicillium* species. Sixty-four of these plasticizers did not support growth of any of the three test fungi. These included primarily esters of phthalic, tartaric, citric and phosphoric acids. The remaining 53 supported growth of at least one of the challenge fungi. These susceptible plasticizers included plant oils and fatty-acid esters. A later report by E. Abrams, United States National Bureau of Standards, discussed growth of *Aspergillus niger* on 83 different plasticizer substrates. His results were similar to those of the previously described studies.

S. Berk, of the United States Army's Frankfort Arsenal, compared the changes in vinyl resin films plasticized with dibutyl sebacate, dioctyl phthalate or butadiene-acrylonitrile after six weeks incubation subsequent to inoculation with *Aspergillus flavus, Penicillium* sp. or *Trichoderma* sp. The sebacate film supported good growth and showed a great increase in tensile strength and decrease in percentage elongation under stress. S. Berk attributed the changes to removal of plasticizer during fungal growth. The films prepared with phthalate or butadiene-acrylonitrile plasticizers showed insignificant growth or physical changes.

J. V. Harvey and F. A. Meloro of United States Army Quartermaster Depot, studied hundreds of experimental films exposed to a mixture of *Penicillium luteum, Aspergillus flavus, Chaetomium globosum, Memnoniella echinata, Penicillium* sp., *Aspergillus fumigatus, Curvularia lunata* and *Myrothecium verrucaria*. These films included almost all combinations of six different plasticizers (two ricinoleic acid derivatives, two phthalate esters, two phosphate esters). Films plasticized with the ricinoleates stiffened during incubation with the test organisms; the others did not. If mixtures of ricinoleate and non-nutritive plasticizers were used, the degree of stiffening was proportional to the amount of the ricinoleate.

During these studies stiffening was noted in some films despite sparse growth of the fungi. Closer examination revealed the presence of *Pseudo-*

monas aeruginosa, which is capable not only of inhibiting growth of fungi, but of utilizing the ricinoleate plasticizers. A review of some of these early studies has been published by Hueck-van der Plas (1960).

These early workers had established that not only could certain plasticizers serve as microbial nutrients when tested as pure chemical compounds, but also when incorporated into plastic formulations. Microbial utilization of these plasticizers effectively removed them from the formulation, resulting in unwanted physical changes.

A variety of 22 commercial plasticizers, representing most of the major chemical types in use at the time, as well as 23 laboratory-synthesized sebacate esters were evaluated for their ability to serve as a nutrient for *Aspergillus versicolor* or *Pseudomonas aeruginosa*. These two organisms were routinely isolated from deteriorated vinyl raincoats (Stahl and Pessen, 1953). Growth was measured quantitatively. The alcohol moiety of the sebacates varied in chain length from 1 to 18 carbons, including some branched-chain alcohols. The commercial plasticizers gave the anticipated results with *Aspergillus versicolor*. Dicarboxylic acid esters and esters of fatty acids supported good growth. In some cases, growth with a plasticizer (i.e. butyl ricinoleate, methyl ricinoleate) was greater than with sucrose or glycerol controls. With *Ps. aeruginosa*, growth was always less than with a glycerol control. Most of the plasticizers that supported fungal growth also supported bacterial growth. Failure to support fungal growth was reflected in the bacterial studies. Some of the ricinoleate esters, however, failed to support growth of the pseudomonad.

These authors also compared growth of purified, laboratory-synthesized sebacate esters with growth on the alcohol moieties of these esters. Although esters with alcohols larger than one and less than 15 carbon atoms supported growth of the fungus, the ethanol esters and esters with chain length of 15 or more carbon atoms supported only scanty growth (Table 3). For the bacterial culture, limited growth was supported by the C_1, C_5 and C_6 esters; all others supported excellent growth. In general, however, it is suspected that most of the growth came at the expense of the sebacic acid rather than from the alcohol moiety. Except for ethanol, none of the alcohols with chain lengths of 11 or less carbons supported growth of *A. versicolor*. Bacterial growth was non-existent, or at best scanty, for all normal alcohols smaller than n-decanol.

In a comprehensive study (Berk *et al.*, 1957), 127 different plasticizers,

Table 3

Growth of *Aspergillus versicolor* and *Pseudomonas aeruginosa* on sebacic acid esters of linear alkane alcohols of different chain lengths and on their corresponding alcohols. From Stahl and Pessen (1953)

| Alcohol chain length | Growth (mg dry wt. of cells from 70 ml of medium) of | | | |
| | *Aspergillus versicolor* | | *Pseudomonas aeruginosa* | |
	Ester	Alcohol	Ester	Alcohol
C_1	0	0	14	0
C_2	58	95	48	0
C_3	153	0	42	0
C_4	240	0	32	0
C_5	174	0	7	0
C_6	242	0	3	0
C_7	243	0	62	0
C_8	231	0	56	6
C_9	256	0	46	6
C_{10}	261	0	61	7
C_{11}	136	0	68	44
C_{12}	111	26	71	98
C_{13}	96	116	61	67
C_{14}	72	104	65	58
C_{15}	26	110	60	58
C_{16}	31	101	63	54
C_{17}	26	—	55	51
C_{18}	33	36	50	47
Glycerol	247	363	79	46

or plasticizer-related compounds, were evaluated for their abilities to serve as nutrients for 24 separate fungi. The plasticizers included all of those tested previously as well as new materials introduced commercially. The fungi represented 13 different genera, and are listed in Table 4. A few of these cultures had been used in previous studies with plasticizers although most had not. The plasticizers were homogenized in mineral salts–agar medium and spot inoculated with spores of the appropriate culture. Susceptibility was determined by the diameter of the fungal colonies after three weeks incubation. A relative susceptibility rating for each plasticizer was derived by averaging the growth of all organisms. The relative ratings of the various groups of plasticizers are shown in Table 5.

Using data from studies on alkyl esters of adipic acid, the authors were able to show a general increase in growth as the carbon chain of the alcohols increased in length from C_3 to C_6. The 1-methyl isomers supported at least as much growth as the ester with the corresponding normal alcohol moieties. More complex branching (2-methyl, 3-methyl, dimethyl or ethyl) decreased the growth attained compared with the normal isomer.

It should not be surprising that so many different fungi utilize plasticizers as nutrients, although the ability to use a specific plasticizer varies from one organism to another. Reese *et al.* (1955) noted that 95% of 358 cultures grew on coconut oil (a mixture of triglycerides of myristic, lauric, palmitic and stearic acids), 90% of 309 cultures metabolized methyl acetyl ricinoleate, and 60% of 82 cultures utilized dihexyl sebacate.

In general, plasticizers came to be regarded as 'susceptible' or 'resistant' with respect to biological utilization as a result of the various tests described above. It was concluded that the types of plasticizers resistant to microbial attack were esters of short-chain aliphatic

Table 4

Organisms utilized in testing plasticizers. From Berk *et al.* (1957)

Micro-organism	Previously studied
Alternaria solani	—
Aspergillus flavus	+
Aspergillus niger	—
Aspergillus oxyzae	+
Aspergillus terreus	—
Aspergillus ustus (three strains)	—
Aspergillus versicolor	+
Curvularia geniculata	—
Fusarium sp.	—
Glomerella cingulata	—
Mucor sp.	—
Myrothecium verrucaria	+
Paecilomyces varioti	—
Penicillium chrysogenum	—
Penicillium citrinum	—
Penicillium frequentans	—
Penicillium funiculosum (two strains)	—
Pullularia pullulans	—
Stachybotrys atra	—
Stemphylium consortiale	—
Trichoderma sp.	+

dicarboxylic acids (e.g. succinates), esters of tricarboxylic acids (citric, aconitic, tricarballylic), phthalic and phosphoric acid esters, derivatives of toluenesulphonic acid, glycol and glycollic acid derivatives having aliphatic moieties with carbon chains less than ten carbon atoms in length and aromatic hydrocarbons.

As so often happens when attempting to ascribe specifics to mutable biological systems, some of the 'susceptible' materials were resistant when used in commercial formulations, and some of the 'resistant' plasticizers could support microbial growth.

For example, several polyester plasticizers previously shown to support excellent microbial growth performed well in soil burial studies lasting four years (DeCoste, 1968). These data were verified (Klausmeier and Jamison, 1973) by incubation of vinyl strips plasticized with butylene glycol polyadipate with either *Ps. aeruginosa* or with fungal spores. This anomaly can be explained if one examines the mobility of the plasticizers in the plastic formulation, and the site of the microbe–ester interaction. The nature and size of polyester molecules severely limit their migratory ability in vinyl formulations, so that they do not migrate to the surface as the surface molecules are removed by microbial action. Since microbial action is limited to the plastic–environment interface, the failure of a susceptible plasticizer to migrate to the inter-

Table 5

Ability of various kinds of plasticizers (the numbers tested are given in parentheses) to support growth of 24 fungi. From Berk *et al.* (1957).

Plasticizer group	Average colony diameter (cm)
Ricinoleic acid esters (7)	6.1
Oleic acid esters (2)	5.9
Polyesters (3)	5.6
Azelaic acid esters (9)	3.9
Sebacic acid esters (12)	3.8
Succinic acid esters (6)	2.8
Esters of polyols (4)	2.7
Adipic acid esters (27)	2.2
Glycollic acid esters (3)	0.6
Phthalic acid esters (9)	0.2
Maleic acid esters (4)	0.1
Phosphoric acid esters (5)	0.0
Dextrose control	6.3
Glycerol control	6.0

face allows that plasticizer to be used successfully in environments that would be avoided on the basis of results obtained with pure plasticizers.

Almost all the above data were derived by use of laboratory cultures. In some instances, the culture had originally been isolated from microbially infested plastics. Whether the culture was metabolizing some component of the plastic formulation, or was fortuitously present growing on contaminating debris, is not known. In other instances, the organisms were those recommended for use in testing textiles for mildew resistance. Berk et al. (1957) noted that *Aspergillus niger* and *A. versicolor*, commonly used in testing plastic materials, produced less growth on plasticizers than many other fungi, and indeed did not utilize some of the plasticizers used by the other cultures.

If, on the other hand, experiments are designed to isolate cultures capable of specific utilization of a plasticizer, the success rate is relatively good. Cultures capable of degrading certain plasticizers (dimethyl, diallyl, dioctyl, dibutyl and octyl decyl phthalates; tributyl citrates; dibutyl tartrate; tricresyl phosphate; sucrose acetate isobutyrate) were isolated by enrichment techniques (Klausmeier and Jones, 1961). Not all isolates could be confirmed as degraders by reincubation with the plasticizer. Twenty-one of 36 dibutyl phthalate degraders were confirmed, as were 5 of 11 isolates from medium containing tributyl citrate, 9 of 20 from media containing sucrose acetate isobutyrate, and 6 of 20 from diallyl phthalate-containing media. The most active phthalate degrader was identified as a member of the genus *Fusarium*; citrate isolates include *Aspergillus terreus* and a black yeast. Further studies with the *Fusarium* sp., including isolation and identification of dibutyl phthalate catabolites, verified beyond doubt the ability of this organism to degrade phthalate esters.

More recently, biodegradation of dibutyl or di-ethylhexyl phthalates by the microflora in fresh water hydrosoil has been reported. Assessment of degradation was verified by radioisotope tracer techniques, utilizing [^{14}C-COOH]-phthalates (Johnson and Lulves, 1975). About 98% of the dibutyl phthalate (1 mg/l) had disappeared from the hydrosoil after five days aerobic incubation and after 30 days of anaerobic culture. Only 53% of the diethylhexyl ester had disappeared after 14 days aerobic incubation and this ester was not degraded anaerobically. Radiorespirometric studies accounted for 78% of the ^{14}C added as butyl ester as $^{14}CO_2$, whereas 60% of the ^{14}C from ethylhexyl ester was collected as $^{14}CO_2$.

butanol. These products accumulate in the cleared area around the colony and persist for some time. In a similar manner, sebacic acid and butanol accumulate during the cometabolism of dibutyl sebacate. The sebacate cometabolizer failed to grow with either butan-1-ol or sodium sebacate as sole organic nutrient.

Similarly, it has been shown that cometabolism of plasticizer occurs with 17 yeast cultures (Osmon et al., 1970), with 28 cultures of actinomycetes (Klausmeier and Osmon, 1976), and with nine out of 19 cultures of thermophilic fungi (Mills and Eggins, 1974).

C. Biochemistry of Plasticizer Degradation

Most plasticizers are esters, and it had been assumed that the deterioration of these esters was initiated by esterases (Reese et al., 1955; Stahl and Pessen, 1953; Berk et al., 1957). Stahl and Pessen (1953; Table 2) noted that various alcohol moieties from sebacate esters would not support growth of either of their test cultures, whereas the complete ester was a very good nutrient. This would suggest a preliminary de-esterification and then utilization of the acid fraction of the substrate.

In their studies, Berk et al. (1957) noted the formation of cleared zones surrounding the colonies of organisms capable of degrading the plasticizer in the opaque homogenized plasticizer agar. These zones varied in size from narrow bands surrounding the colonies to clearings of almost the entire agar plates. This clearing suggests that the cultures produced extracellular esterases which diffuse beyond the colony and convert insoluble esters into water-soluble components. Unpublished studies in the laboratories at the Naval Weapons Support Center by the authors have supported this hypothesis by demonstrating that an aseptic transfer of a square of agar free from viable organisms, excised from a zone of clearing, to agar containing the same plasticizer produces further clearing on the new substrate.

Mycelia of organisms capable of degrading plasticizers have been shown to contain esterases capable of degrading dibutyl sebacate (Williams et al., 1969; Mills and Eggins, 1974). The lipase (an esterase) activity of six fungi (Neurospora sitophila, Aspergillus amstelodami, A. niger, Chaetomium globosum, Penicillium cyclopium and Paecilomyces varioti, all used in at least one test method for plastics) varied from medium to high when polyoxyethenesorbitan monolaurate was used as substrate (Soru *et al.*, 1965). The activity was generally higher when the organisms were cultured on malt extract–agar than when they were grown on vinyls formulated with dioctyl sebacate. Klausmeier and Jamison (1973) have demonstrated weight losses exceeding 18% in dibutyl sebacate-plasticized polyvinyl films incubated with the cell-free spent medium of a strain of *Pseudomonas aeruginosa*. This study also suggested the presence of an extracellular esterase.

Esterolytic cleavage is suggested as the means of degradation of dibutyl phthalate from studies demonstrating accumulation of the monobutyl ester of phthalic acid in the spent dibutyl phthalate-containing growth medium (Klausmeier and Jones, 1961; Johnson and Lulves, 1975). Klausmeier and Jones (1961) have also identified production of monopropyl phthalate by organisms grown on dipropyl phthalate-containing medium. Further, both monobutyl phthalate and butanol persist in the growth medium of a culture cometabolizing dibutyl phthalate (Klausmeier, 1966). Sebacic acid and butanol were shown to accumulate during cometabolism of dibutyl sebacate.

Because an organism has synthesized an esterase that enables it to degrade one plasticizer, it does not necessarily follow that this organism has the capability of degrading all ester plasticizers. For example, the *Fusarium* species capable of de-esterifying dibutyl phthalate, with the liberation of butan-1-ol, was capable of growth at the expense of free butanol. The same culture did not hydrolyse (or grow on) tributyl citrate or dibutyl tartrate (Klausmeier and Jones, 1961), although it did grow on the butyl esters of oleic and sebacic acids and the mixed ester, butyl isodecyl phthalate.

Klausmeier (1966) evaluated 32 bacterial isolates (capable of degrading dipropyl phthalate only in the presence of added organic nutrient) for their abilities to cometabolize a variety of other phthalate esters. Phthalate esters having more than four carbon atoms in either alcohol fraction (butyl isodecyl-, di-n-octyl-, n-octyl n-decyl-, iso-octyl isodecyl phthalates) were not degraded. However, the ability to degrade phthalate esters may vary from a culture that degrades only the phthalate ester from which it was isolated to one that will apparently degrade any diester of phthalic acid having less than four carbon atoms in each alcohol moiety (Table 8).

Once a plasticizer has been de-esterified, the biochemistry of catabolism is well understood. Fatty acids, dicarboxylic acids, phthalic acid, most alcohols and the many aromatic compounds are degraded

Table 8

Cometabolism of phthalate esters by 32 microbial isolates from dipropyl phthalate–yeast extract medium. From Klausmeier (1966)

Plasticizers cleared	Length of side chain	Cometabolizing organisms (%)
Dipropyl phthalate	C_3	3
Dipropyl and diallyl phthalates	C_2 and C_3	13
Diallyl, dipropyl and dibutyl phthalates	C_2–C_4	13
Diallyl, diethyl and dipropyl phthalates	C_2 and C_3	38
Dimethyl, diethyl, diallyl and dipropyl phthalates	C_1–C_4	28
Diallyl, diethyl, dipropyl and dibutyl phthalates	C_2–C_4	3
Dimethyl, diethyl, diallyl, dipropyl and dibutyl phthalates	C_1–C_5	3

by micro-organisms by well-established metabolic pathways and need not be discussed here.

However, the fate of the plasticizer in the presence of extraneous nutrient need not be as straightforward as already suggested. A strain of *Aspergillus terreus* that grows with dibutyl sebacate, but only slowly and after several days acclimatization, readily degrades the plasticizer (as seen by a zone of clearing) within 24 hours in the presence of yeast extract (Klausmeier *et al.*, 1971). These studies showed that growth was negligible during the first three days in either yeast-extract broth or dibutyl sebacate–mineral salts broth; however, growth was already significant after 24 hours in a yeast extract–sebacate broth. Radioactive tracer studies revealed that ^{14}C-labelled material from either butanol- or sebacic acid-labelled dibutyl sebacate was incorporated into cell material during the first 24 hours of growth in a yeast extract–plasticizer broth. Labelled $^{14}CO_2$ did not appear until later. The specific activity of the cells increased with age between 24 and 72 hours. The sebacate was hydrolysed faster than it could be further metabolized; sebacic acid accumulated in the medium during the first two days, then rapidly disappeared. On the basis of these and similar data, the authors proposed that, in the presence of yeast extract, both dibutyl sebacate and yeast extract are metabolized concurrently during early growth stages. The nutrients provided by the yeast extract serve primarily as a source of energy, but may supply some carbon for cellular synthesis, whereas the sebacate serves primarily as a major source of carbon for synthesis. When the energy-yielding metabolites of yeast extract are exhausted, the sebacate ester must serve as both a source of energy and carbon. The plasticizer catabolites that accumulate in the medium during early stages of growth are further metabolized during later stages of growth.

In summary, most of the plasticizers currently in use are capable of being microbiologically degraded under given conditions. Those that serve as nutrients for a wide variety of organisms are the least practical for use in plastics under any but carefully controlled conditions unless protected by biocides. Others are degraded only in the presence of excess nutrient. Their use can be generally recommended, but use in organic-rich environments (compost systems, food-processing functions and fermentation industry, for example) should be avoided.

IV. PRESERVATIVES AND ANTIMICROBIALS

It is obvious that plastic formulations that afford almost no nutrients for micro-organisms can be selected. Therefore the problem of microbial deterioration of plastics should be of historical interest only; its practical impact should be negligible. Granted, compared with the 1940s and 1950s, the problems are few and surmountable.

Unfortunately, there are a variety of engineering problems that do not lend themselves to solution by selection of microbially inert formulations. Surface growth, especially on contaminating organic debris which may accumulate during service life, is one such problem. Such growth may be unsightly, or contribute musty odours in closed environments. Thin films that retain a high degree of flexibility are best prepared from polyvinyl chloride plasticized with fatty-acid esters.

To prevent growth in these and similar situations the plastics engineer resorts to incorporating various antimicrobial compounds into formulations. A short representative list of such biocidal agents is presented in Table 9. Many authors have discussed the properties of the ideal antimicrobial for use in plastic systems. A consensus of such characteristics dictates that the ideal biocide should be: (a) effective against a wide variety of organisms, including fungi, bacteria and even algae;

Table 9

Some representative commercial biocides used in plastic formulations

---

Mercury-containing compounds
    Phenyl mercury salicylate
    Phenyl mercury phthalate
    Phenyl mercury saccharin
Other metal-containing compounds
    Copper-8-quinolinolate
    Tributyl tin oxide
    Arsine-epoxy-soya complex
    Diphenyl antimony ethylhexoate
Sulphur-containing compounds
    Zinc dimethyldithiocarbamate
    Tetramethylthiuram disulphide
    Dithiopyridine dioxide
Quaternary ammonium compounds
    Dodecyl dimethylbenzyl ammonium naphthenate
Halogenated compounds
    Dibromosalicylanilide

(b) highly toxic to organisms at low concentrations; (c) non-toxic to humans at the concentrations used, even after repeated exposures; (d) environmentally safe when used properly; (e) low in cost; (f) capable of storage under normal ambient conditions for long periods of time without loss of effectiveness; (g) easy to apply; (h) usable without causing harm to the appliers or the equipment; (i) effective in the plastic formulation for the life of the product; (j) compatible with all ingredients in all plastic formulations without loss of activity; (k) colourless and odourless. (l) 'non-blooming' (do not exhibit migration to the surface) at effective doses. Needless to say, the ideal in biocides is very difficult to achieve.

The problem of selection of a biocide is not a simple one. Kaplan (1968) suggested that at least six groups of people play a role in the efficiency of a biocide treatment. They include the biocide producer, the biocide formulator, the treater, the end-product manufacturer and the user who establishes the requirements.

The biocide producer is usually a chemical manufacturer whose prime interest is not necessarily production of biocides for plastic formulations. The biocide may have other uses. For example, organotin compounds are used as stabilizers for plastic formulations and as rodent repellants, as well as fungicides. The fungicide, then, may have its chemical or physical properties defined by its use in another role, since the biocide role is secondary from the manufacturer's point of view. The biocide producer is also limited by external forces beyond his control. For example, the ecological impact of the product may lead to regulations banning its production or use, no matter how effective it may be. Organic mercurials, for example, are no longer available, since mercury is a persistent and cumulative toxicant.

Many biocides are not usable generally in the form they come from the producer. The formulator blends the fungicide into a product usable for a specific application. This blending may be as much an art as a science. For example, a variety of techniques have been developed for incorporating copper-8-quinolinolate (Darby, 1953a, b; Malone, 1951; Field, 1951) or organic mercurials (Smith, 1949; Smith and Walker, 1949) into plastic formulations. The formulator must consider such problems as heat or light sensitivity of the biocide, tendency to crystallize or 'bloom' (rapidly migrate to the surface), tendency to leach from the system when water (such as rainfall) is involved, and then must formulate compositions that overcome these problems.

The roles of the treater and the end-item manufacturer are often combined, but not necessarily so. Some moulding powders may be provided that already are treated with biocides. Textiles to be coated may be treated before use by the end-item producer. On the other hand, specific biocide formulations may be incorporated directly into the final processing mix by the ultimate processor. The manufacturer is concerned about the economy and simplicity of manufacturers. If incorporation of a biocide requires special precautions, special equipment or additional steps, he may look for a 'comparable' antimicrobial which will not cause additional expenditures. Unfortunately, such a substitution may not meet end-use requirements.

The user, who makes the final judgment as to the adequacy of the biocidal treatment for the end-product, has the greatest challenge. There are many factors which must be considered by the user in reaching such a decision. Economic factors (cost of treatment compared to possible adverse consequences of failure to treat) often play a primary role. The ultimate use of the material must also be a priority consideration. In situations in which the material is worn or frequently handled by consumers, or comes into contact with foodstuffs, the problem of human toxicity must be considered. If specific colours or odours are intolerable in the end-product, the use of coloured copper-containing biocides, as well as the malodorous alkyl tin compounds, may be unsatisfactory.

Of special importance to the user in determining the type and amount of antimicrobial protection required is a consideration of the environment in which the material is to be used. For example, materials to be used outdoors, such as tents and awnings, required a preservative resistant to leaching by rainfall.

It should also be recognized that fungi are the predominant biodegrading organisms when moisture is largely in the vapour phase or only intermittently as condensate. However, when liquid water is constantly present (e.g. in soil, irrigation ditch liners, plastic swimming pools), bacteria, and perhaps even algae, can develop as slimes that can rapidly degrade plasticized vinyl liners (Yeager, 1968; Beiter, 1973).

Further, Ross (1957) reported that a plasticized vinyl barrier material could be protected from fungal attack by several biocides, but not protected against deterioration when exposed to bacterial growth, or when buried in the soil. Obviously the nature of the water environment must be considered.

Other environmental factors which should be considered are the effects of ultraviolet radiation and heat from sunlight. It has been found, especially during outdoor weathering of vinyl formulations, that no matter how durable the fungicides these products ultimately lose their protective qualities, even though 70–85% of the fungicide remains intact and chemically unchanged in the formulation (Yeager, 1968). It is believed that, under the effects of heat and ultraviolet radiation, hydrochloric acid is split from the polyvinyl chloride polymer, allowing subsequent cross-linking of the polymer chains. This entraps the remaining biocide within the formulation, retarding its movement to the surface, the site of biodegradation.

Another factor to be considered involves the exact degree of protection to be afforded. For example, in vinyl systems, the biocide is dispersed in the plasticizer, which is usually the most biodegradable component. As the surface film which is present in all new vinyl systems is removed, fresh plasticizer will migrate to the surface, carrying with it a constant source of biocide. However, in developing plasticizers that are permanent (non-migratory), the industry also eliminated the method of replenishing the supply of biocide. This rate of replenishment can become important. Girard and Koda (1959) and Yeager (1962) reported staining of vinyl materials as a result of the growth of pigment-producing microbes on organic debris that accumulated on the plastic. The polymer surface must always maintain a level of biocide sufficient to prevent growth on the accumulated debris.

It can be seen that preservation of plastics by use of antimicrobials is not a simple undertaking. It requires a thorough knowledge, appreciation and understanding of the role of all of those concerned with the design of the plastics material as well as those involved with the production and utilization of the treatment. It may happen that, in certain situations, the requirements imposed by the use cannot be met, and compromises may have to be made. The ideal biocide is not yet available and, until it is, compromise appears to be the best solution to the problem.

V. TEST METHODS

After plastic formulations were found to support and be deteriorated by microbial growth, it became obvious that methods would have to be developed to assist in selection of materials that would be less vulnerable to the effects of micro-organisms. The urgencies and priorities of the Second World War led to development of many of the methods that are still, with many improvements, being used today. Many of these methods were originally used for evaluating cellulosic textiles, but, in principle, were found to be equally effective in testing other polymeric materials.

In devising test methods, at least five factors must be considered. These are the nature of the organisms used, the provision of an environment conducive to microbial growth, the physical characteristics of the sample to be tested, any pretreatment of the sample and the nature of the data desired.

Most early studies on microbial degradation of materials were conducted by mycologists, probably because the most obvious forms of microbial growth on damaged material are aerial hyphae of microscopic fungi. Accordingly, most test methods to evaluate the microbial degradation of plastics utilized fungi. However, as reported in early sections of this chapter, a variety of types of micro-organism (bacteria, yeasts and actinomycetes) also can cause deterioration of plastic formulations. The test inoculum should include those types of organisms most likely to cause problems under conditions of use of the end-product.

The test environment must be at a temperature suitable for growth of the test inoculum. Water must be present (as humidity for fungi, liquid water for bacteria and yeasts). Oxygen is required for almost all deteriorative microbes. Minerals to supply growth requirements other than carbon may be added in some tests, depending on test philosophy.

The physical nature of the sample may require the use of certain test conditions. For example, an electronic package measuring 30 cm on each side and weighing 18 kg does not lend itself to testing in a culture dish. Samples for measurement of strength or elasticity may be 15 cm long or longer.

Materials to be exposed to weathering during use frequently are exposed to artificial or accelerated sunlight and/or rainfall prior to microbial testing. Materials may be sterilized, or non-destructive physical measurements performed, before exposure to micro-organisms. These and similar factors must be considered in developing a test plan.

Finally, in designing a test, the nature of the final data requirements must be considered. Questions like the following must be answered. Will visual assessment of growth meet the demands? Are physical measurements necessary to reflect change and what should be the nature of these physical measurements? Must measurements be performed on the sample as it comes from test complete with microbial growth (i.e. electrical measurements) or can the sample be cleaned before functional testing?

Tests generally have one of two functions. One is to test materials or items during their research or development phase. The second is to test end-item acceptability as manufactured.

The research and development test is intended to accumulate data which would allow the design engineer to make the best selection of materials for the item being designed. The primary function of this kind of test is to determine whether a material will support microbial growth in the item's end-use environment, and whether that growth affects the desired physical characteristics.

The acceptance test, on the other hand, is essentially a quality check on the manufacturer. It is not intended to serve as a research and development tool, and it is unwise to use it as such. In general, it is not as severe as research and development testing. In reality, it serves to distinguish between the supplier who provides a good product and one who sells a poor-quality item. It is possible that a supplier with criminal intent could readily sell products that could support significant deteriorative microbial growth but which would meet the requirements of the more commonly used acceptance test methods.

There are four general types of tests used for assessing microbial deterioration of materials. Two of these, namely field exposure and humidity-chamber tests, will be described briefly. They are primarily used for complete items of equipment, composed of many materials, rather than just for testing plastics.

Field testing means exactly what is implied by the terminology. The test is performed in a natural environment similar to that in which it is supposed to function (tropical rain forest, for example), and is evaluated for functionality after various periods of exposure. This test measures deterioration caused not only by micro-organisms but by rain, sunlight, insects and other environmental influences. This kind of testing obviously provides the greatest insight into what can happen to an item in use. However, it definitely is not an accelerated test.

The humidity-chamber or tropical-chamber test involves placing the item to be tested in a chamber at approximately 30°C and 90–100% relative humidity. The sample is usually inoculated by spraying with a mixture of fungal spores suspended in distilled water or in a dilute mineral salts solution. In some chambers, the temperature is changed cyclically to induce moisture condensation on and/or in the sample for several hours each day. Normally the test is completed at the end of 28 days, although longer test periods are possible. Most acceptance tests are humidity-chamber tests.

The other two methods, culture-dish and soil-burial tests, are more adaptable to testing materials rather than completely fabricated items. These tests are used most commonly in the research and development phases of materials or item design.

Culture-dish tests are the more commonly used of the two in assessing biodeterioration of plastics. A more detailed discussion of these types of tests is available (Osmon and Klausmeier, 1977). The use of this type of test to determine the biodeterioration of plasticizers, either by direct metabolism or by cometabolism (Berk et al., 1957; Klausmeier, 1966), has been discussed in the section on plasticizers and will not be further mentioned.

When this test is used for plastic formulations, a layer of mineral salts–agar is placed in a culture vessel of sufficient size. The sample to be tested is then placed on the agar. In one such type of test, a second layer of agar, seeded with the inoculum, is imposed over the top surface of the sample. The sample itself is inoculated with suitable test organisms, incubated for 7 to 28 days, and evaluated.

The composition of the mineral-salts medium used is not critical. Many are listed in test methods. Most of them support adequate microbial growth in the presence of a plastic sample containing organic nutrients. Klausmeier *et al.* (1963) evaluated 18 such prescribed media for their abilities to support growth of 18 commonly used test fungi and concocted a formula for a medium that would provide elements for adequate growth for all 18 organisms.

Osmon *et al.* (1972) studied factors in culture-dish tests that limit the rate of deterioration of plasticized vinyl formulations. In all probability, the same factors could apply to other plastic formulations. They noted that, when a plastic sample is placed on the surface of an agar plate, microbial growth would first occur only at such points on the sample at which organisms, sample, inorganic nutrients, water and sufficient oxygen were all present. It was shown that growth was effectively limited to a zone 1–3 mm wide around the sample perimeter due to the oxygen limitation under the sample and inadequate amounts of mineral salts and water on the upper surface. When samples of identical composition and area, but with different perimetric measurements, were tested in the same manner, deterioration proceeded faster with an increase in perimeter size. As a result, these authors suggested use of narrow (approximately 1 mm wide) specimens in such tests. If wider strips were needed to make physical measurements, they recommended immersing the sample in seeded agar to ensure uniform degradation of the upper surface of the sample.

Considerable attention has been given to the inocula used in culture-dish tests. Two types of inocula are generally used, namely filamentous fungi and non-filamentous micro-organisms. A series of interlaboratory experiments conducted under the aegis of the Organization for Economic Cooperation and Development investigated the inoculum used in fungus tests. Three groups of fungi (Table 10) were evaluated for their ability to degrade plasticized polyvinyl samples (Hazeu, 1967). These inocula are representative of organisms used in a variety of international test methods. It should be noted that culture strains, as designated by culture numbers, are specified in Group A, whereas strain numbers are not so designated in Groups B and C. The results of this five-laboratory (from five countries) co-operative experiment indicated that no differences could be noted among the three inocula, even though they consisted of different groups of organisms.

These studies were extended by utilizing individually the three fungi common to all groups in Table 10. Deterioration caused by these individual organisms was compared to that caused by a mixture of the three, and to that caused by the six-member composite of Group A (Klausmeier, 1972a). In essence, weight losses caused by the six-member or three-member mixtures were not substantially greater than those effected by any of the three individually. When these same inocula were used to challenge susceptible vinyl formulations that had been protected with biocides, they showed preferential deterioration of formulations, depending on biocide susceptibility (Klausmeier, 1972b).

Theden and Schultze-Motel (1960) studied seven fungi frequently used as a mixed-spore inoculum for testing plastics. These included *Aspergillus amstelodami*, *Aspergillus niger*, *Penicillium brevi-compactum*, *Penicillium cyclopium*, *Paecilomyces varioti*, *Stachybotrys atra* and *Chaetomium globosum*. Two of them, *P. brevi-compactum* and *S. atra*, failed to grow at the test temperature of 30°C, and growth of *P. cyclopium* was severely restricted. The optimum growth temperature for all three organisms

Table 10

Fungal inocula used in co-operative experiments. From Hazeu (1967)

Group A	Group B	Group C
Aspergillus niger A.T.C.C.[a] 9642	*Aspergillus niger*	*Aspergillus niger*
Aspergillus flavus A.T.C.C. 9643	*Aspergillus flavus*	
Aspergillus versicolor A.T.C.C. 11730		
Penicillium funiculosum A.T.C.C. 9644	*Penicillium funiculosum*	*Penicillium funiculosum*
Trichoderma sp. A.T.C.C. 9645	*Trichoderma* T-1	*Trichoderma* T-1
Pullularia pullulans A.T.C.C. 9348		
	Aspergillus amstelodami	
	Penicillium brevi-compactum	
	Penicillium cyclopium	
	Paecilomyces varioti	*Paecilomyces varioti*
	Chaetomium globosum	*Chaetomium globosum*
	Myrothecium verrucaria	
	Stachybotrys atra	
	Memnoniella echinata	

[a] American Type Culture Collection strain designation

is approximately 22°C. When this fungus mixture was brought into contact with nutrients (unfortunately, none of which were plastics or ingredients in plastic formulations), only one, two or at most three species developed, depending on the nutrient provided. Additions of biocides altered the nature of the developing species, allowing growth of the more biocide-resistant strains.

The number of different chemicals which conceivably could be formulated into a plastic is large. The ability to degrade a plasticizer, for example, varies from species to species (Berk *et al.*, 1957). Even various strains of the same species do not all metabolize the same plasticizer (Berk *et al.*, 1957; Klausmeier *et al.*, 1976). It seems obvious therefore that, except in very specific situations, use of a single pure culture as a means of assessing whether or not a plastic formulation is degraded microbially is not feasible. The number of fungi to be used in the mixed inoculum is a matter of choice, convenience and economics. In some situations, the microflora obtained from suspensions of rich soil is used to supplement laboratory cultures.

On the other hand, selection of a bacterial inoculum for use in testing plastics deterioration is relatively simple. The only bacterium utilized in a test method is *Pseudomonas aeruginosa*. It was selected solely because of its ability to degrade plasticizers (Klausmeier, 1968). The plastic samples are placed between two layers of mineral salts–agar seeded with the test organism. After Bejuki (1960) compiled a series of suggestions from medical microbiologists and pathologists that *Ps. aeruginosa*, an opportunistic pathogen, was too dangerous to use as a routine test culture, Osmon and Klausmeier (1971) suggested that using *Candida lipolytica*, a non-pathogenic yeast, would give almost identical results under the same test conditions as *Ps. aeruginosa*. To date *C. lipolytica* has not been incorporated into any test procedures.

Burial in soil is generally considered the most severe biodegradative test to which a material can be subjected. Obviously, the material is confronted by a larger group of organisms than could be present in other types of testing. Rich soils may contain 10^{10} or more microorganisms per gram. The 'soil', usually composed of a well-composted mixture of manure, loam and sand, contains sufficient nutrients to support good microbial growth, thus allowing cometabolism. Further, the nature of the soil may cause some minor chemical changes in the sample which may render the material more amenable to microbial dissimilation. Ross (1957) reported that biocide-protected vinyl formulations that were rated acceptable when tested by a fungus culture dish test, lost flexibility during soil burial. Diamond *et al.* (1975) demonstrated that the deterioration patterns of polyester plastics were different when tested by a fungus culture-vessel test and by soil burial (see Table 2).

Booth and Robb (1968), not content with its normal microflora, supplemented soil with *Pseudomonas* and *Brevibacterium* species for their studies. They did not compare the enriched with normal soil, nor did they determine the persistence of the added cultures in the test. Pitis (1964) reported that supplementing soil with fungal spores and/or with peptone enhanced deterioration of plasticized vinyl formulations. These data are suspect, however, since the authors reported that the deteriorated specimens lost strength. Plasticized polyvinyl chloride formulations normally gain strength after biodegradation.

The methods used for evaluating the results of the various type tests also vary significantly. For many years it has been our opinion that deterioration measurements should be those that reflect the effect of the micro-organisms on the material, not the reverse. Growth of the test organisms is the most common method of assessing deterioration. Visual assessments are aided by development of some type of rating scheme (e.g. 0 = no growth; 1 = traces of growth, <10% covered; 2 = light growth, 10–30% covered; 3 = medium growth, 30–60% covered; 4 = heavy growth, 60–100% covered). Over the years, many laboratories have expressed dissatisfaction with visual growth assessment which is very subjective and is particularly difficult to do with non-filamentous microbes. Hazeu (1967), reporting on studies performed simultaneously by six laboratories using identical samples, cultures and test procedures, noted that in some laboratories the visual ratings were the same for all samples, resistant or susceptible. In others, the ratings were higher for degradable than for resistant materials and, in some, the resistant materials had the higher ratings. Hitz *et al.* (1967) also noted that the extent of fungal growth (by visual assessment) is not necessarily related to biodeterioration.

Other assessment methods, developed during World War II, included measurements of changes in flexibility, tensile strength and elasticity modulus. Burgess and Darby (1964, 1965) established that the loss of weight, especially in plasticized vinyl samples, was a simple rapid method for assessing deteriorative changes. Wendt *et al.* (1970) demonstrated that this method is applicable to soil burial as well as to culture-vessel testing. Klausmeier (1972b) has shown that weight losses approximate plasticizer losses in vinyl formulations. At present, weight losses are probably the most reported quantitative assessment of biodeterioration.

Hitz *et al.* (1967) compared weight losses with measurements of tensile strength at break and stress at 33% elongation (which they called Biological Value). They concluded that elongation at break was relatively insensitive and relatively imprecise. At low extensions, a distinct difference could be noted between inoculated specimens and sterile controls. Biological Values agreed well with weight losses.

In addition to tests normally used, a number of other tests have been described. Burgess and Darby (1964, 1965) described a manometric method for measuring respiration of the test organisms growing on plastic samples mounted on mineral salts–agar. Pankhurst and Davies (1968) utilized manometry to measure respiration of resting cells in contact with vinyl formulations, as did Cavett and Woodrow (1968). The last authors noted that respirometric techniques were complicated by the occurrence of endogenous respiration of the organisms. Endogenous respiration could be suppressed, enhanced or unchanged during the presence of exogenous substrates. Endogenous respiration in non-growing Gram-negative bacteria (*Ps. aeruginosa*, in this case) usually occurred at the expense of nitrogenous cellular reserves, and was accompanied by release of ammonia. They suggested that ammonia liberation during manometric studies could serve as an index of endogenous metabolism. Further studies (Sharpe and Woodrow, 1972) have shown that endogenous ammonia production decreased as the concentration of exogenous nutrient available was increased. Comparison of ammonia liberation from resting cells exposed to various vinyl formulations for 48 hours with weight losses from the plastics during the same period showed a linear relationship.

Kaplan (1964) expressed mistrust of manometry as a routine tool for measuring biodeterioration indicating that this technique was extremely sensitive and required extreme precision and technology.

In studies by Hitz and Zinkernagel (1967), growth of *Ps. aeruginosa* in a mineral-salts solution containing plasticized vinyl strips was measured nephelometrically and was compared with weight losses of the samples. In general, there was a correlation between the two results (greater growth equated to greater weight loss). However, when these studies were attempted in three separate laboratories using identical cultures, plastics and test methods, the results were somewhat contradictory (Hitz *et al.*, 1967).

Klausmeier and Jamison (1973) reported that pregrown cells of *Ps. aeruginosa*, when incubated for 20 to 28 hours in the spent medium with plasticized vinyl strips, resulted in weight losses of the strips equivalent to one week incubation with the same organism in a culture-dish method. Commercial plastic formulations incubated for four hours with the pregrown cells had weight losses ranging from 23 to 36% of the weight losses noted in conventional 28-day culture-dish tests. These four-hour tests are adequate to assess deterioration.

Osmon *et al.* (1972) has suggested a non-biological approach to measuring biodeterioration of plasticized vinyl formulations. Essentially, the relative biosusceptibilities of all plasticizers are known. Degradation is a function of the diffusibility of the plasticizers to the surface. Methods for measuring diffusion rates of plasticizers, involving implantation in activated charcoal at elevated temperatures, are available. Weight losses from plastic formulations due to microbial action and to diffusion correlate well. Therefore, it was suggested that diffusion studies, coupled with available knowledge of plasticizer reactions with microbes, could supply information on the degradability of the entire formulation. All of these test methods are capable of furnishing data. The challenge to the user is to be able to select those methods or that method which would provide data having direct application to the use for which the end-product is intended.

VI. SUMMARY

The contents of this chapter can be summarized in a few simple statements.

(1) Very few synthetic polymers can be degraded microbially. The exceptions are very low-molecular-weight polyethylenes, some polyesters and some polyester polyurethanes.

(2) The most serious problem in biodeterioration of plastics is the use of biosusceptible additives. The most significant of these additives are plasticizers in vinyl formulations.

(3) Biodeterioration of plastics can usually be avoided by selection of formulation components or by use of biocides.

Palei, M. I., Trepelkova, L. I., Akopdovzhanyan, E. A. and Golodnaya, S. L. (1965). *Soviet Plastics 1965*, 67.

Pitis, I. (1964). *Industrie des Plastiques Modernes* **16**(2), 109.

Potts, J. E., Clendinning, R. A. and Ackart, W. B. (1972). National Technical Information Service (U. S. Department of Commerce) Report PB 213 488.

Reese, E. T., Cravetz, H. and Mandels, G. R. (1955). *Farlowia* **4**, 409.

Ross, S. H. (1957). U. S. Army Frankford Arsenal (Philadelphia, PA) Report R-1396.

Sharpe, A. N. and Woodrow, N. M. (1972). *In* 'Biodeterioration of Materials' (A. H. Walters and E. H. Hueck-van der Plas, eds.), vol. 2, p. 233. Applied Science Publisher Ltd., London.

Smith, H. E. (1949). United States Patent 2490100.

Smith, H. E. and Walker, H. (1949). United States Patent 2491287.

Spencer, L. R., Hiskins, M. and Guillet, J. E. (1976). *In* 'Proceedings of the Third International Biodegradation Symposium' (J. M. Sharpley and A. M. Kaplan, eds.), p. 753. Applied Science Publishers Ltd., London.

Stahl, W. H. and Pessen, H. (1953). *Applied Microbiology* **1**, 30.

Soru, E., Savulescu, A., Istrati, M. and Lazar, V. (1965). *Revue Roumaine Biologie—Botanique* **10**(5), 419.

Theden, G. and Schultze-Motel, M. (1960). *Angewandte Botanik* **34**, 133.

Wendt, T. M., Kaplan, A. M. and Greenberger, M. (1970). *International Biodeterioration Bulletin* **6**, 139.

Wiles, D. M. (1973). *Polymer Science and Engineering* **13**(1), 74.

Williams, P. L., Kanzig, J. L. and Klausmeier, R. E. (1969). *Developments in Industrial Microbiology* **10**, 177.

Yeager, C. (1962). *Plastics World* (*December*), 14.

Yeager, C. C. (1968). *In* 'Biodeterioration of Materials' (A. H. Walters and J. J. Elphick, eds.), p. 161. Elsevier, Amsterdam.

AUTHOR INDEX

Numbers in italics are those on which References are listed

A

Abbot, E. M., 24, 25, *60*
Abbott, B. S., 266, *301*
Ackart, W. B., 437, 438, 439, 442, 446, *474*
Adams, M. E., 354, *385*
Agarwal, P. N., 275, *301*
Aharonowitz, Y., 8, *17*
Ahearn, D. G., 398, 415, *428, 429*
Ajello, L., 368, *380*
Akbar, S. A., 172, *199, 201*, 297, *304*
Akehurst, B. C., 237, 240, *256*
Akopdovzhanyan, E. A., 454, *474*
Albertsson, A. C., 440, *472*
Aleshina, W. A., 264, *305*
Alexander, M., 439, *472*
Alexander, P., 82, *128*
Al-Haidary, N. K., 281, 283, 285, *302*
Allen, F. H., 264, *301*, 363, 364, 375, *380*, 397, *426*
Allsopp, D., 4, *18*
Alquist, H. E., 274, *303*
Alteras, I., 368, *380*
Al-Zubaidy, T. S., 296, 297, 298, *301, 303*
American Cyanamid Company, 377, *380*
Anderson, K. F., 407, *427*
Anderson, P. J., 238, 246, 255, *256*
Andrews, A. V., 413, *427*
Angert, L. G., 330, 331, 332, 333, 334, 335, 336, 344, 347, 372, 373, *381*
Anon, 20, 22, 41, 42, 45, 48, *56*, 90, 92, 94, 97, 99, 102, 114, 116, 117, *128*, 265, *301*
Araktcheyena, D. Z., 227, *234*
Arens, P., 328, *380*
Ashton, S. A., 190, *199*

Ashworth, J. M., 8, *17*
Association of British Pharmaceutical Industry, 423, 425, *426*
Atwood, H., 437, *472*
Auquiere, J. P., 266, *304*
Austrian, R., 401, *429*
Awad, Z. A., 403, 405, 407, 416, 417, 420, *426*
Ayerst, G., 206, *233*, 248, *256*
Ayliffe, G. A. J., 397, 416, *426*
Ayres, J. C., 3, *17*, 394, *429*
Azova, L. G., 99, *130*

B

Baars, J. K., 179, 198, *199*
Bacot, A. M., 236, *256*
Badzoing, W., 182, *199*
Baecker, A. A. W., 25, *56*
Bailey, C. A., 276, *304*
Bailey, D. G., 145, *147*
Bailey, I. W., 35, *56*
Bailey, W. J., 437, *472*
Baillie, A. J., 25, *58*
Baines, E. F., 26, 29, 32, 33, 34, *56, 59*, 60
Baird, R. M., 403, 404, 407, 410, 411, 413, 416, 417, 418, 420, *426*
Bakanauskas, A., 372, 373, 375, *380*
Bakanauskas, S., 265, *301*
Baker, J. M., 23, *56*
Baker, P. G., 253, *257*
Banaszak, E. F., 298, *301*
Banks, W. B., 25, *56*
Bansleben, D., 437, *472*

Barbiers, A. R., 397, *427*
Barkin, S. M. 116, *129*
Barr, A. R. M., 100, *128*
Barr, M., 424, *428*
Barrowcliff, D. F., 416, *426*
Barry, D. R., 397, *426*
Bartäkova, B., 356, *384*
Bassett, D. C. J., 401, *426*
Bassi, M., 73, *79*, 208, 222, *233*
Bathershall, R. D., 371, *385*
Batson, D. M., 107, *128*
Bavendamm, W., 28, *56*
Baxter, R. M., 278, *303*, 364, *382*
Bean, H. S., 414, 424, *426*
Becker, G., 4, *18*
Becker, H., 325. 345. *380*
Beckwith, J. D., 364, *385*
Beeby, M. M., 409, *428*
Beech, J. C., 48, *56*
Beeley, F., 325, *380*
Beijerinck, M. W., 179, *199*
Beiter, C. B., 462, *472*
Bejuki, W. M., 468, *472*
Bell, A. A., 371, *380*
Bell, E. J., 252, *256*
Bell, J. P., 437, *472*
Bell, W. H., 75, *78*
Bemelmans, J. M. H., 237, *256*
Beng, C. G., 393, *427*
Bengough, G. D., 179, *199*
Bennett, E. O., 290, 291, 292, 296, *301*, *302, 303*
Bennett, J. V., 396, *428*
Benrud, N. C., 142, *147*
Berger, A. J., 252, *256*
Berk, S., 441, 449, 451, 452, 453, 456, 465, 468, *472*
Berman, N., 83, *130*
Bernstein, I. A., 319, *320*
Bernstein, L., 6, *17*
Berry, D. R., 8, *18*
Berry, H., 276, *301*
Betz Handbook of Industrial Water Conditioning, 161, *199*
Beveridge, E. G., 391, 394, 403, 410, *426*, *427, 428*
Bieri, B., 372, *380*
Bignell, J. L., 397, *427*
Binford, C. H., 368, *381*
Binnington, C. D., 295, *305*

Biodeterioration Information Centre, 325, *380*
Bitritto, M., 437, *472*
Blackwell, B., 395, *427*
Blake, J. T., 337, 345, 355, 356, *380*
Blanchard, G. C., 171, *199*, 268, *301*
Blank, F., 83, *129*
Bletchly, J. D., 22, 23, *56*
Block, S. S., 9, 10, 16, *17*
Blow, C. M., 324, 327, 329, *380*
Bluestone, H., 371, *385*
Bobilioff, W., 328, *380*
Boggs, W. A., 276, 278, *301*
Bohdanowicz-Murek, K., 295, *301*
Born, M. J., 298, *301*
Booth, G. H., 171, 182, 183, 189, 190, 193, 194, 197, *199*, *202*, 454, 469, *472*
Borecki, Z., 224, *233*
Borghard, W., 437, *472*
Bothast, R. J., 250, *256*
Boud, M. R., 401, *427*
Boustead, W., 212, 215, *233*
Boutelje, J. B., 25, 50, *56*
Bravery, A. F., 25, 28, 36, 37, 50, *56, 57*
Brazin, J. G., 293, *304*
Breitbarth, W., 368, *383*
Brent, M. M., 319, *320*
British Cotton Industry Research Association, 375, *380*
British Standard, 327, 363, *380*
Brown, D. S., 374, *380*
Brown, P. M., 299, *303*
Brown, W. R. L., 403, 411, 413, 418, *426*
Bryce, D. M., 389, *427*
Buchanan, R. E., 142, *147*
Bucherer, H., 391, *427*
Buck, A. C., 407, *427*
Bultman, J. D., 186, *199*
Buono, F., 308, *320*
Bunker, H. J., 354, *385*
Burges, A., 63, *79*
Burgess, R., 105, *128*, 469, 470, *472*
Burkholder, W. H., 238, 239, *257*
Burns, R. P., 298, *303*
Bush, L. P., 239, *257*
Bushnell, L. D., 264, *301*
Butcher, J. A., 52, 54, *57, 58*
Butler, N. J., 393, 398, *427*
Byrom, D., 292, 293, 294, *301, 303*

C

Calandra, J. C., 239, 257
Calder, M. W., 396, 414, 428
Calderon, O. H., 362, 380
Callahan, L. T., 298, 304
Cameron, J. L., 172, 201, 266, 268, 278, 304
Campbell, J., 20, 57
Campbell, W. G., 28, 57
Cantwell, S. G., 344, 380
Carey, J. K., 43, 48, 49, 51, 52, 57, 60
Carlile, M. J., 25, 58
Carroll, M. T., 177, 200, 274, 302
Carrie, M. S., 90, 128
Cartwright, K. St. G., 45, 57
Castleberry, J. R., 159, 199
Catley, B. J., 319, 320
Cavalcante, M. S., 25, 57
Cavallo, J. J., 293, 304
Cavett, J. J., 470, 472
Chamberlain, E. A. C., 358, 380
Chandler, A. C., 290, 303
Chapin, T., 437, 472
Chaplin, J. F., 243, 257
Cheong, S. F., 368, 380
Chiatante, D., 73, 79
Chican, M., 376, 384
Chidester, C. G., 410, 429
Chong, Y. H., 393, 427
Christian, J. H. B., 6, 17, 248, 257
Churchill, A. V., 166, 171, 178, 187, 199
Ciach, T., 62, 63, 79
Clapp, A. F., 230, 233
Clarke, P., 16, 18
Clendinning, R. A., 437, 438, 439, 442, 446, 474
Clothier, C. M., 414, 427
Clubbe, C. P., 24, 25, 28, 51, 52, 57
Cluff, L. E., 395, 429
Cole, J. S., 238, 242, 256
Colin, G., 438, 439, 440, 472
Collins, R. N., 398, 428
Commins, B. T., 63, 79
Conen, P. E., 407, 427
Connolly, R. A., 357, 358, 380
Cook, A. S., 368, 370, 380
Cooke, E. M., 399, 400, 407, 427, 429
Cooney, J. D., 438, 439, 440, 472

Cooney, J. J., 266, 301
Cooper, A. W., 182, 189, 194, 199, 454, 472
Cooper, D. R., 142, 145, 147, 148
Cooper, P. M., 182, 189, 194, 199
Cope, W. A., 320, 320
Coppier, O., 66, 79
Corbaz, R., 253, 256
Corbett, N. H., 27, 29, 34, 35, 36, 57
Corry, J. E. L., 5, 17
Cosmetic, Toiletry and Fragrance Association Report, 404, 422, 423, 427
Costello, J. A., 192, 199
Cousminev, J. J., 331, 385
Cowap, D., 293, 305
Cowen, R. A., 419, 427
Cowling, E. B., 25, 33, 40, 57, 58, 60
Cozenda, B., 24, 25, 59
Cragg, J., 413, 427
Cravetz, H., 451, 456, 474
Crompton, D. O., 397, 407, 427
Cropper, D. G., 319, 320
Croshaw, B., 414, 427
Cross, T., 244, 256, 325, 348, 350, 352, 353, 354, 355, 382
Crossley, A., 28, 34, 35, 36, 37, 38, 39, 57
Crowden, C. A., 403, 420, 426
Crum, M. G., 172, 200, 267, 269, 271, 302, 304, 365, 380, 384, 445, 446, 472
Cserjesi, A. J., 40, 60, 237, 256
Culver, S. C., 172, 200
Cundell, A. M., 337, 339, 340, 341, 342, 343, 344, 345, 346, 347, 348, 349, 350, 352, 353, 354, 355, 371, 374, 375, 376, 377, 379, 380, 381
Cundliffe, E., 392, 427
Cunha, D. G., 225, 228, 233
Cunha, G. M., 225, 228, 233
Curtis, R. F., 237, 256
Czajnik, M., 229, 233
Czerwińska, E., 224, 233
Czerwonka, M., 76, 78, 79, 232, 233

D

Dachselt, E., 364, 381
Dale, J. K., 397, 427
Dallyn, H., 5, 17

Darby, A. E., 469, 470, *472*
Darby, J. R., 461, *472*
Darby, R. T., 266, *301*, 441, 446, 447, *472*, *473*
David, J., 100, *128*
Davidenko, W. N., 358, *383*
Davies, D. J. G., 397, *428*
Davies, I., 260, 261, 267, 293, 295, *303*
Davies, L., 169, *200*
Davies, M. J., 470, *473*
Davies, P. E., 292, *303*
Davis, D. L., 255, *256*
Davis, J. B., 187, *199*
Davis, P. A., 255, *256*
Dayal, H. M., 275, *301*, 360, 377, *385*
Dearling, T. B., 25, *56*
Deasy, C., 140, *148*
De Coste, J. B., 452, *472*
De Gray, R. J., 265, 276, *301*
Delsol, G., 95, *128, 129*
De Mare, J., 290, *304*
De Mele, M. F. L., 172, *199*
Dempsey, M., 136, *147*
Dennis, C., 237, *256*
De Szocs, J., 390, *427*
De Vries, O., 328, *381*
Diamond, M. J., 443, 444, 469, *472*
Dickenson, P. B., 343, 348, 353, *381*
Dickinson, D. J., 26, 33, 34, 38, 41, 44, 45, 46, 47, 49, 52, 53, 54, 55, *56, 57, 58, 60*
Dimond, A. E., 333, 372, 373, *381*
Din, 363, *381*
Dittmer, C. K., 191, *200*
Dobb, M. G., 82, *129*
Doelle, H. W., 7, *17*
Doerner, M., 205, *233*
Dogadkin, B. A., 324, *381*
Doi, S., 37, *58*
Dolezel, B., 371, *381*
Domaslowski, W., 62, 74, *79*
Dow Chemical Company, 376, *381*
Drysdale, J. A., 54, 55, *57*
Dubok, N. N., 330, 331, 332, 333, 334, 335, 336, 344, 347, 372, 373, *381*
Dubrovin, G. I., 374, 375, *381*
Duff, R. B., 72, *79*
Duffer, N. D., 290, *301*
Duncan, C. G., 24, 25, *59*
Duncan, C. O., 318, *320*
Dunkley, W. E., 348, *381*
Dunleavy, J. A., 25, 43, *57*

Dunlop Rubber Company, 375, *381*
Dunnigan, A. P., 404, *427*
Durân-Grande, M., 239, *256*
Durrel, L. W., 76, *79*
Dwyer, G., 26, 33, *59*

E

Eadie, G. R., 193, *199*
Eaton, B. J., 326, 328, 329, 370, *381*
Eaton, R. A., 25, 35, *57, 59*
Ebbon, G. P., 8, *18*
Ebeling, W., 83, *129*
Ebert, H., 441, 449, 451, 452, 453, 456, 465, 468, *472*
Ecklund, B. A., 24, *57*
Eckstein, Z., 224, *233*
Ederer, G. M., 397, *427*
Editorial, 398, *427*
Eggins, H. O. W., 3, *17*, 106, *129*, 439, 454, 456, *473*
Eirich, F. R., 324, *381*
Eisman, P. C., 409, *427*
Eldridge, E. F., 400, 417, *428*
Elek, S. D., 407, *427*
Elford, L., 190, *199*
Elhag, K. M., 410, *426*
Ellis, L. F., 291, *301*
Ellis, W. J., 93, *129*
Ellwood, E. L., 24, *57*
Elphick, J. J., 166, 167, 174, 178, *200*, 363, 364, *381*
Elphick, J. S., 3, *18*
Eltringham, S. K., 22. *58*
Emmons, C. W., 368, *381*
Enebo, L., 25, *57*
Engel, W. B., 265, *301*
Englehard, W. E., 296, *304*
Engler, R., 16, *17*
English, C. F., 252, *256*
English, M. P., 368, *381*
Ernerfeldt, F., 397, 398, 399, 402, 407, 413, 421, *428*
Esuruoso, O. F., 328, *381*
Evans, D. A., 169, *200*, 260, 261, 267, 295, *303*
Evans, J. R., 404, *427*
Everette, G. A., 255, *256*
Eykyn, S., 396, 410, *429*

F

Fabian, F. W., 292, *304*
Fajfr, M., 373, *382*
Falkow, S., 392, *427*
Fall, R. R., 344, *380*
Fancher, O. E., 239, *257*
Farmer, J. J., 298, *303*
Farquhar, G. B., 187, *200*
Farrer, T. W., 188, *200, 202*
Farwell, J. A., 416, 417, 420, *426*
Fass, E., 344, *384*
Feinstein, R. R., 397, *429*
Feisal, E. V., 290, *302*
Feist, W. C., 49, *57*
Felts, S. K., 396, *427*
Fennell, D. I., 248, *257*
Ferguson, R., 146, *147*
Fialkow, P. J., 401, *428*
Field, W., 461, *472*
Fields, R. D., 442, 443, *472*
Findlay, G. W. D., 36, *57*
Findlay, W. P. K., 35, 36, 45, 47, *57, 58*
Finefrock, V. H., 274, 278, *304*
Fink, J. N., 298, *301*
Finlay, A. J., 358, *380*
Finn, R. K., 442, *472*
Fiore, J. V., 253, *257*
Fisher, R. C., 23, *58*
Fitzgerald, G. P., 78, *79*, 232, *233*
Fitzgibbons, W. O., 276, *301*
Fletcher, J. T., 248, *258*
Florczak, K., 248, *256*
Floyd, J. D., 309, *320*
Fol, J. G., 329, *384*
Folkes, B. F., 25, *58*
Fore, D., 363, 364, 375, *380*
Forkner, C. E., 397, *427*
Formisano, M., 142, *147*
Forsyth, T. J., 171, 174, 176, *201*, 266, 271, *305*
Fountaine, W. B., 253, *257*
Fowle, T. I., 281, *302*
Fox, A., 5, *17*
France, D. R., 399, 407, *429*
Francke-Grossmann, H., 23, *58*
Frankenburg, W. G., 250, 251, *256*
Fraser, I. E. B., 84, 85, *129*
Frazier, W. C., 3, *17*
Freedman, B., 443, 444, 469, *472*
Freeman, G., 25, *58*

Freifeld, M., 308, *320*
Friedman, S. D., 116, *129*
Fudales, D. S., 339, 342, 358, 375, 379, *382, 384, 385*
Fudales, P. S., 339, 342, *382, 385*
Fukazawa, K., 37, *58*
Fuller, W. H., 24, *59*
Fullerton, R. G., 328, 329, 370, *381*

G

Gage, S., 263, 279, *303*
Gaines, R. H., 179, *200*
Gale, E. F., 392, *427*
Gallo, F., 207, 224, 228, *233*
Gallo, P., 224, 228, *233*
Galloway, A. L., 145, *147*
Galloway, L. D., 218, *233*
Galvanek, T. E., 16, *17*
Gangarosa, E. J., 401, *428*
Ganser, P., 264, *302*
Gardner, D. A., 413, *429*
Gargani, G., 71, *79*, 207, 221, 231, *233*
Garibaldi, J. A., 443, 444, 469, *472*
Garner, W., 102, *129*
Garrard, S. D., 298, *302*
Garratt, G. A., 10, *17*
Garrett, S. D., 26, *58*
Gaya, H., 399, 407, *429*
Gayed, S. K., 238, 239, 241, 243, 255, *256*
Gee, J. M., 237, *256*
Gee, M. G., 237, *256*
Geiger, W. B., 82, *129*
Genner, C., 271, 282, 286, 288, 289, 293, 294, 295, *302, 303*
Gentles, J. C., 368, *382*
George, E., 401, *428*
Gershenfeld, L., 424, *428*
Getz, M. E., 368, *380*
Giacobini, C., 206, 208, 218, 221, 222, 228, 232, *233, 234*
Gibbon, O. M., 292, *303*
Gibbons, N. E., 142, *147*
Gibson, M. D., 368, *381*
Gilbert, P. T., 194, *200*
Gilbert, R. J., 4, *17*
Gill, J. W., 309, *320*
Gillespie, J. M., 324, *382*
Gilmartin, J. N., 170, 171, *200*, 365, *382*
Gipson, I. K., 298, *303*
Girard, T. A., 441, 462, *472*

Gledhill, W. E., 266, *301*
Glencross, E. J., 398, 407, *427*
Gold, S. H., 290, *301*
Goley, A. G., 401, *428*
Goll, M., 308, 309, 313, 318, 319, *320, 321*
Golodnaya, S. L., 454, *474*
Gorog, J., 268, *302*
Gorter, K., 326, *382*
Goucher, C. R., 171, *199*, 268, *301*
Grabherr, H., 125, *130*
Graham, K., 23, *58*
Graham Jones, J., 293, *303*
Grant, C. W., 337, *385*
Grant, D. J. W., 390, *427*
Grantham, J., 326, *381*
Grantham, P., 328, *381*
Greaves, H., 24, 25, 26, 33, 53, *58, 59*
Green, G. H., 96, *129*
Green, H., 16, *18*
Green, N. B., 38, *58*
Green, R. H., 263, *302*
Greenberger, M., 446, 469, *473, 474*
Greig, B. J. W., 42, *58*
Grenn, F. P., 400, 417, *428*
Griffiths, N. M., 237, *256*
Groenewege, J., 326, *382*
Gross, H., 325, 345, *380*
Grossman, J. P., 198, *200*
Grunwald, C., 239, *257*
Grzesikowa, H., 206, *233*
Guide to Good Pharmaceutical Manufac-
   turing Practice, 423, *427*
Guillet, J. E., 439, 440, *473, 474*
Gustan, E. A., 263, *302*
Guynes, G. J., 290, *302*

H

Haas, H. F., 264, *301*
Haines, B. M., 145, 146, *147, 148*
Haines, J. R., 439, *472*
Haley, D. E., 252, *257*
Halkias, J. E., 170, 171, *200*, 268, *302*
Hall, S. W., 398, 415, *429*
Halter, H. M., 253, *257*
Hamai, J., 392, *429*
Hammersley, V. L., 171, *200*, 466, *473*
Harbison, G. A., 410, *428*
Hannan, P. J., 279, *304*
Harbord, P. E., 426, *427*

Hardy, P. C., 397, *427*
Harke, H. P., 299, *302*
Harmsen, L., 25, *58*
Harris, M., 82, *129*
Harris, W. V., 22, *58*
Hart, A., 397, *427*
Hart, W. C., 293, *304*
Harter, L. L., 241, *258*
Hartill, W. F. T., 238, *256, 257*
Harvan, D., 239, *257*
Harvey, W. R., 253, *257*
Hashimoto, K., 392, *429*
Hawker, L. E., 25, *58*
Hayman, R. H., 85, *129*
Hayward, A. C., 397, *428*
Hazeu, W., 466, 467, 469, *472*
Hazzard, G. E., 166, *200*
Hazzard, G. F., 265, 278, *302*
Heap, W. M., 371, *383*
Hedrick, H. G., 170, 171, 172, 177, 195,
   200, 267, 268, 269, 271, 274, *302, 304*,
   365, *380, 382, 384*, 445, 446, *472*
Heinisch, K. F., 328, 370, *382*
Heinrichs, T. F., 293, *302*
Heiss, F., 404, *427*
Hejmanek, M., 24, *58*
Henderson, A. E., 84, 88, 100, *129*
Henderson, M. E. K., 64, 71, 72, *79, 80*
Hendey, N. I., 268, 280, *302*
Hennrich, N., 83, *129*
Hepworth, A., 119, *129*
Hercules Powder Company, 377, *382*
Herron, W. C., 172, *201*, 266, 268, 278,
   304, 365, *383*
Heskins, M., 439, *473*
Hesseltine, C. W., 250, *256*
Hickin, N. E., 22, 23, *58*
Highley, T. L., 38, *58*
Hildebrand, J. E., 268, *302*
Hildebrand, J. F., 170, 171, *200*
Hill, E. C., 169, 173, 174, 178, *200, 202*,
   260, 261, 262, 266, 267, 268, 270, 271,
   273, 274, 275, 276, 277, 280, 281, 282,
   283, 284, 285, 286, 287, 288, 289, 291,
   292, 293, 294, 295, 296, 297, 298, *301*,
   302, 303, 305, 366, *384*
Hills, D. A., 344, 354, 373, 374, 375, 376,
   379, *381, 382*
Hills, S., 398, *428*
Hirsch, M. M., 298, *302*
Hiskins, M., 440, *474*

Hitz, H. R., 469, 470, 471, *472*
Hitzman, D. O., 274, 275, *303*
Hoadley, A. W., 298, *303*
Hoar, T. P., 188, *200*
Hocking, A. D., 248, *257*
Hodgson, G., 298, *303*
Hofmann, W., 376, *382*
Holdeman, Q. L., 238, 239, *257*
Holdom, R. S., 295, *303*
Holmes, W. F., 308, *320*
Holtzman, G. H., 290, *304*
Honeywell, G. E., 410, *429*
Hope, I. A., 403, 410, *426*
Hopkins, J. C. F., 241, *257*
Hopkins, W. J., 145, *147*
Horsfall, J. G., 333, 372, 373, *381*
Horvath, J., 188, *200*
Horvath, R. S., 319, *320*
Hosteler, H. F., 276, 279, *303*
Houghton, D. R., 263, 279, *303*
Huang, S. J., 437, *472*
Hudson, R. F., 82, *128*
Hueck, H. J., 228, *233*, 347, *382*
Hueck-Van Der Plas, E. H., 3, *18*, 72, 74, 78, *79*, 224, 232, *233*, 325, 372, 374, 376, *382*, 449, *473*
Hughes, I. R., 145, *148*
Hugo, W. B., 391, 394, *427*, *428*
Hulme, M. A., 54, *58*
Hummer, C. W., 186, *199*
Hunt, G. M., 10, *17*
Hunter, S. K. P., 166, *200*
Hutchinson, M., 325, 348, 350, 352, 353, 354, 355, *382*

I

Iglewski, B. H., 298, *301*, *303*
Ignatescu, M., 368, *380*
Imagawa, H., 37, *58*
Imschenezki, A. H., 24, *58*
International Electrotechnical Commission Publication, 363, *382*
International Organization for Standardization Recommendation, 363, *382*
Ioachimescu, M., 330, 331, 332, 333, 334, 335, 336, 374, 376, 378, 379, *383*
Isenberg, D. L., 290, *303*
Isquith, I. R., 319, *321*

Istrati, M., 456, *474*
Iverson, W. P., 171, 187, *200*

J

Jacobson, J. A., 298, *303*
Jamison, E. I., 452, 454, 456, 457, 458, 466, 468, 471, *473*
Jaton, C., 66, *79*
Jawetz, E., 415, *429*
Jay, S. M., 3, *17*
Jen-Hao, L., 438, *473*
Jennings, D., 46, *58*
Jerusalimsky, N. D., 266, *303*
Jirsak, K., 373, *382*
John, C. K., 326, 327, 344, 368, 369, *382*
Johnson, B. T., 453, 454, 457, *473*
Johnson, E. L., 237, *256*
Johnson, J., 237, 238, 239, *257*
Johnson, R. L., 16, *17*
Johnston, R. R. M., 192, 193, *200*, *201*
Jones, D. B., 398, *428*
Jones, E. B. G., 22, *58*
Jones, P. H., 439, *473*
Jones, W. A., 171, *200*, 453, 457, 466, *473*
Julian, A. J., 398, *429*
Jung, H. D., 368, 375, 379, *384*
Jung, R., 368, *383*

K

Kaarik, A., 27, *58*
Kaben, U., 368, 375, 379, *384*
Kalinenko, V. O., 329, *382*
Kallings, L. O., 397, 398, 399, 402, 407, 413, 421, *428*
Kallio, R. E., 266, *304*
Kaminski, E. J., 239, *257*
Kanzig, J. L., 171, *200*, 456, 458, 466, *473*, *474*
Kapica, L., 83, *129*
Kaplan, A. M., 3, *18*, 265, 266, 267, 273, *304*, 441, 446, 447, 469, 470, *472*, *473*, *474*
Karkanis, A. G., 27, *58*
Karnop, G., 24, *58*, *59*
Kathpalia, Y. P., 219, *233*
Katsuya, N., 266, *303*
Kauffmann, J., 66, 70, *79*

Keall, A., 397, *428*
Keczkes, K., 299, *303*
Kedzia, W., 390, *428*
Kellaway, C. H., 397, *428*
Keller, C. J., 239, *257*
Kelly, B. J., 158, *200*
Kelman, A., 25, *58*
Kelsey, J. C., 409, *428*
Kempner, E. S., 162, *200*
Kennare, M. A., 407, *427*
Kennedy, H., 348, *382*
Kennedy, M. E., 328, *385*
Kereluk, K., 278, *303*, 364, *382*
Kerner-Gang, W., 348, 352, 354, 379, *382*, *383*
Keuringinstuut voor Waterleidingartik-flen, 348, 349, 354, *383*
Kewell, C. B., 358, *380*
Kieslinger, A., 65, *79*
Killian. L. N., 265, 274, 278, *301*, *304*
King, B., 25, 33, *56*, *58*
King, R., 284, 285, *303*
King, R. A., 158, 168, 169, 184, 185, 190, 191, 192, *199*, *200*, *201*
Kinsel, N. A., 265, *303*
Klausmeier, R. E., 171, *200*, 441, 452, 453, 454, 455, 456, 457, 458, 465, 466, 467, 468, 470, 471, *473*, *474*
Klemme, D. E., 265, 279, *303*, *304*, 437, *473*
Klockow, M., 83, *129*
Knox, J. R., 437, *472*
Knuth, D. T., 24, 25, *59*
Kobrin, G., 186, *200*
Kitchin, D. W., 337, 345, 355, 356, *380*
Koch, H. A., 368, *383*
Koda, C. F., 441, 462, *472*
Koenig, M. G., 396, *427*
Koenigs, J. W., 38, *59*
Kohn, S. R., 424, *428*
Komagata, T., 266, *303*
Koppenhoefer, R. M., 140, *148*
Korhonen, J., 186, *202*
Kost, A. N., 374, 375, *383*
Kowalik, R., 207, 224, *233*, 287, *304*
Kozhova, E. I., 99, *130*, 329, *383*
Krapivina, I. G., 27, *58*
Kreger-Van Rij, N. J. W., 23, *56*
Krigrens, A. G., 172, *201*, 266, 268, 278, *304*, 365, *383*
Kritzinger, C. C., 141, *148*

Krumbein, W. E., 62, 64, 66, 70, 71, 72, 74, *79*
Krumperman, P. H., 215, 218, *233*
Krupka, L. R., 390, *428*
Kuczera, M. H., 339, *382*
Kuehne, J. W., 398, 415, *428*, *429*
Kuhr, P., 328, *382*
Kula, T. J., 266, *301*
Kull, F. C., 409, *427*
Kulman, F. E., 356, *383*
Kumar, P., 399, 407, *429*
Kundsin, R. B., 415, *428*
Kunz, L. J., 398, *428*
Kuo, W. C., 437, *472*
Kuritzina, D. S., 214, 221, *233*
Kuster, E. C., 278, *302*
Kuznetsova, V. M., 329, *384*
Kwiatkowska, D., 339, 340, 341, 343, 352, *383*
Kwon-Chung, K. J., 368, *381*

L

Laidlaw, R. A., 23, *56*
Lajudie, J., 66, 76, *79*
Laker, M., 396, 410, *429*
Lalk, R. H., 309, *320*
Lamanna, C., 83, *129*
Lamont, D., 371, *385*
Lancaster, E. B., 250, *256*
Land, D. G., 237, *256*
Landsdown, A. R., 272, *303*
Lang, H., 83, *129*
Lang, J. M., 75, *78*
Langguth, R. P., 116, *129*
Laskownicka, Z., 367, *383*
Last, P. M., 410, *428*
Lau, C. M., 369, *382*
Lau, E. P., 344, *380*
Lawrie, G. M., 396, 414, *428*
Lazanas, J. C., 239, *257*
Lazar, V., 330, 331, 332, 333, 334, 335, 336, 374, 376, 378, 379, *383*, 456, *474*
Leathen, W. W., 265, *303*
Lee, J. C., 401, *428*
Lee, M., 290, *303*
Leeflang, K. W. H., 337, 339, 348, 352, 354, 374, 375, *383*
Leesment, H., 441, *473*
Lefevre, M., 75, *79*

Le Gall, J., 183, *200*
Lehnert, Z., 229, *233*
Leightley, L. E., 35, *59*
Lennox, F. G., 89, 93, 94, 98, *129*
Lennox-King, S. M. J., 407, *428*
Leonard, J. M., 265, 279, *303*
Leong, K. W., 437, *472*
Lepard, C. W., 397, *428*
Lerczynski, S., 229, *233*
Le Roux, N. W., 163, *200*
Levenstein, I., 403, 404, *429*
Levi, J. D., 8, *18*
Levin, R. A., 223, *233*
Levy, J. F., 24, 25, 26, 29, 32, 33, 34, 35, 37, 38, 43, 53, 55, *56, 57, 58, 59, 60*
Lewis, J., 82, 111, 118, 119, 120, 123, 124, *129*
Lewis, P. F., 29, *59*
Lewon, J., 390, *428*
Lhuede, E. P., 146, *148*
Liberthson, L., 290, *303*
Liese, W., 24, 25, 27, *59*
Liggett, S., 278, 279, *304*
Limpel, K. L. E., 371, *385*
Lingappa, Y., 319, *320*
Linstedt, G., 391, *428*
Linton, A. H., 25, *58*
Linton, K. B., 401, *428*
Lipson, D., 287, *304*
Lithander, A., 397, *428*
Llanos, M. C., 238, *257*
Lloyd, A. O., 102, 109, *129*
Lloyd, G., 296, *304*
Lloyd, G. I., 294, 296, *304, 305*
Logan, A. G., 197, *200*
Logvin, V. V. 358, *383*
London, S. A., 274, 278, *304*
Long, K. D., 33. *58*
Lowbury, E. J. L., 397, 401, 416, *426, 428*
Lowe, J. S., 327, *383*
Lucas, G. B., 237, 238, 239, 242, 248, *257, 258*
Lulves, W., 453, 454, 457, *473*
Lutz, J. F., 24, 25, *59*

M

MacCallum, P., 397, *428*
Machemer, W. E., 16, *18*
Mack, D., 146, *147*

Mackel, D. C., 396, *428*
MacLachlan, J., 371, *383*
MacPeak, M. D., 24, *59*
MacPherson, R., 197, *201*
Macura, A., 367, *383*
Maeda, S., 248, *257*
Main, C. E., 243, *257*
Maki, D. G., 396, *428*
Malik, K. A., 106, *129*
Malizia, W. F., 401, *428*
Mallette, M. F., 83, *129*
Malone, R. W., 461, *473*
Mandels, G. R., 451, 456, *474*
Mandels, M., 37, *59*
Mandelstam, J., 7, *18*
Mara, D. D., 190, 191, *200*
Mark, M., 146, *148*
Markovetz, A. J., 266, *304*
Marouchoe, S. R., 16, *18*
Marsh, G. A., 192, *201*
Marsh, J. A., 410, *428*
Marsh, J. T., 126, *129*
Marsh, P. B., 105, *129*
Marshall, K. C., 320, *320*
Martin, K., 137, *148*
Martin, D. R., 400, *427*
Martin Walker, W., 397, *426*
Matsen, J. M., 397, *427*
Maurer, I. M., 401, 410, 413, *428, 429*
Maxwell, M. E., 93, 95, *129*
May, M. E., 366, *383*
May, R., 179, *199*
Mayer, R. L., 409, *427*
Mazhar, M. M., 396, 414, *428*
McCail, C. E., 398, *428*
McCallan, S. E. A., 106, *130*
McCormack, K. R., 415, *427*
McCoy, E., 24, 25, *59*
McCready, R. G. L., 64, *79*
McCreary, M., 24, 25, *59*
McCullough, J. C., 397, *428*
McGlown, D. J., 319, *320*
McInnes, C. A., 25, *58*
McKenzie, P., 172, *201*, 284, 285, 297, *303, 304*
McKinstry, D. W., 252, *257*
McMullen, A. I., 325, 326, *383*
McQuillen, K., 7, *18*
McQuire, A. J., 25, 43, *57, 59*
Meakin, B. J., 397, *428*
Mecey, L. W., 116, *129*

Meek, I. R., 358, *380*
Meers, P. D., 396, 414, *428*
Melly, M. A., 396, *427*
Mendelow, M., 291, *301*
Mercer, E. H., 82, *129*
Mercer, P., 40, *59*
Merianos, J., 16, *18*
Merdinger, E., 319, *320*
Merrill, W., 33, *57*
Merz, A., 469, 470, 471, *472*
Metz, H., 83, *129*
Meyer, K. K., 10, *18*
Michaels, I., 276, *301*
Midtvedt, T., 391, *428*
Mikhailov, W. A., 358, *383*
Millbank, J. W., 26, 33, *56*, *59*, *60*
Miller, C. E., 170, 171, *200*, 268, *302*
Miller, J. D., 297, *304*
Miller, J. D. A., 158, 161, 168, 169, 172, 182, 183, 184, 185, 187, 190, 191, 192, 194, 195, 197, *199*, *200*, *201*, *202*
Miller, R. N., 172, *201*, 266, 268, 278, *304*, 365, *383*
Miller, W. G., 308, 309, *320*
Mills, J., 439, 454, 456, *473*
Miraz, E. A., 49, *57*
Mitchell, R., 75, *78*
Mitchell, R. G., 397, *428*
Mitchell, T. G., 244. *257*
Mixer, R. Y., 293, *304*
Miyoshi, M., 261, *304*
Mizell, L. R., 82, *129*
Molyneux, G. S., 82, 85, *129*
Monsanto Chemical Company, 372, 373, 376, *383*
Montgomery, R. A. P., 37, 38, 39, *59*
Moore, K. E., 397, *427*
Moreau, C., 6, *18*
Morgan, M. H., 439, *473*
Moroney, J. P., 25, 43, *57*
Morse, L. J., 400, 417, *428*
Moshy, R. J., 253, *257*
Mozokhina-Porshnyakova, N. N., 329, *384*
Mulcock, A. P., 84, 85, 88, *129*, 337, 339, 340, 341, 342, 343, 344, 345, 346, 347, 348, 349, 350, 352, 353, 354, 355, 371, 374, 375, 376, 377, 379, *380*, *381*
Mundt, J. O., 3, *17*
Muracka, J. S., 356, *383*
Murphey, T. P., 117, *130*

Muse, R. R., 371, *380*
Muthukuda, D. S., 370, *382*
Myers, G. E., 64, *79*, 403, 404, 415, 417, *428*

N

Nadarajah, M., 328, 369, 370, *382*
Nakase, T., 266, *303*
Narita, T., 437, *472*
Nasroun, T. A. H., 28, 36, 37, *59*
Natarjan, K., 437, *472*
National Association of Corrosion Engineers, 197, *201*
Neihof, R. A., 276, 279, *303*, *304*, 366, *383*
Nelson, J., 280, *305*
Nelson, J. H., 401, *428*
Nelson, L. A., 247, *258*
Nepumuceno, J., 437, *472*
Nette, I. T., 329, 374, 375, *383*
Newman, P. L., 48, *56*
Nicholls, C. H., 123, 126, 127, *130*
Nicklen, D. K., 313, *321*
Niel, C. B., van, 329, *384*
Nigam, S. S., 275, *301*
Nippon Oil Company, 376, *383*
Nissen, T. V., 25, *58*
Noble, W. C., 399, 407, 416, *426*, *428*
Nook, M. A., 397, *427*
Norkans, B., 37, *59*
Norman, A. G., 24, *59*
Norris, D. M., 23, *59*
North, A. A., 163, *200*
Norton, D. A., 397, *428*
Nuksha, J. P., 219, *233*
Nyns, E. J., 266, *304*

O

Obold, W. I., 24, 25, *60*
O'Brien, T. E. H., 370, *383*
O'Connell, J., 344, *382*
O'Farrell, S. M., 400, 403, 420, *426*, *427*
Okamoto, Y., 437, *472*
Old, G., 319, *320*
Olin, G., 397, *428*
Oliver, A. C., 22, *59*
Olofinboda, M., 33, *59*
Olsen, H., 406, *428*

Olson, R. L., 263, *302*
Ong, C. O., 368, *380*
Onions, W. J., 99, *130*
Orth, H. D., 83, *129*
Osmon, J. L., 454, 456, 458, 465, 466, 468, 471, *473*
Ouchterlony, O. T., 398, *428*
Owen, H. P., 177, *200*
Owens, R. G., 239, *257*
Oxley, T. A., 3, 4, *17, 18*, 33, *58*

P

Pacitti, J., 371, *383*
Padgett, D., 319, *321*
Paigen, K., 7, *18*
Palei, M. I., 454, *474*
Pankhurst, E. S., 470, *473*
Parameswaran, N., 55. *59*
Parbery, D. G., 167, 171, *201*, 268, 280, *304*
Park, P., 273, *304*
Parker, M. T., 399, 407, *429*
Parker, W. D., 197, *200*
Parkin, E. A., 23, *59*
Pasutto, F. M., 403, 404, 415, 417, *428*
Patel, N., 399, 407, *429*
Patouillet, P. J., 279, *304*
Pattee, H. E., 239, *257*
Patterson, W. I., 82, *129*
Patterson, W. S., 186, *201*
Payne, H. F., 215, 232, *233*
Payne, R. A., 16, *17*
Payne, R. W., 412, *428*
Peberdy, J. F., 8, *18*
Peel, J. L., 237, *256*
Penberthy, I. O., 294, *303*
Penkala, B., 70, *79*
Pepper, M., 326, *384*
Pero, R. W., 239, *257*
Pessen, H., 449, 450, 454, 456, *474*
Peters, G. A., 55, *59*
Petrocci, A., 16, *18*
Petty, J. A., 53, *59*
Phillips, I., 396, 401, 410, *428, 429*
Phillips, L. C., 372, *384*
Philpott, M. W., 327, *384*
Pietrykowska, S., 208, 215, 218, *233, 234*
Pietrykowska-Leznicka, S., 219, *234*
Pilgrim, A. J., 263, *302*

Piper, E., 410, *429*
Pitis, I., 374, 376, 379, *384*, 469, *474*
Pitt, J. I., 248, *257*
Pivnick, H., 291, 292, 296, *301, 304*
Plenderleith, H. J., 225, 226, 228, 230, *234*
Plotkin, S. A., 401, *429*
Plumbe, W. J., 230, *234*
Pochon, J., 62, 63, 66, 68, 71, *79*
Pollock, M., 298, *304*
Pomorsteva, N. V., 329, 374, 375, *383*
Popova, T. A., 357, 374, *384*
Porges, N., 107, *128*
Postgate, J. R., 180, 181, 183, 187, 197, 198, *200, 201*
Potts, J. E., 437, 438, 439, 442, 446, *474*
Pounds, J. R., 237, *257*
Powelson, D. M., 265, *304*
Powers, E. J., 276, 279, *303*
Prasad, D., 439, *473*
Pratt, O. S., 355, *380*
Pressley, T. A., 98, *130*
Preston, R. D., 36, 53, *59*
Prince, A. E., 166, *201*, 265, *304*, 372, 373, 375, *380*
Prindle, B., 84, *130*
Pritchard, J. A. V., 293, *303*
Proceedings of the Second International Conference, 198, *201*
Prokop, J. V., 261, *305*
Provasoli, L., 75, *79*
Przedpelski, Z., 62, *79*
Public Health Laboratory Service, 398, 403, 404, 414, 420, *429*
Purkiss, B. E., 159, 161, 162, 186, 196, *201*
Pursiainen, M., 186, *202*

Q

Quinby, H. L., 182, 186, *202*

R

Rabotnova, I. L., 329, *384*
Race, E., 83, 84, 99, *130*
Racle, F. A., 390, *428*
Raistrick, M., 83, *130*
Rampling, A., 418, *429*
Ranby, B., 440, *472*
Raper, K. B., 248, *257*

Raw, F., 63, *79*
Rawlings, D. E., 142, *148*
Reddie, R. N., 123, 126, 127, *130*
Reese, E. T., 37, *59*, 451, 456, *474*
Reid, J. J., 252, *257*
Reinecke, J., 238, *257*
Remberger, U., 342, *384*
Report: Chemistry Research, 163, *201*
Reszka, J., 339, *384*
Reszka, K. R., 339, *384*
Rettger, L. F., 83, *130*
Reynolds, P. E., 392, *427*
Reynolds, R. J., 172, *200*, 267, 269, *302*, 365, *380*, *384*, 445, *472*
Rhame, F. S., 396, *428*
Richardson, N. E., 397, *428*
Richmond, J. B., 298, *302*
Richmond, M. H., 392, *427*
Ridgway, J. W., 325, 348, 350, 352, 353, 354, 355, *382*
Ringertz, O., 397, 398, 399, 402, 407, 413, 421, *428*
Rishbeth, J., 42, *59*
Ritzinger, G. B., 372, *384*
Robb, J. A., 190, *199*, 454, 469, *472*
Roberts, G. A. H., 187, *201*
Robins, C. S., 49, *60*
Robinson, D., 237, *256*
Robinson, E. P., 403, *429*
Roby, M., 437, *472*
Rodriguez, F., 442, 443, *472*
Roff, J. W., 40, *60*
Rogers, H. J., 25, *60*
Rogers, M. R., 265, 266, 267, 273, *304*, 446, *473*
Rogers, P. J., 192, 193, *200*, *201*
Romanoff, M., 186, *201*
Ronay, D., 268, *302*
Rook, J. J., 337, 339, 341, 348, 352, 372, 373, *384*
Roper-Hall, M. R., 397, *426*
Rosa, H., 219, 225, 230, *234*
Rose, A. H., 7, *18*
Rosenberg P., 309, *320*
Ross, E. O., 373, 375, 376, *384*
Ross, R. T., 210, 212, 214, 218, *234*. 308, 318, *320*, *321*
Ross, S. H., 462, 468, *474*
Rossell, S. E., 24, 25, 43, *57*, *60*
Rossmore, H. W., 290, 292, 293, 296, 300, *302*, *304*

Rotta, J. R., 400, 417, *428*
Round, F. E., 74, *79*
Rousseau, R., 415, *429*
Rowe, L. C., 268, *304*
Roznerska, M., 231, *234*
Ruban, G. I., 330, 331, 332, 333, 334, 335, 336, 344, 347, 372, 373, *381*
Rubber and Plastic Biodeterioration, 325, *384*
Rubidge, T., 178, *201*, 274, *304*
Rudakova, A. K., 357, 374, *384*
Rudniewski, P., 208, 209, *234*
Russell, D. R., 271, *304*
Ruth, L., 291, *301*
Rychtera, M., 356, *384*
Rypacek, V., 24, *58*
Rytych, B. J., 339, 342, 358, 371, 375, 379, *384*, *385*

S

Sabina, L. R., 291, *304*
Sadiff, J. C., 298, *301*
Sadurska, I., 207, *233*, 287, *304*
Saleh, A. M., 197, *201*
Salvarezza, R. C., 172, *199*
Samuel-Maharajah, R., 291, *301*
Sanders, A. C., 426, *429*
Sandine, W. E., 3, *17*
Saunders, J. C., 276, *304*
Savin, J. A., 399, *428*
Savory, J. G., 35, 36, 43, 44, 48, *58*, *60*
Savulescu, A., 456, *474*
Schacht, H., 35, *60*
Schade, A. L., 328, *384*
Schaffner, W., 396, *427*
Schaschl, E., 192, *201*
Schmid, R., 27, *59*
Schmitt, J. A., 319, *321*
Schmitt, R. L., 140, *148*
Schofield, J., 296, *304*
Schonbeck, L. E., 400, *428*
Schönborn, Ch., 368, 375, 379, *384*
Schulze-Motel, M., 467, *474*
Schwartz, A., 330, 331, 332, 347, *384*, 438, *473*
Scobie, D., 45, *60*
Scott, J., 337, *384*

Scott, J. A., 171, 174, 175, 176, 178, *201*, *202*, 261, 266, 271, 272, 273, 278, *304*, *305*, 366, *384*, 415, *428*
Scott, W. R., 187, *202*
Scroggie, J., 146, *148*
Sekhar, K. C., 327, 370, *380*, *384*
Sergent, C., 95, *128*, *129*
Seubert, W., 344, *384*
Shaposhnikov, N. V., 99, *130*, 329, *384*
Sharp, R. F., 26, 33, *60*, 106, *129*
Sharpe, A. S., 470, *474*
Sharpell, F. H., 11, 12, 16, *18*
Sharpley, J. M., 3, *18*
Shaw, E. J., 410, *426*
Sheffer, T. C., 24, 25, 40, *59*, *60*
Shennan, J. L., 8, *18*
Sheridan, J. E., 167, *202*, 280, *305*
Shields, L. M., 76, *79*
Shigo, A. L., 24, 25, 40, 43, *59*, *60*
Shinn, P. M., 189, *199*
Shiroki, C., 248, *257*
Shnyrev, I. M., 358, *383*
Shoemaker, N., 437, *473*
Shooter, R. A., 399, 400, 403, 407, 411, 413, 416, 417, 418, 420, *426*, *427*, *429*
Shotton, J. A., 274, *303*
Sikorski, J., 119, *129*
Silverstolpe, L., 397, 398, 399, 402, 407, 413, 421, *428*
Simmons, E. G., 266, *301*
Simmons, N. A., 413, *429*
Sinclair, A., 293, *303*
Singh, V., 298, *305*
Sisowa, T. P., 214, *233*
Siu, R. G. H. 24, 25, *60*, 105, *130*
Skinner, C. E., 222, *234*
Skryabin, G. K., 266, *303*
Sladen, J. B., 210, 212, *234*
Slansky, B., 212, 227, 230, *234*
Smart, R., 389, 393, 413, *427*, *429*
Smirnova, B. J., 230, *234*
Smith, D. W., 5, *18*
Smith, G., 110, *130*, 249, *257*
Smith, G. A., 23, *56*
Smith, H. E., 461, *474*
Smith, J. E., 8, *17*, *18*
Smith, J. S., 191, *200*, *202*
Smith, T. H. F., 290, 299, *304*, *305*
Smyth, H. F., 24, 25, *60*
Snoke, L. R., 356, *384*
Snow, J. P., 239, *257*

Soane, G. E., 48, *60*
Soboslai, J., 437, *473*
Societe des Usines Chemiques Rhone-Poulenc, 373, *384*
Söhngen, N. L., 329, *384*
Soimajarvi, J., 186, *202*
Sokoll, M., 76, 78, *79*, 222, 223, *234*
Sokolski, W. T., 410, *429*
Solti, M., 188, *200*
Somer, G. L., 140, *148*
Sorkhoh, N. A. A. H., 52, 53, 54, 55, *57*, *60*
Sorokin, T., 239, *257*
Soru, E., 456, *474*
Southwell, C. R., 186, *199*
Sowa, F. J., 372, *384*
Spence, D., 329, *384*
Spencer, G., 401, *428*
Spencer, L. R., 440, *474*
Sperry, J. A., 83, *130*
Spooner, D. F., 393, 413, *429*
Spruit, C. J. P., 188, *202*
Spurr, H. W., 243, *257*
Stachowiak, E., 227, *234*
Staffeldt, E. E., 362, *380*
Stahl, W. H., 449, 450, 454, 456, *474*
Stanescu, E., 376, *384*
Starkey, R. L., 164, *202*
Staszewska, G., 78, *79*, 232, *234*
Stauber, P. C., 244, *257*
Steiger, B., 419, *427*
Steinbach, H. L., 415, *429*
Stelka, J. P., 16, *17*
Stephen, R. C., 238, *257*
Stephenson, M., 179, 180, *202*
Stewart, R. B., 395, *429*
Stewart, W. J., 308, *320*
Stickland, L. H., 179, 180, *202*
Stokes, K. J., 401, *426*
Strzelczyk, A., 63, *79*, 205, 208, 215, 218, 219, 225, 226, 230, 232, *233*, *234*
Stuart, L. S., 140, *148*
Sturges, W. S., 83, *130*
Sturgis, M., 416, 417, 420, *426*
Surmatis, J. D., 252, *257*
Sussman, A. S., 319, *320*
Suzukii, T., 331, *385*
Svanberg, M., 417, *429*
Swan, G. W., 40, *60*
Swanepoel, O. A., 119, 125, *130*
Swart, N. L., 326, *382*
Swateck, F. E., 265, *301*

Algae
  growth in cooling-water systems, 157
  in destruction of rocks and stones, 74–77
  on mural paintings, 222
    control, 232
Algicides, treatment of stone objects with,
    78
Alcaligenes spp., growth on degraded poly-
    ethylene, 440
Alcaligenes faecalis, in hand cream, health
    hazards, 400
Alkanes, microbiology, 438–439
Alstonia scholaris, wood colonization by, 54
Alternaria spp.
  growth on Thiokol latex MX linings for
    aviation fuel tanks, 364
  in microbial degradation of paintings
    on canvas, 216
  in subsonic aircraft fuel, 266
  in tobacco, biodeterioration by, 245
  on paint films, 318
  watercolour degradation by, 218
Alternaria alternata, in tobacco curing, bio-
    deterioration by, 241, 243, 258
Aluminium
  corrosion, sulphate-reducing bacteria
    and, 194
  fuel tanks, microbial corrosion, 268
  rolling oils, microbial contamination,
    294
Aluminium alloys, microbial corrosion,
    171
Ambrosia beetles, wood degradation by,
    22
Amidase, skin protein digestion by, 135
Amines, inhibition of persulphate activity
    by, 312
Aminoglycosides
  bacterial resistance to, 392
  biodegradation, 391
Ammonia catabolite repression, 8
Ammonia, for *Hevea* latex preservation,
    368
Amorphotheca resinae
  Cladosporium resinae and, 167
  in aviation fuel tank corrosion, 166
Aniline, 2,6-dichloro-4-nitro-, in control
    of tobacco biodeterioration, 242
Anisole, chloro-, formation in tobacco,
    237
Anobiideae, wood decay by, 211

Anobium punctatum—*See* Common furniture
    beetle
Anodic reaction in electrochemical corro-
    sion, 150–151
Antibiotics, algal, in destruction of rocks
    and stones, 75
Antimicrobial agents in plastics, 459–463
Anti-oxidants for rubber, microbiological
    resistance, 333, 335
Anti-ozonants for rubber, microbiological
    resistance, 333, 335
Anti-ozonant waxes for rubber, microbio-
    logical resistance, 336
Antiseptics
  in control of mildew on wool, 110
  microbial contamination, health haz-
    ards, 400–401
Applicators, drugs and cosmetics, micro-
    bial contamination, 417
Aprophylite, fungi destruction of, 72
Aspirators, microbial contamination, 417
Arachis oil, liquid, microbial contamina-
    tion, 409
Archaeological excavations, microflora
    control in, 77
Armillaria mellea, in standing commercial
    timbers, 42
Aromatic compounds, *Aureobasidium pul-
    lulans* growth on, 319
Arthrobacter spp., growth on degraded
    polyethylene, 440
Asbestos pipes, in control of corrosion by
    sulphate-reducing bacteria, 196
Ascomycetes, wood degradation by, 27
Aspergillus spp.
  biodeterioration by, 11
  development on painting surfaces, 208
  in aircraft tank sludge, 166
  in destruction of rocks and stones, 72
  in lubricating oils, 284
  in microbial degradation of paintings
    on canvas, 216
  in raw natural rubber, 329
  in subsonic aircraft fuel, 266
  in tobacco, biodeterioration by, 246
    storage and, 248
  in tobacco curing, biodeterioration and,
    285
  on paint films, 318
  processed wool damaged by, 99
  skin deterioration by, 146

watercolour degradation by, 218
wood decay by, 211
Aspergillus amstelodami, plasticizer degradation by, 456
Aspergillus clavatus, marble deterioration, by, 73
Aspergillus echinulatus, a_w value and, 6
Aspergillus flavus
  aflatoxins from, in drugs and cosmetics, 393
  growth, on high-density polyethylene, 438
    on polycaprolactone, 442
    on polyesters, 443
    on polyethylene, 437
  in tobacco curing, biodeterioration and, 258
Aspergillus fumigatus
  in fuel-water systems, 268
  in supersonic aircraft fuel systems, 174, 270
  marble deterioration by, 73
Aspergillus glaucus
  in tobacco, storage and, 248
  on cigars, 253
Aspergillus niger
  aspirin degradation by, 390
  growth, on high-density polyethylene, 438
    on plasticizers, 453
    on polycaprolactone, 442
    on polyesters, 443
    on polyethylene, 437
  in tobacco, storage and, 248
  on tobacco, 249
  plasticizer degradation by, 456
Aspergillus penicilloides
  on cigars, 253
  on cigar tobacco, 249
Aspergillus repens
  in tobacco, storage and, 248
  marble deterioration by, 73
  growth, on dibutyl sebacate, 458
    on plasticizers, 453
    on polyesters, 443
Aspergillus ustus, marble deterioration by, 73
Aspergillus versicolor
  growth, on plasticizers, 448-449, 453
    on polyesters, 443
    on sebacic acid esters, 450

Aspirin, biodegradation, 390
Atropine, biodegradation, 390
Aureobasidium spp.
  in microbial degradation of paintings on canvas, 217
  on paint films, 318
Aureobasidium pullulans, biodeterioration by, 11
  effect on paints and varnishes, 49
  on paint films, 318
  paint degradation by, 214
Autolysis, deterioration of hides and skins by, 133-136
Aviation fuel
  microbial contamination, subsonic aircraft, 264
  microbial growth on, 261, 262
Aviation fuel tanks
  microbial corrosion, 165-179
  polyurethane coating, microbiology, 445
  polyurethane foam in, microbiology, 446
  subsonic aircraft, corrosion, mechanism, 171-173

B

Bacillus spp.
  biodeterioration by, 11
  growth on plasticizers, 454
  in aircraft tank sludge, 166
  in hydraulic oils, 300
Bacillus asterosporus, wood degradation by, 24
Bacillus cereus
  in supersonic aviation fuel tanks, 174
  microbial slime and, cooling-water systems, 158
  processed wool damaged by, 99
  wood degradation by, 24
Bacillus licheniformis, in detergents, effect on wool, 116
Bacillus mascerans, wood degradation by, 24
Bacillus megatherium
  marble deterioration by, 73
  on polysulphide rubber linings of aviation fuel tanks, 363
  wood degradation by, 24
Bacillus mesentericus vulgatus, processed wool damaged by, 99

plasticizer degradation by, 456

Chalk as pipeline surround, in control of corrosion by sulphate-reducing bacteria, 196

Chilling in curing skins, 146

Chlamydobacteriaceae in cooling-water systems, 159

Chloramphenicol acetyltransferase from bacteria, 391

Chloramphenicols, biodegradation, 391

Chlorella spp.
  in Lascaux Caves, 76
  on ancient buildings at St-Cloud, 76

Chlorination, water, natural rubber microbial deterioration and, 353

Chlorobotrys sp., in Lascaux Caves, 76

Chlorococcus spp., microbial slime and, cooling-water systems, 158

Chloroprene, rubber, microbial activity, 331

Chromates
  as biocide in fuel systems, 273, 274
  in control of aviation fuel tank contamination, 178

Chrome yellow, paint, microbial degradation, 212

Chromium compounds in control of mildew on wool, 113

Chromobacterium prodigiosum, protein breakdown by, 83

Chromococcus spp., microbial slime and, cooling-water systems, 158

Chrysosporium fastidium, in prunes, biodeterioration by, 248

Chrysosporium xerophilum, in prunes, biodeterioration by, 248

Churches, microbial deterioration of paintings and sculptures in, 206

Cigars, microbial biodeterioration, 253

Citric acid in microbial corrosion of aviation fuel tanks, 172

Citrobacter spp., shampoo degradation by, 392

Cladosporium spp.
  growth, on acrilonitrile-butadiene rubber coating for aviation fuel tanks, 364
    on Thiokol latex MX linings for aviation fuel tanks, 364
  in aircraft tank sludge, 166
  in destruction of rocks and stones, 72

in jet fuel systems, 265

in microbial degradation of paintings on canvas, 216

in tobacco, biodeterioration by, 245

on paint films, 318

watercolour degradation by, 218

aluminium alloy corrosion by, 172

in aviation fuel tank corrosion, 166, 167–171

in fuel-water systems, 268

in marine fuels, 279

in polyurethane coatings for aircraft fuel tanks, 445

in subsonic aircraft fuel, 266

in supersonic aviation fuel tanks, 174

natural history, 280

on equipment for manufacturing drugs and cosmetics, 410

Clay loams, microbial pipe corrosion and, 194

Cleaning equipment in drugs and cosmetic manufacture, microbial contamination, 413

Cleaning solutions, contact lenses, microbial contamination, standards, 421

Clean-room technology in drug and cosmetic manufacture, 411

Clostridium spp.
  in aircraft tank sludge, 166
  toxins from, in drugs and cosmetics, 393
  wood degradation by, 24

Clostridium histolyticum, skin deterioration by, 137

Clostridium omelianski, wood degradation by, 24

Clostridium tetani
  in powders, health hazard, 407
  in raw material for drugs and cosmetics, 409
  in talcum, health hazard, 398

Closures, drug and cosmetic contamination and, 412–413

Coalescers for removal of water from aviation fuel, 176

Coatings
  fuel tanks, microbial corrosion and, 277–278
  aviation fuel tanks, microbial degradation, 363–366
  polyurethanes, microbiology, 445

Coconut oil, microbiology, 451

Coleoptera, wood deterioration by, 22

Collagenase in skin deterioration, 139

Colonization, wood, effect of preservatives, 51–56

Colour
drugs and cosmetics, microbial growth and, 393
tobacco, biodeterioration and, 239

Colouring materials for rubber, microbiological resistance, 333, 334

Cometabolism
of plasticizers, by bacteria, 455
by filamentous fungi, 455
phthalate esters, 458

Common furniture beetle, wood degradation by, 22

Compounding ingredients in rubber, microbiological resistance, 333–337

Concentration, nutrients, in cooling-water systems, 157

Concentration cells
in microbial metal corrosion, 155–162
metal corrosion by, microbial growth and, 156–159

Condensation, microbial deterioration of paintings and sculptures and, 205

Coniophora cerebella
microbial deterioration of paintings and sculptures by, 205
wood decay by, 211

Coniophora puteana
in window and door frames, 48
wet rot, 47

Contact lenses
cleaning solutions, microbial contamination, health hazard, 397
microbial contamination, standards, 421

Containers, drug and cosmetic contamination and, 412–413

Contamination
drugs and cosmetics, during use, 414–418
health hazards, 406–407
sources, 408–418

Control, drug and cosmetic microbial contamination, 420–424

Cooling-water systems
concentration cell corrosion in, microbial growth and, 156

corrosion, control by biocides, 196
microbial corrosion, 186
microbial growth, control, 159–162

Copper, corrosion, sulphate-reducing bacteria and, 194

Copper naphthenate in control of mildew on wool, 113

Copper-8-quinolinolate as stabilizer for plastics, 461

Copper sulphate in prevention of microbial biodeterioration, 10

Coriolus versicolor, in window and door frames, 48

Cork stoppers for disinfectant solutions, microbial contamination, 401

Corrosion
aviation fuel tanks, microbial, 165–179
subsonic aircraft, mechanism, 171–173
by concentration cells, microbial growth and, 156–159
by sulphate-reducing bacteria, control, 195–198
iron, by sulphate-reducing bacteria, mechanism, 187–194
metals, microbial growth on hydrocarbons and, 264
sulphate-reducing bacteria and, 179–198
prevention, 195–198
by thiobacilli, 163–164
concentration cells and, 155–162
microbial growth in fuel and, 268
microbial growth in fuels and, 263
non-ferrous metals, by sulphate-reducing bacteria, 194–195

Corrosion cells in iron corrosion by, sulphate-reducing bacteria, 188

Corynebacterium spp., biodegradation of atropine by, 390

Corynebacterium humiferum, wood degradation by, 24

Cosmetics, biodeterioration, 387–429

Cracking, emulsions, microbial degradation and, 394

Creams
cracking, microbial activity and, 394
microbial contamination, health hazard, 399

Crenothrix spp., in cooling-water systems, 159

in intravenous fluids, nosocomial sepsis and, 396

in tobacco curing, biodeterioration and, 241, 259

Erwinia aroideae, in tobacco curing, biodeterioration and, 259

Erwinia carotovora, in tobacco curing, biodeterioration and, 259

Escherichia coli
in drugs and cosmetics, standards, 421
in emulsions, 296
in hand creams, 416

Escherichia intermedia, in hand cream, health hazards, 400

Esterases in plasticizers, degradation, 456

Esters, fatty acids, microbiology, 448

Ethylene glycol alkyl ethers as biocide in fuel systems, 274

Ethylene glycol monomethyl ether
as biocide in fuel systems, 273
in control of aviation fuel tank contamination, 177

Ethylene oxide
for disinfection of paintings on paper, 230
in control of mildew on wool, 113
in disinfection of paintings on wood and sculptures, 228
in sterilization of latex, 326

Extenders for rubber, microbiological resistance, 333

Extractives in heartwood, decay and, 30

Eye cosmetics
microbial contamination, in use, 415
health hazard, 397

Eye drops, microbial contamination, health hazards, 397

Eye infections
contaminated eye drops and, 397
emulsion spoilage and, 296

Eye ointment, microbial contamination, risk level, 402

Eyes, drugs and cosmetics, microbial contamination, health hazards, 408

F

Farmers lung, 298

Fats, microbial polypeptide digestion and, 83

Fatty acids
esters, microbiology, 448
formation in emulsions by micro-organisms, 297
in latex, 327

Fellmongering, 90

Fences, decay, 50

Fermentation, tobacco, 251

Ferrobacillus ferrooxidans, sulphuric acid production by, metal corrosion and, 162–163

Field tests, plastics, microbial deterioration, 464

Fillers for rubber, microbiological resistance, 333, 334

Filling containers, drugs and cosmetics, microbial contamination, 413

Filters
blocking, microbial growth on fuels and, 263
for removal of water from aviation fuel, 176

Filtration, fuel, contamination and, 272

Fire, wood degradation by, 21

Flavobacterium spp.
biodeterioration by, 11
growth on degraded polyethylene, 440
microbial slime and, cooling-water systems, 158
on paint films, 318
wood degradation by, 25

Flavobacterium diffusum, on polysulphide rubber linings of aviation fuel tanks, 363

Flavobacterium fulvum, growth on Thiokol latex MX linings for aviation fuel tanks, 364

Flavobacterium meningosepticum, in drugs, health hazards, 406

Fleece wool, micro-organisms in, 84

Flexibility, plastics, microbial deterioration and, 469

Flindersia pimenteliana, wood colonization by, 54

Flue-curing
tobacco, 240
barn rot, economics, 255

Fomes annosus, butt-rot of conifers, 41

Food microbiology, 3

Formaldehyde
for *Hevea* latex preservation, 368

in control of mildew on wool, 113
wool treatment by, effect of biological detergents and, 123
Fragilaria spp., on ancient buildings at St-Cloud, 76
Fragilaria capucina, on stone monuments and plasters, 76
Freezing in curing skins, 146
French chalk, microbial contamination, 409
Frescos
   biodeterioration, 204
   microbial deterioration, 220
FSII—*See* Fuel system icing inhibitor
Fuel baffles, microbial deterioration, 269
Fuel pumps in subsonic aircraft, microbial growth on fuel and, 264
Fuel system icing inhibitor in control of aviation fuel tank contamination, 177
Fuel systems, microbial infections, detection, 271–272
Fuel tanks
   aluminium, microbial corrosion, 268
   aviation, coatings, microbial degradation, 363–366
   linings, microbial degradation, 363–366
   coatings, microbial corrosion and, 277–278
Fuels
   biodeterioration, 259–305
   microbial growth, conditions for, 261
Fullers earth, microbial contamination, 409
Fumago vagans, skin deterioration by, 147
Fumigation
   microbial biodeterioration and, 9
   paintings and sculptures, 224–255
Fungi
   effect on marble, pigeon excrement and, 73
   growth in aviation fuel tanks in supersonic aircraft, 173–174
   in aircraft fuel, 265
   in destruction of rocks and stones, 71–74
   oil painting deterioration by, 209
Fungi imperfecti, wood degradation by, 27
Fungicidal vapours in treatment of paint-

ings and sculptures, 225
Fungicides
   in control of micro-organisms on paintings and sculptures, 223
   in disinfection of paintings and sculptures, 226–233
   in vulcanizates, fungal resistance, 378
Furs, processing, 133
Fusarium spp.
   biodeterioration by, 11
   growth, on plasticizers, 453
      on Thiokol latex MX linings for aviation fuel tanks, 364
   in aircraft tank sludge, 166
   in cutting oils, 290
   in destruction of rocks and stones, 72
   in rubber pipe joint rings, 352
   in subsonic aircraft fuel, 266
   in tobacco curing, biodeterioration and, 258
   watercolour degradation by, 218
Fusarium oxysporum, marble deterioration by, 73
Fusarium solari, in mascara, health hazard, 398

G

Gallionella spp., in cooling-water systems, 159
Gamella spp., growth on degraded polyethylene, 440
Gasoline
   for subsonic aircraft, microbial contamination, 264
   microbial growth on, 261
Gas masks, mould growth on, 360
Gas production in latex paints, bacterial growth and, 317
Geotrichum spp., in microbial degradation of paintings on canvas, 217
Geranyl-coenzyme A carboxylase in *Pseudomonas* spp., 344
Gleocapsa spp.
   on ancient buildings at St-Cloud, 76
   on stone monuments and plasters, 76
Gleocapsa sanguinea, effect on granite, 75
Gliocladium roseum, wood degradation by, 28

M

Machine tools, soluble oil emulsions, ecology, 288

Mackinawite in iron corrosion by sulphate-reducing bacteria, 191

Magnesium, algal growth and, 75

Manufacture
  drugs and cosmetics, microbial control, 423–424
  tobacco, biodeterioration in, 251–253

Marble, effect of fungi on, pigeon excrement and, 73

Marine borers, 22

Marine diesel engines, lubricating oils, microbial growth and, 284

Marine fuels, microbial growth on, 263, 279

Marine timber, decay, 50

Marketing, tobacco, biodeterioration in, 246–250

Mascara, microbial contamination, health hazard, 398

Mechanical injury, tobacco, bacterial decay and, 242

Membrane separators in oil pollution control, 366

Mercuric chloride in prevention of microbial biodeterioration, 10

Metals, deterioration, 149–202

Methyl cellosolve—See Ethylene glycol monomethyl ether

Methylhydroxyethyl cellulose, effect on cellulosic ethers, 311

Methylhydroxypropyl cellulose, effect on cellulosic ethers, 311

Microbial attack in biodeterioration, methods, 8–9

Microbial biodeterioration
  history and scientific basis, 1–18
  prevention, 9–17
  physiological aspects, 4–9

Microbial contaminants, survival, factors for, 418–420

Microbial corrosion, metals, classification, 153–155

Microbial deterioration
  mural paintings, 220–223
  painted sculptures, 210–215
  paintings and sculptures, moisture and, 205–207

paintings, on canvas, 215–218
  on paper, 218–220
  on wood, 210–215

Microbial growth
  concentration cell corrosion and, 156–159
  in cooling-water systems, control, 159–162

Microbicides for rubber products, 371–377

Micrococcus spp.
  in aircraft tank sludge, 166
  on paint films, 318
  wood degradation by, 24

Micrococcus aurantiacus, in latex, 326

Micrococcus candidus, in latex, 326

Micrococcus luteus, fatty acids in latex and, 327

Microflora on stone, origin, 63–65

Micromonospora spp., wood degradation by, 25

Micro-organisms
  entry into aviation fuel tanks, prevention, 176–177
  on painting surfaces, development, 207–209
  wood degradation by, 23–28

Micropolyspora faeni, in tobacco manufacture, 253

Milan Cathedral, marble deterioration on, 73

Mildew
  on processed wool, 98
  on wool, chemical control, 110–114
  damage detection, 101–103
  development, 99–101
  eradication, 114

Mining cables, chloroprene-rubber, microbial degradation, 358

Mink, curing, 144

Mixed-culture techniques in testing of textile materials for resistance to microbial attack, 106

Moisture
  effect of microbial deterioration on paintings and sculptures, 205–207
  wood decay and, 31–32

Moisture content
  tobacco, microbial deterioration and, 236
  wool, mildew development and, 99

Molecular weight, polycaprolactones, microbiology and, 443
Monascus bisporus, a_w value and, 6
Monilia spp., in subsonic aircraft fuels, 264
Monitoring, drug and cosmetic contamination, 420–424
Moulds
  in drugs and cosmetics manufacture, risk level, 402
  primary, wood degradation by, 26–27
  secondary, wood degradation by, 28
  skin deterioration by, 146–147
Mucor spp.
  development on painting surfaces, 208
  in destruction of rocks and stones, 72
  in microbial degradation of paintings on canvas, 217
  on paint films, 318
  watercolour degradation by, 218
Mucor hiemalis, marble deterioration by, 73
Multidose stock pots, contamination in use, 416
Mummification, 9
Mural paintings
  disinfection, fungicides for, 231–233
  microbial deterioration, 220–223
Mycotoxins in tobacco, biodeterioration and, 239
Mystox B in control of mildew on wool, 111

N

National Health Service, drug expenditure, microbial contamination and, 425
Natural rubber
  preservation, 370–371
  vulcanizates, microbial degradation, 340
  microbiological resistance, 337–344
Navicula spp.
  microbial slime and, cooling-water systems, 158
  on ancient buildings at St-Cloud, 76
Nebulizers, microbial contamination, 417
Neurospora sitophila, plasticizer degradation by, 456
Nipagin as fungicide for disinfection of

canvas paintings, 227
Nitrifying bacteria, stone deterioration and, 70–71
Nitrogen, wood decay and, 33–34
Nitrogen fixation
  by algae on rocks and stones, 75
  in wood, 26
Nocardia spp.
  growth, on plasticizers, 454
    on polyethylene, 438
  in rolling mill emulsions, 294
Nosocomial infections, microbial contamination of drugs and, 406
Nosocomial sepsis, intravenous fluid contamination and, 396
Nutrients
  in fuels, micro-organism growth on, 262
  in microbial corrosion of metals, 153
  in wool, mildew development and, 100
  microbial biodeterioration and, 7–8
  wood decay and, 33–34
Nystatin, disinfection of mural paintings, 231

O

Oak, decay, structure and, 29
Off-odours, tobacco, 237
Oil emulsions, soluble, microbial growth in, 288–300
Oil industry, microbial metal corrosion in, 186–187
Oil-in-water emulsions
  microbial contamination, 294–295
  health hazard, 399
Oil paintings, microbial deterioration, 209
Oil pollution, membrane separators for, 366
Oil wells, metal corrosion in, control with biocides, 196–197
Oils, biodeterioration, 259–305
Ointments
  cracking, microbial activity and, 394
  microbial contamination, health hazard, 399
Olivine, fungal destruction of, 72
Onychomycosis, rubber products and, 367
Ophthalmic preparations

microbial contamination, health hazards, 397
non-sterile, microbial contamination, health hazard, 397–401
Organoborates
    as biocide, in fuel systems, 273, 274, 276
    in marine fuels, 279
Organoborinates in control of aviation fuel tank contamination, 177
Organoboron compounds in control of aviation fuel tank contamination, 177
Organomercury compounds
    as stabilizer for plastics, 461
    in control of mildew on wool, 113
Organotin compounds as stabilizer for plastics, 460
Oscillatoria spp.
    microbial slime and, cooling-water systems, 158
    on ancient buildings at St-Cloud, 76
    on mural paintings, 222
Over-absorbancy, water-storing timber and, 42
Oxalic acid, formation by lichens on rocks and stones, 77
α-Oxoglutaric acid in microbial corrosion of aviation fuel tanks, 172
Oxygen
    in latex paints, bacterial growth and, 313
    in microbial corrosion of metals, 153

P

Pads, drugs and cosmetics, microbial contamination, 417
Paecilomyces spp., in raw natural rubber, 329
Paecilomyces varioti, plasticizer degradation by, 456
Paint films on wood, exterior, biodeterioration, 318–320
Paint on wood, decay, 212–215
Painting in wool removal from sheepskin, 90
Paintings
    biodeterioration, 203–233

disinfection, 224–226
microbial deterioration, 209–210
    moisture and, 205–207
micro-organisms on, control, 223–233
on canvas, microbial deterioration, 215–218
on paper, disinfection, fungicides for, 230–231
    microbial deterioration, 218–220
on wood, microbial deterioration, 210–215
on wood and sculptures, disinfection, fungicides for, 228–230
surfaces, microbial development on, 207–209
Paints, microbiology, 49
Palmelococcus spp., in Lascaux Caves, 76
Pancreatin extracts, microbial contamination, 409
Pancreatin powder, microbial contamination, health hazard, 398
Papain in wool removal from sheepskin pieces, 98
Paper
    paintings on, biodeterioration, 204
    disinfection, fungicides for, 230–231
    microbial deterioration, 218–220
Paper-making industry, microbial corrosion, 186
Papularia spp., watercolour degradation by, 218
Paraffin, liquid, microbial contamination, 409
Pastels
    biodeterioration, 204
    disinfection, 225
    microbial deterioration, 219–220
Pathogenic bacteria in emulsions, 296–297
Pathogenic micro-organisms, colonization of rubber products by, 367–368
Paxillus pannoides, wood decay by, 47
Pellicle, drugs and cosmetics, 393
Penicillinases from bacteria, 391
Penicillins, biodegradation, 391
Penicillium spp.
    biodeterioration by, 11
    development on painting surfaces, 208
    in aircraft tank sludge, 166
    in destruction of rocks and stones, 72

in microbial degradation of paintings on canvas, 216
in raw natural rubber, 329
in subsonic aircraft fuel, 266
in tobacco, biodeterioration by, 246
in tobacco curing, biodeterioration and, 258
on paint films, 318
paint degradation by, 212
pastel deterioration by, 220
processed wool damaged by, 99
skin deterioration by, 146
watercolour degradation by, 218
wood decay by, 211
Pencillium cyclopium
marble deterioration by, 73
plasticizer degradation by, 456
Pencillium funiculosum
growth, on high-density polyethylene, 438
on polycaprolactone, 442
Penicillium luteum, growth on polyethylene, 437
Peniophora gigantea
control in packaged timber, 44
in controlling pine crop infection by *Phomes annosum*, 41
Peptidases, 83
Peppermint water, microbial contamination, health hazard, 399
Perfusion techniques in testing of textile materials for resistance to microbial attack, 106–107
Peronospora tabacina, in tobacco curing, biodeterioration and, 258
Persulphate, effect on cellulosic ethers, 311
Peyronellaea glomerata
black fungus tip and, 88
on fleece wool, 87
pH value
effect on algae in destruction of rocks and stones, 75
mildew development on wool and, 99
Pharmaceuticals, active ingredients, biodegradation, 390–393
Phellinus sp., in window and door frames, 48
Phenol, 4-nitro-, as preservative for natural rubber, 370
Phenol, *o*-phenyl-, for disinfection of textiles and leather, 228

Phenol, pentachloro-
as fungicide for disinfection of canvas paintings, 227
biodegradability, 115
effect on micro-organism colonization of wood, 51
for mildew-proofing of wool, 115
in disinfection of paintings on wood and sculptures, 229
toxicity, 115
Phenylmercuric acetate
as fungicide for disinfection of canvas paintings, 227
disinfection of mural paintings, 231
for disinfection of paintings on paper, 230
in disinfection of paintings on wood and sculptures, 229
Phoma spp.
growth on polyethylene, 441
on paint films, 318
Phoma pigmentovora, paint degradation by, 215
Phosphate esters, microbiology, 448
Photodegradation, polymers, 439
Phthalate esters, cometabolism, 458
Phthalic acid esters, microbiology, 448
Phycomycetes, wood degradation by, 27
Phytobacterium spp., in tobacco curing, biodeterioration and, 259
Phytophthora parasitica, in tobacco curing, biodeterioration by, 241, 258
Pichia spp.
growth, on hydrocarbons, 266
on plasticizers, 454
on paint films, 318
Pieing, 97
Pigeon excrement, effect on fungal decomposition of marble, 73
Pink rot, fleece wool, 84, 86
Pinus radiata
packaged, control of *Peniophora gigantea* in, 44
sapwood, treatability, 30
Pinus sylvestris—See Scots pine
Pipelines
blocking, microbial growth on fuels and, 263
rubber ring joints in, microbial degradation, 348–355

Pyrite, oxidation by *Thiobacillus ferro-oxidans*, 164

Pythium spp., in tobacco curing, biodeterioration and, 258

Q

Quaternary amonium compounds as biocides in fuel systems, 277

Quercus robur—*See* Oak

R

Radiation in control of micro-organisms in cutting oils, 293

Railway sleepers, decay, 50

Raincoats, vinyl, microbial degradation, 449

Raw materials for drugs and cosmetics, microbial contamination, 408–410

Red heat on skins, prevention, 143

Red heat in salting skins, 142

Redox potential in bacterial growth, 313

Redrying tobacco, 235

Reducing agents, sulphur-containing, viscosity of cellulosic ethers and, 310

Resins—*See* Polymers

Respiration, micro-organisms, plastic degradation and, 470

Respiratory malfunctions
  emulsion spoilage and, 296
  microbial contamination of cutting oil emulsion and, 297

Resuscitators, microbial contamination, 417

RH 886 as biocide in fuel systems, 277

Rhizopus spp.
  in microbial degradation of paintings on canvas, 217
  on paint films, 318
  watercolour degradation by, 218

Rhizopus arrhizus, in tobacco curing, biodeterioration by, 241, 258

Rhizopus nigricans, in tobacco curing, biodeterioration and, 258

Rhodotorula spp.
  growth, on hydrocarbons, 266
  on plasticizers, 454
  on paint films, 318

Ricinoleic acid
  derivatives, microbiology, 451
  esters, microbiology, 448

Rocket propellants, microbial growth on, 263

Rocks
  decomposition, chemolithotrophic bacteria and, 65–71
  destruction, algae in, 74–77
    heterotrophic bacteria and fungi in, 71–74
  microflora on, 64–65

Rolling mill emulsions, microbial growth on, 294

Ropey strands, drugs and cosmetics, 393

Rotting, tobacco, 237

Rubber
  biodeterioration, 323–383
  raw natural, composition, 327
  micro-organisms, 327–329
  synthetic, microbial resistance, 329–333

Rubber goods, microbial deterioration, 347–367

Rubber insulation, microbial degradation, 355

Rubber products
  colonization by pathogenic micro-organisms, 367–368
  protection, 371–379

Rubber ring joints in pipelines, microbial deterioration, 348–355

S

Saccharomyces spp., growth on plasticizers, 454

Saccharomyces cerevisiae, growth on plasticizers, 454

Saccharomyces rouxii, a_w value and, 6

Safes, microbial deterioration of paintings and sculptures in, 206

Salting in skin curing, 141–143

Salicylanilide
  as fungicide for disinfection of canvas paintings, 227
  in control of mildew on wool, 113

Salmonella spp.
  in drugs and cosmetics, standards, 421
  in emulsions, 296

Salmonella agona, in raw materials for drugs and cosmetics, 409

Salmonella bareilly
  in raw materials for drugs and cosmetics, 409
  in thyroid tablets, risk level, 402

Salmonella muenchen
  in raw materials for drugs and cosmetics, 409
  in thyroid tablets, risk level, 402

Salmonella oranienburg, a_w value and, 6

Sandocide liquid in control of mildew on wool, 111

Sandstones, microbial deterioration, 66

Sap stain, post-harvest deterioration of wood and, 43

Saponite, fungi destruction of, 72

Sapwood
  decay and, 30
  in soft-wood timber from small trees, 42

Sarcina spp., wood degradation by, 24

Savlon, microbial contamination, health hazards, 401

Scenedesmus spp., microbial slime and, cooling-water systems, 158

Schizothrix spp., on stone monuments and plasters, 76

Sclerotinia sclerotiorum
  in tobacco, biodeterioration by, 246
  in tobacco curing, biodeterioration by, 241, 258

Scopulariopsis spp., on cured tobacco, 250

Scots pine
  colonization by micro-organisms, preservatives and, 51
  decay, nutrients and, 33

Sculptures
  disinfection, 224–226
  microbial deterioration, 209–210
    moisture and, 205–207
  micro-organisms on, control, 223–233
  painted biodeterioration, 203–233
    microbial deterioration, 210–215
  paintings on, disinfection, fungicides for, 228–230
  wooden, microbial degradation, 211

Scytonema spp.
  on mural paintings, 222
  on stone monuments and plasters, 76

Sedimentation, drugs and cosmetics, 393

Sepedonium spp., watercolour degradation by, 218

Septicaemia, microbial contamination of drugs and, 406

Serpula lacrimans
  dry rot, 46–47
  wood decay by, 45

Serratia spp.
  in contact-lens cleaning solutions, health hazard, 397
  wood degradation by, 25

Serratia marcescens
  in hand creams, health hazards, 400
  penetration in sandstone, 65

Shampoos
  biodegradation, 392
  medicated, microbial contamination, 400
  microbial contamination, health hazards, 400

Sheepskin
  pieces, wool removal from, 97–98
  structure, 90
  wool removal from, proteolytic enzymes in, 89–98

Shigella spp., in emulsions, 296

Shipping tobacco, biodeterioration in, 246–250

Ships, microbial corrosion, 186

Shipworm—*See Teredo navalis*

Shrink-proofed wools, effect of biological detergents on, 118

Sienna, microbial degradation and, 212

Siliceous rocks, microbial deterioration, 66

Silicon rubber, microbial resistance, 360–362

Sistotrema spp., in window and door frames, 48

Skin, infection from drugs and cosmetics, 408

Skin care, hand creams, microbial contamination, health hazards, 400

Skins
  deterioration, 131–148
  dewoolled, use, 90

Slime
  drugs and cosmetics, 393
  in cooling-water systems, 157
  in stone destruction, 72

Slimicides in cooling-water systems, 160

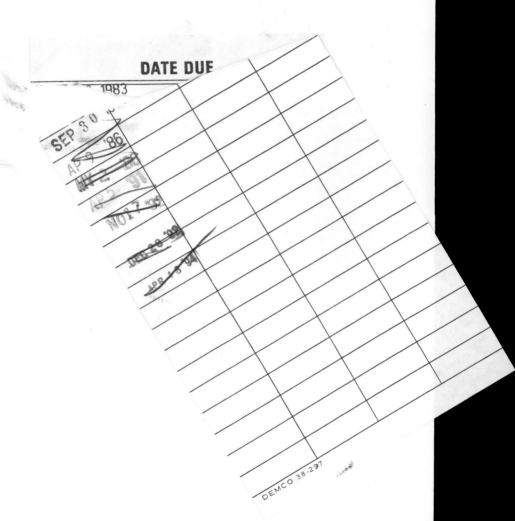

DATE DUE

1983

SEP 30 '86

AP 7

MY 2 '91

AP 2

NO 17

APR

DEMCO 38-297